NCCER

Field Safety

Upper Saddle River, New Jersey
Columbus, Ohio

NCCER

President: Don Whyte
Director of Curriculum Revision and Development: Daniele Dixon
Director of Safety and Management Education: Gary Wilson
Safety Project Manager and Editor: Tara Cohen
Production Manager: Debie Ness
Quality Assurance Coordinator: Jessica Martin
Desktop Publishers: Laura Parker and Rachel Ivines

The NCCER would like to acknowledge the contract service provider for this curriculum:
Topaz Publications, Liverpool, New York.

This information is general in nature and intended for training purposes only. Actual performance of activities described in this manual requires compliance with all applicable operating, service, maintenance, and safety procedures under the direction of qualified personnel. References in this manual to patented or proprietary devices do not constitute a recommendation of their use.

18
ISBN-0-13-106256-5

Preface

This volume was developed by the National Center for Construction Education and Research (NCCER) in response to the training needs of the construction, maintenance, and pipeline industries. It is one of many in NCCER's *Contren™ Learning Series*. The program, covering training for close to 40 construction and maintenance areas, and including skills assessments, safety training, and management education, was developed over a period of years by industry and education specialists.

NCCER also maintains a National Registry that provides transcripts, certificates, and wallet cards to individuals who have successfully completed modules of NCCER's *Contren™ Learning Series*, when the training program is delivered by an NCCER Accredited training Sponsor.

The NCCER is a not-for-profit 501(c)(3) education foundation established in 1995 by the world's largest and most progressive construction companies and national construction associations. It was founded to address the severe workforce shortage facing the industry and to develop a standardized training process and curricula. Today, NCCER is supported by hundreds of leading construction and maintenance companies, manufacturers, and national associations, including the following partnering organizations:

PARTNERING ASSOCIATIONS

- American Fire Sprinkler Association
- American Petroleum Institute
- American Society for Training & Development
- American Welding Society
- Associated Builders & Contractors, Inc.
- Association for Career and Technical Education
- Associated General Contractors of America
- Carolinas AGC, Inc.
- Carolinas Electrical Contractors Association
- Citizens Democracy Corps
- Construction Industry Institute
- Construction Users Roundtable
- Design-Build Institute of America
- Merit Contractors Association of Canada
- Metal Building Manufacturers Association
- National Association of Minority Contractors
- National Association of State Supervisors for Trade and Industrial Education
- National Association of Women in Construction
- National Insulation Association
- National Ready Mixed Concrete Association
- National Utility Contractors Association
- National Vocational Technical Honor Society
- North American Crane Bureau
- Painting & Decorating Contractors of America
- Plumbing-Heating-Cooling Contractors National Association
- Portland Cement Association
- SkillsUSA
- Steel Erectors Association of America
- Texas Gulf Coast Chapter ABC
- U.S. Army Corps of Engineers
- University of Florida
- Women Construction Owners & Executives, USA

Some features of NCCER's *Contren™ Learning Series* are:

- An industry-proven record of success
- Curricula developed by the industry for the industry
- National standardization providing portability of learned job skills and educational credits
- Credentials for individuals through NCCER's National Registry
- Compliance with Apprenticeship, Training, Employer, and Labor Services (ATELS) requirements for related classroom training (CFR 29:29)
- Well-illustrated, up-to-date, and practical information

The Contren™ Safety Learning Series

Welcome to the *Contren™ Safety Learning Series*. This systematic approach to safety education and training provides a standardized curriculum in a modularized form that allows organizations to design custom programs to fit their specific needs. The series is composed of four independent titles: *Safety Orientation*, *Field Safety*, *Safety Technology*, and *Safety Management*. While none of these books serve as mandatory prerequisites for the others, they are listed in a logical education path.

NCCER offers individual recognition for successful completion of all programs, ranging from certificates of completion to Safety Instructor Certification. NCCER's *Contren™ Safety Learning Series* provides unique recognition, the education necessary to gain opportunities for career advancement, and the skills needed to build a safe workplace.

In addition to NCCER's certifications, the *Contren™ Safety Learning Series* has been recognized by the Council on Certification of Health, Environmental, and Safety Technologists (CCHEST) as exam preparation for their Safety Trained Supervisor (STS) and Construction Health and Safety Technician (CHST) certifications. CCHEST is a joint venture of the Board of Certified Safety Professionals and the American Board of Industrial Hygiene.

For more information about courses, certifications, or any other related issue, please contact NCCER Customer Service at (888) 622-3720; visit us online at www.nccer.org; or email your questions to info@nccer.org.

Features of This Book

Capitalizing on a well-received campaign to redesign our textbooks, NCCER is publishing select textbooks in a two-column format. *Field Safety*, a part of the *Contren™ Safety Learning Series*, incorporates the design and layout of our full-color books along with special pedagogical features. The features augment the technical material to maintain the participants' interest and foster a deeper appreciation of the trade.

The **Think About It** feature uses "what if?" questions to help participants apply theory to real-world experiences and put ideas into action.

The **Case History** feature demonstrates the significance of adhering to safety guidelines and best practices by focusing on the importance of job-site safety. Participants read about causes and repercussions of real-life job-site hazards, incidents, accidents, and near-misses.

The **What's Wrong with This Picture**? feature includes photographs of standards violations and failures to adhere to best practices. This feature enables participants to study an image, identify the unsafe condition or act, and discuss how to recognize, avoid, and prevent such situations on the job site.

The **Profile in Success** feature presents one-page biographies of successful industry professionals and shares their related experiences and advice.

We're excited to be able to offer you these improvements and hope they lead to a more rewarding learning experience. As always, your feedback is welcome. Please let us know how we are doing by visiting NCCER at www.nccer.org or e-mailing us at info@nccer.org.

Acknowledgments

This curriculum was developed as a result of the vision and leadership of NCCER's National Safety Committee and those who served as subject matter experts:

Jim Humphry, Quanta Services
Art Deleon, Underground Construction Co.
Frank McDaniel, Casey Industries

For the long-term professional services provided to NCCER on all safety-related subjects, including the development of the *Contren™ Safety Learning Series*, a sincere thanks is extended to **Steven Pereira, Professional Safety Associates**, Denham Springs, LA.

Contents

Module 75101-03

Introduction to Safety

COURSE MAP

This course map shows all of the modules in Field Safety. The suggested training order begins at the bottom and proceeds up. The local Training Program Sponsor may adjust the training order.

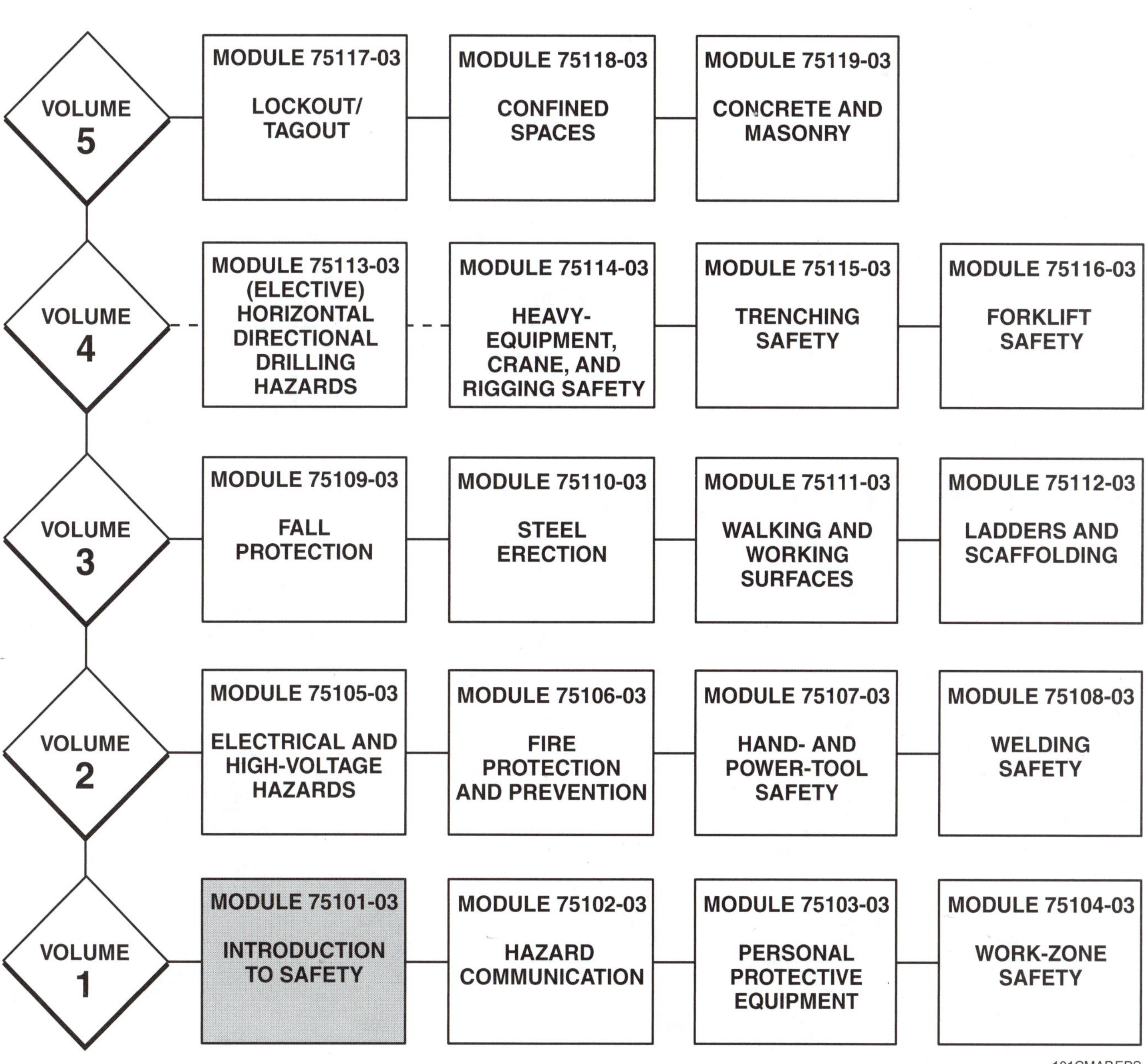

MODULE 75101-03 CONTENTS

Figures

Tables

MODULE 75101-03

Introduction to Safety

Objectives

When you have completed this module, you will be able to do the following:

1. Explain the difference between compliance and best practices.
2. Describe the purpose and function of the Occupational Health and Safety Administration (OSHA).
3. Explain how accident costs affect everyone on a job site.
4. Describe proper materials handling procedures and safeguards.

Prerequisites

There are no prerequisites for this module.

Required Materials

1. Pencil and paper
2. Appropriate personal protective equipment

1.0.0 ◆ INTRODUCTION

Construction work can be dangerous; however, it doesn't have to be deadly. Most **accidents** on a construction site are caused by worker carelessness, poor safety planning, lack of training, or failure on the part of the employer or employee to follow safety regulations.

To help prevent accidents, your company must have a safety program. This program will provide you with the rules and safeguards you need to work safely. Safety must be incorporated into all phases of the job and involve all employees at every level, including management.

Some safety programs require that a company-appointed **competent person** be on site before you start any job. A competent person is someone who has experience and training for the job and knows the hazards of the job. He or she knows the rules and regulations associated with the job and has the authority to stop a job if work is not being done safely.

In addition to your company's rules, there are many regulations that companies have to comply with to meet local, state, and national guidelines. Safety regulations are created to make work sites safe and accident-free. Safety policies and procedures are available to help you and your company comply with these regulations. It is important to remember that there is a good reason why each regulation exists. Following good safety practices helps to save lives.

2.0.0 ◆ TEN BASIC RULES OF SAFETY

The following ten basic rules of safety should become second nature to everyone on the job site:

1. Use your tools, equipment, and personal protective equipment the way they were designed.
2. Wear your hard hat and safety equipment at all times where required.
3. Inspect your equipment daily.
4. Only perform tasks for which you have been trained.
5. Understand your company safety rules and policies.
6. Take responsibility for yourself and your co-workers.
7. Correct or report all unsafe conditions to your supervisor immediately.
8. Accept the "zero accident" philosophy.
9. Get involved with your company safety program.
10. When in doubt, stop and ask!

3.0.0 ◆ COMPLIANCE VERSUS BEST PRACTICES

There is a difference between **compliance** and best practices. The **Occupational Safety and Health Administration (OSHA)** determines compliance; individual company policies determine best practices. OSHA sets minimum safety requirements, investigates serious accidents, and inspects work sites. A company is in compliance when it meets all minimum safety regulations.

In addition to specific safety guidelines, OSHA regulations include a general duty clause. Each employer has a general duty to provide employees with a workplace free from recognized hazards that may cause death or serious harm. Companies are required to set practices and procedures to make sure all workers are safe. A safety program that includes the best available practices and equipment to ensure worker safety is known as a **best practices plan**. Best practices often exceed the minimum safety requirements. Best practices can change as new safety equipment and procedures are developed.

Keep in mind, however, that if you get injured and you are not following the company's safety program or **OSHA standards**, you can be held responsible for the accident.

DID YOU KNOW?

Occupational Hazards

In 1999, there were 5.7 million occupational injuries and illnesses among U.S. workers. Approximately 6.3 of every 100 workers experienced a job-related injury or illness and 6,023 workers lost their lives on the job.

Source: The Occupational Safety and Health Administration (OSHA)

3.1.0 OSHA

OSHA, part of the U.S. Department of Labor, is a government agency that helps protect millions of workers each year. Their mission is to ensure a safe and healthy environment in the workplace. Since the agency was created in 1971, workplace deaths have been cut in half and injury and illness rates have declined 40 percent. At the same time, there has been an increase in the number of people working. In fact, since 1971, the number of people working in the United States has doubled from 56 million workers at 3.5 million work sites to 111 million workers at 7 million sites.

OSHA adopts and enforces safety regulations known as **standards**. Some, like the hazard communications standard, apply to all industries. Others only apply to a specific industry. There is a series of additional OSHA standards for construction, in addition to the general industry standards.

Twenty-six states have their own OSHA programs. The state regulations must be at least as strict as the national standards. Find out which regulations apply to your job site. The federal and state OSHA programs keep workplaces safe through accident investigation, inspections, and outreach.

3.1.1 Accident Investigation

If there is a serious accident at a job site in which three or more workers are hospitalized or someone is killed, OSHA must be notified. OSHA will then investigate the accident. Accident investigation is OSHA's tool for uncovering hazards that were missed during inspections. Accident investigation, however, is really about accident prevention. A lot can be learned from an accident, most importantly how to avoid another one. Investigations are very useful when they discover every contributing factor to the accident. This helps correct the unsafe condition or activity and helps prevent it from happening again.

3.1.2 OSHA Inspections

OSHA inspects workplaces to ensure compliance with minimum safety standards. If OSHA compliance officers find any violations on a site, they may issue a citation and a penalty. Citations

"Get Out of That Trench!"

That's the order an OSHA inspector gave to a worker in an unshored, unsloped trench. It's a good thing the inspector was paying attention because 30 seconds after the worker left the trench, the wall near where he had been standing collapsed. The worker was spared serious injury or death because he listened to the warning and left the trench.

The Bottom Line: Be aware of your surroundings, and always listen to OSHA inspectors.

Source: The Occupational Safety and Health Administration (OSHA)

inform the employer and employees of the regulations and standards that have been broken. Penalties are fines the company must pay for each violation. Even though the company is paying the fine, you will still feel the effects. Fines affect you because the company will have less money for raises, bonuses, and hiring more workers to replace those who may have been injured.

3.1.3 OSHA Resources

OSHA plays an important role in preventing on-the-job injuries and illnesses. This is accomplished through outreach, education, and compliance assistance. OSHA provides various types of training materials to assist employers and workers in understanding the standards.

The OSHA Web site (www.osha.gov) has nearly 450,000 pages of safety information. It includes a special section dedicated to construction standards. The Web site also offers assistance and interactive e-tools to help both employers and employees to understand and comply with the standards.

In addition to their Web site, OSHA offices provide employers and employees with information and training to help them understand and comply with safety regulations. OSHA also offers free workplace consultations. These consultations are available to any business that needs on-site help in establishing safety and health programs, as well as identifying and correcting workplace hazards.

3.2.0 Best Practices

Best practices are the rules and safety procedures that the industry considers to be the safest method to get a job done. Best practices take a pro-active approach to safety, identifying possible safety issues in advance and establishing procedures to avoid potential accidents. Although your company sets best practices, they are often based on OSHA regulations. Best practices are designed to help you and your co-workers comply with required safety standards.

Confined-space safety is a good example of best practices and compliance. Industrial sites are required to follow the confined-space standard. Construction sites are exempt; however, all employers have a general duty to keep workers safe while working in confined spaces. A best practices plan at a construction site would follow confined-space entry rules because it is the safest way to do the work.

You have a responsibility to follow your company's rules and safety procedures. These are some general safety rules that promote a safe workplace:

- Follow all rules, policies, and procedures for the job you are doing. Rules, policies, and procedures can be both written and spoken.
- Never start a job until you have reviewed the appropriate procedure for the job and understand how to do the task properly.
- Never do a job that you have not been authorized to perform.
- Never start a job that appears unsafe or that will unnecessarily put you at risk.
- Never start a job that you are incapable of doing without assistance.
- Never put yourself or co-workers at risk by engaging in any unsafe behaviors.
- Stay focused on the job at all times. Think about what could happen and protect yourself and your co-workers.
- Immediately report all job-related **incidents**, near misses, and injuries to your supervisor.

OSHA Inspector Helps Prevent a Fatality

While investigating the death of an aerial lift operator, an OSHA inspector helped prevent another serious injury or death. The first incident involved an aerial lift that rolled off the side of a flatbed truck and threw the operator to the ground. The worker died from his injuries.

During his investigation, the OSHA inspector noticed a tow truck operator preparing to bring the lift upright. The inspector realized that by doing this, the aerial lift could shift and possibly swing into the driver's side of the tow truck. The inspector told the tow truck driver to use the passenger side controls to winch up the aerial lift instead of using the driver's side. This act saved the driver's life because the basket of the aerial lift did hit the driver's side of the truck. Had the worker been there, he would likely have been killed.

The Bottom Line: OSHA inspectors are experienced safety officers. Following their advice prevents accidents and saves lives.

Source: The Occupational Safety and Health Administration (OSHA)

- Immediately report all hazards to your supervisor. Correct those conditions that are within your control.
- Tell co-workers to stop what they are doing when they are being unsafe or creating a hazardous situation.

4.0.0 ◆ CAUSES OF ACCIDENTS

Accidents and injuries result in needless pain and suffering, as well as financial hardship for employees, their families, and the company. Every employer and worker on the site has moral, legal, and financial obligations to prevent accidents and injuries. To do this, it is important to understand that accidents are caused by unsafe acts and unsafe conditions. Unsafe acts are things you do or do not do that can cause an accident. Unsafe conditions are external factors that make the work area dangerous.

DID YOU KNOW?

Eye Injuries

The average cost of an eye injury is $1,463. That includes both the direct and indirect costs of accidents, not to mention the long-term effects on the health of the worker; that's priceless.

Source: The Occupational Safety and Health Administration (OSHA)

Accidents and Incidents

Accidents and incidents are different. An accident is an unplanned event preceded by an unsafe act, unsafe condition, or both. Accidents disrupt activities and often result in personal injury or property damage. An incident is any undesirable loss of resources or undesired event that , under slightly different circumstances, could have resulted in personal harm or property damage.

4.1.0 Unsafe Acts

Unsafe acts often lead to serious injury and sometimes death. You can stop unsafe acts by changing your behavior. It is your responsibility to recognize unsafe acts and stop them immediately. This can mean stopping what you are doing or stopping your co-worker(s).

Examples of the most common unsafe acts are:

- Using defective equipment
- Disabling a safety device
- Operating equipment at improper speeds
- Servicing equipment while it is in motion or energized
- Loading or placing equipment or supplies improperly or in a dangerous way
- Using equipment improperly
- Working while impaired by alcohol or legal or illegal drugs
- Horseplay (*Figure 1*)

An unsafe act also includes not doing the correct thing. Many accidents are caused or worsened by workers who fail to use proper personal protective equipment, fail to follow safety procedures, or do

101F01.EPS

Figure 1 ◆ Horseplay is dangerous.

Ask Questions

Find out about all potential job hazards before you begin work. Ask your supervisor or the competent person on site to answer the following questions before you start a job.

- What are the hazards associated with this job?
- Where is the energy source and how can it be shut down or controlled?
- What could possibly go wrong?
- What must I do to protect myself and my co-workers from personal injury?

not warn co-workers of potentially hazardous conditions. Keep yourself healthy and safe and look out for the safety of others. Always follow the rules and use the right equipment for the job.

4.2.0 Unsafe Conditions

Unsafe working conditions are also a major cause of accidents and injury. Unsafe conditions can be caused by environmental factors such as noise, extreme heat or cold, poor lighting, or poor air circulation. These conditions impair your reactions or limit your movements. Poor housekeeping can also create unsafe conditions. Clutter in walkways, spills, and improper waste disposal can make merely walking on a construction site dangerous. To work safely, you need to be able to hear, see, breathe, and maintain your balance. Correct unsafe working conditions before you start the job.

5.0.0 ◆ COSTS OF ACCIDENTS

Accidents are very costly. When they happen, everyone involved loses, including the company and its employees. You may not believe that all accidents will cost you money. Think about it this way. If there is an accident on a job site, the company will have to pay higher insurance rates, costs for medical care and/or repairs, downtime, and the cost of investigations or fines. If the company is paying for these things, there is less money to spend on raises and performance bonuses for workers or hiring replacement workers. What that means to you is more work without an increase in pay. You are a part of the company (*Figure 2*) and you will be affected by job-site accidents, even if you aren't directly involved.

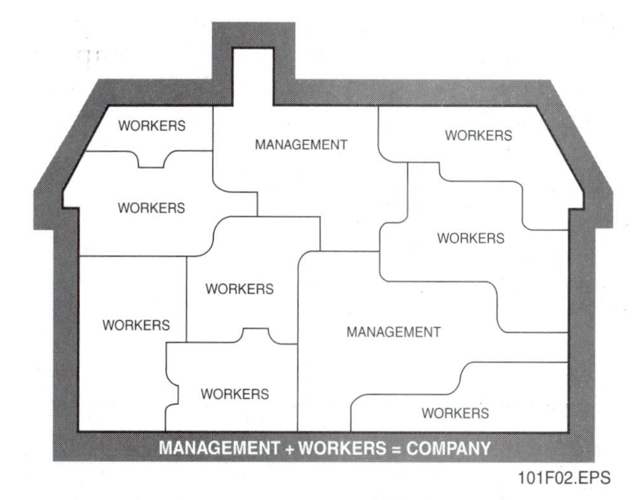

Figure 2 ◆ Companies are a combination of management and workers.

DID YOU KNOW?

Alcohol and Drug Abuse on Construction Sites

Most experts agree that at least one out of five construction workers has an alcohol or drug abuse problem. That's 20% of a company's workforce. The impact of this problem results in increased workers' compensation and health care costs; decreased productivity; increased tool and materials costs; and increased absenteeism.

Accident costs can be classified as direct (insured) and indirect (uninsured). The costs associated with accidents can be compared to an iceberg, as shown in *Figure 3*. The tip of the iceberg represents the direct costs, such as medical bills, compensation, and insurance premiums. On the other hand, the larger indirect costs are underwater or unseen. These include property and equipment damage, production delays, and lost time. The indirect costs of accidents are usually two to seven times greater than the direct costs. *Table 1* shows the costs of accidents that are not covered by insurance.

6.0.0 ◆ CONSTRUCTION ERGONOMICS AND MATERIALS HANDLING

Back injury accidents are some of the most common and costly in the construction industry, accounting for 25% of all construction-related injuries annually. Back injuries result in an average of six days of lost work time, but often require even more.

A variety of unsafe acts and conditions, mainly related to **materials handling** and ergonomics, can cause such accidents. The personal and financial costs related to job-related accidents make it essential that every employee on the job site be familiar with and practice good materials handling procedures. A thorough understanding of proper ergonomics is a good place to start.

6.1.0 Construction Ergonomics

Ergonomics is the study of how people are physically affected by work-related movements, motions, and postures. It is a critical factor to your long-term health. Back injuries and repetitive motion illnesses are a major concern for the construction industry. Many of these injuries or illness show up years later and can limit your quality of life. Stretching prior to work can help reduce injuries and improve performance. Construction work can be as strenuous as many athletic events. As in athletics, you should warm up before beginning any activity.

Adhere to the following guidelines when you perform tasks that involve constant repetitive motion or exposure to vibration for long periods of time:

- Take a break for approximately 15 minutes every two hours to get your blood circulating.
- Shake your hands and arms frequently to stimulate blood flow.
- Rotate with co-workers as often as possible.
- Use anti-vibration gloves when needed.

6.1.1 Back Safety

Back injuries can leave you unable to work and can cause lifelong, painful impairments. It is your responsibility to ensure that you do everything possible to avoid these and other injuries. To reduce back injuries caused by lifting, follow these safeguards:

- Always assess the object you need to lift before attempting to lift it.
- Plan your lift; know where it is to be unloaded and whether there are any hazards in your path.
- Make sure you have firm footing.
- Bend your knees.

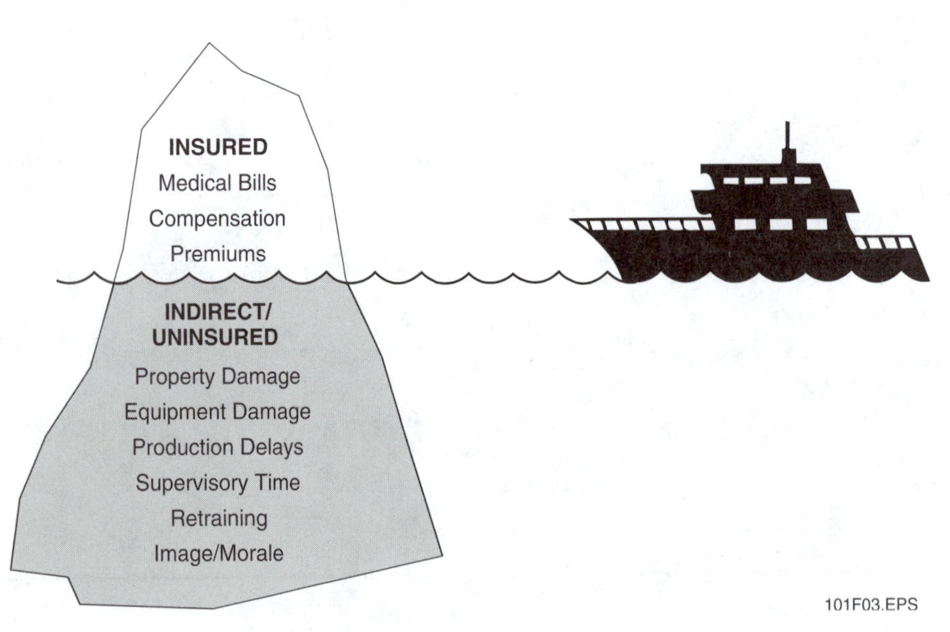

Figure 3 ◆ Hidden costs of accidents.

Table 1 Costs Not Covered by Insurance

INJURIES	WAGE LOSSES	PRODUCTION COSTS	OFF-THE-JOB ACCIDENTS	INTANGIBLES	ASSOCIATED COSTS
First aid expenses	Idle time of workers whose work is interrupted	Product spoiled by accident	Cost of medical services	Lowered employee morale	Difference between actual losses and amount recovered
Transportation costs	Time spent cleaning up accident area	Loss of skill and experience	Time spent on injured worker's welfare	Increased labor conflict	Rental of equipment to replace damaged equipment
Costs of investigations	Time spent repairing damaged equipment	Lowered production of replacement worker	Loss of skill and experience	Unfavorable public relations	Surplus worker for replacement of injured worker
Cost of processing reports	Time lost by workers receiving first aid	Idle machine time	Training replacement worker		Wages or other benefits paid to disabled worker
			Decreased production of replacement worker		Overhead costs while production is stopped
			Benefits paid to injured worker or dependents		Loss of bonus or payment of forfeiture for delays

101T01.EPS

- Get a good grip; take the time to set down your load and re-lift if you do not have a good grip.
- Lift with your legs, keep your back straight, and keep your head up.
- Keep the load close to your body.
- Never turn or twist until you are standing straight, then pivot your feet and body.
- Ask for assistance when lifting heavy loads or break the load down into manageable parts.
- Use mechanical lifting devices when available.

6.1.2 Lifting Procedures

Once you know the rules for lifting safely, you can start practicing the steps. With one fourth of all occupational injuries happening during materials handling, it is essential that you know how to lift heavy objects the right way. Follow these steps (see *Figure 4*):

Step 1 Move close to the object you are going to lift. Position your feet in a forward/backward stride, with one foot at the side of the object.

Step 2 Bend your knees and lower your body, keeping you back straight and as nearly upright as possible.

Step 3 Place your hands under the object, wrap your arms around it, or grasp the handles. To get your hands under and object that is flat on the floor, use both hands to lift one corner. Slip one hand under that corner. With one hand under, tilt the object to get the other hand under the opposite side.

Step 4 Draw the object close to your body.

Step 5 Lift by slowly straightening your legs and keeping the object's weight as much as possible over your legs.

Step 6 Pick the object up facing the direction you are going to go, to avoid twisting your knees or back.

These steps let you use your strongest muscles (those in your legs) instead of your weakest ones (those in your back) to lift. Practice with light objects. Once you've got it down, move on to heavier ones.

Figure 4 ◆ Proper lifting procedures.

6.2.0 Materials Handling

Materials handling includes moving any materials, such as wood, bricks, or other supplies, on the construction site. You will encounter manual and mechanical materials handling methods on the job site. Though both methods are everyday activities, handling materials either way can be dangerous. It's easy to hurt yourself by tripping and/or falling when lifting materials by hand. As well, you are particularly at risk for back injuries during this work. When lifting mechanically, you can easily hurt others if you are not acutely aware of your surroundings. In either instance, unsafe acts or conditions can cause damage to the materials you are handling.

During any type of materials handling, you must ensure that you and your co-workers, materials, and equipment are safe from unexpected movements such as falling, slipping, tipping, rolling, blowing, or other uncontrolled motions. Follow these general guidelines during all materials handling:

Are You a Safety-Conscious Worker?

A safety-conscious worker is one who has the ability to recognize a hazardous condition or behavior and a willingness to do something about it. A safety-conscious worker is one who sees a spill and cleans it up, spots a frayed cord on a tool and turns it in for repair, or notices a co-worker lifting a heavy box and offers to help.

Now, think about a worker who is not safety-conscious. Have you ever observed a worker who sees a spill and avoids it, but does nothing to clean it up? This worker was safe, but not safety conscious. This worker took care of himself but did not protect his co-workers. Have you ever seen a worker who just walks past a spill without noticing it at all? A safety-conscious worker looks for hazards and does something about them. Safety-conscious workers look after their co-workers. Does that describe you?

- Use a safety harness or positioning belt with lanyard as required.
- Protect the area below you.
- Salt or sand icy walking and working areas immediately.
- Clean up all grease and oil spills immediately.
- Chock all material and equipment, such as pipe, drums, tanks, reels, trailers, and wagons, as necessary to prevent shifting or rolling.
- Tie down or band all light, large surface area material that might be moved by the wind.
- When working at heights, secure all tools and equipment from falling.

6.2.1 Manual Materials Handling

During manual materials handling, it is very easy to hurt yourself. Unsafe acts and conditions during materials handling activities are particularly common causes of back injuries. Therefore, when lifting by hand, it is critical that you always adhere to the following appropriate lifting procedures:

- Use gloves whenever cuts, splinters, blisters, or other injuries are possible.
- Know the weight of any object to be handled. Get help for heavy or awkward loads.
- Warm up, stretch, and flex.
- Keep loads close to the body.
- Squat down to the load, keep your back straight, and use your legs to lift the load.
- Do not extend the load out from your body.
- If you must turn with a load, change the position of your feet. Don't twist your back.
- If you are carrying a load with a partner, let your partner know when you are going to release the load.
- Avoid awkward or tight positions while handling loads.
- Get enough help to safely handle the load.

6.2.2 Mechanical Materials Handling

When lifting materials mechanically, you must carefully secure the materials you are loading. You must also ensure that you are aware of any obstructions, such as people or equipment, that are in your path so that you may to avoid potential accidents. When handling materials with a machine, such as a dolly or forklift, always follow these guidelines:

- Know the weight of the object to be handled/moved.
- Know the capacity of the handling device (crane, forklift, chain-fall, dolly, come-along, etc.) that you intend to use and never exceed it.
- Use tag lines to control loads.
- Get rigging instructions from a qualified employee before beginning.
- Ensure that your handling equipment is in good working order and free of damage.

Summary

Working in construction can be hazardous. You have a responsibility to be aware of the hazards associated with your job and know how to protect yourself. Protecting yourself means following the rules, safeguards, and regulations established by your company and OSHA. It is also important to understand that you are part of the company and therefore will be affected by accident costs.

Review Questions

1. A *competent person* for safety purposes is _____.
   a. appointed by the company
   b. an OSHA inspector
   c. a worker who has received safety training
   d. any foreman or supervisor

2. Best practices for safety _____.
   a. are minimum standards
   b. are developed by OSHA
   c. are required by law
   d. often exceed minimum safety standards

3. OSHA's mission is to protect _____.
   a. the economy
   b. employers
   c. the environment
   d. workers

4. OSHA must investigate an accident _____.
   a. if it involves an injury
   b. if someone is killed
   c. to levy a fine
   d. to issue citations

5. The two types of materials handling are _____.
   a. heavy and light
   b. manual and mechanical
   c. single and team
   d. safe and unsafe

6. OSHA citations and penalties affect _____.
   a. company management
   b. everyone on the job site
   c. safety technicians
   d. site supervisors

7. Ergonomics is the study of how people are physically affected by _____.
   a. accidents and incidents on the job site.
   b. temperature changes during work
   c. work-related movements, motions, and postures
   d. long-term back injury complications

8. Using legal drugs, such as prescription painkillers, does not constitute an unsafe act.
   a. True
   b. False

9. Cold weather can cause unsafe working conditions.
   a. True
   b. False

10. The cost of accidents only affects the company.
    a. True
    b. False

PROFILE IN SUCCESS

Don McLellan, Black & Veatch

Field Safety and Health Manager

To newcomers in the Safety Field, Don McLellan says, "Don't ever give up! Be a sculptor. Keep chipping away until you have that finished piece of art."

What has been your greatest career-related accomplishment?
My greatest achievement to date was earning the First OSHA Voluntary Protection Program (VPP) STAR Award for construction on a project in South Carolina. I've participated in four other projects that earned the VPP Star Award. I also completed an SCR retrofit project in 2003 with approximately 350,000 manhours worked, zero injuries, no OSHA recordables, and no lost-time accidents.

What types of training have you been through?
After earning my Master's of Science Degree in Industrial Safety Management in 1990, I began working in construction safety and have been there ever since. Since starting in the field, I have completed the following training: OSHA 10-Hour Construction, Compliance with the New Steel Erection Standard, OSHA Recordkeeping, Fall Protection, Substance-Abuse Supervisor Training, Performance Evaluation and Feedback Training, Construction Supervisor Training, Safety & Health Program Administration, Lockout/Tagout, Confined Spaces, Emergency Response, Hazard Communication, Excavation, Risk Management, First Aid/CPR, and Sexual Harassment Prevention. I have taken refreshers in several of these courses numerous times since starting my career.

What kinds of work have you done in your career?
One of my proudest moments in life was being hired for my first job at age 14 as a soda fountain employee at a neighborhood drug and pharmacy store. In high school, I worked retail sales and later worked retail buying. I served in the military, and upon my release, earned a Bachelor of Arts Degree. I owned a Pest Control Services business for over 14 years and pursued my Master's Degree while working as a Licensed Pest Control Operator.

Tell us about your present job.
I'm currently working as a Field Safety and Health Manager at a coal-fired power plant retro-fitting two Selective Catalytic Reduction units and a Baghouse. I am working an SCR retrofit project with approximately 92,000 manhours to date, zero first-aid cases, zero recordable injuries, and zero lost-time injuries.

What factors have contributed most to your success?
Pre-planning is the most important factor contributing to success in construction work. I have a very strong commitment to use of Job Hazard Analyses and daily Safety Task Assignments on my projects. These tools provide for detailed pre-planning if used correctly. Additionally, craft involvement in the safety and health program is extremely important. If the crafts are involved in the safety program, they get a sense of ownership of the program and participation just naturally follows. Resolution to safety concerns, issues, or problems on a project are a shared commitment by craft and management. Working together for a resolution gets better results than dictating a resolution. You must be innovative and creative in order to develop a positive safety attitude on each project.

GLOSSARY

Trade Terms Introduced in This Module

Accident: An unplanned event preceded by an unsafe act, unsafe condition, or both, which disrupts activities and often results in personal injury or property damage.

Best practices plan: A safety plan that includes the best available practices and equipment to ensure worker safety.

Competent person: A company-appointed person who can identify working conditions or surroundings that are unsanitary, hazardous, or dangerous to employees and who has authorization to correct or eliminate these conditions promptly.

Compliance: A safety plan that meets the minimum OSHA safety standards.

Incident: An undesirable loss of resources or undesired event that, under slightly different circumstances, could have resulted in personal harm or property damage.

Materials handling: The act of manually or mechanically moving materials, such as wood, bricks, or other supplies, on or around the construction site.

Occupational Safety and Health Administration (OSHA): The division of the U.S. Department of Labor mandated to ensure a safe and healthy environment in the workplace.

Standard: A practice or procedure that is widely recognized or employed, especially because of its excellence.

OSHA standard: A standard adopted by OSHA, which requires conditions, or the adoption or use of one or more practices, means, methods, operations, or processes, reasonably necessary or appropriate to provide safe or healthful employment and places of employment [*29 CFR 1910.2(f)*].

NCCER CURRICULA — USER UPDATE

NCCER makes every effort to keep its textbooks up-to-date and free of technical errors. We appreciate your help in this process. If you find an error, a typographical mistake, or an inaccuracy in NCCER's curricula, please fill out this form (or a photocopy), or complete the online form at **www.nccer.org/olf**. Be sure to include the exact module ID number, page number, a detailed description, and your recommended correction. Your input will be brought to the attention of the Authoring Team. Thank you for your assistance.

Instructors – If you have an idea for improving this textbook, or have found that additional materials were necessary to teach this module effectively, please let us know so that we may present your suggestions to the Authoring Team.

NCCER Product Development and Revision

13614 Progress Blvd., Alachua, FL 32615

Email: curriculum@nccer.org

Online: www.nccer.org/olf

❑ Trainee Guide ❑ AIG ❑ Exam ❑ PowerPoints Other ______________________________

Craft / Level: Copyright Date:

Module ID Number / Title:

Section Number(s):

Description:

____

____

____

____

Recommended Correction:

____

____

____

____

Your Name:

Address:

____

Email: Phone:

Module 75102-03

Hazard Communication

COURSE MAP

This course map shows all of the modules in Field Safety. The suggested training order begins at the bottom and proceeds up. The local Training Program Sponsor may adjust the training order.

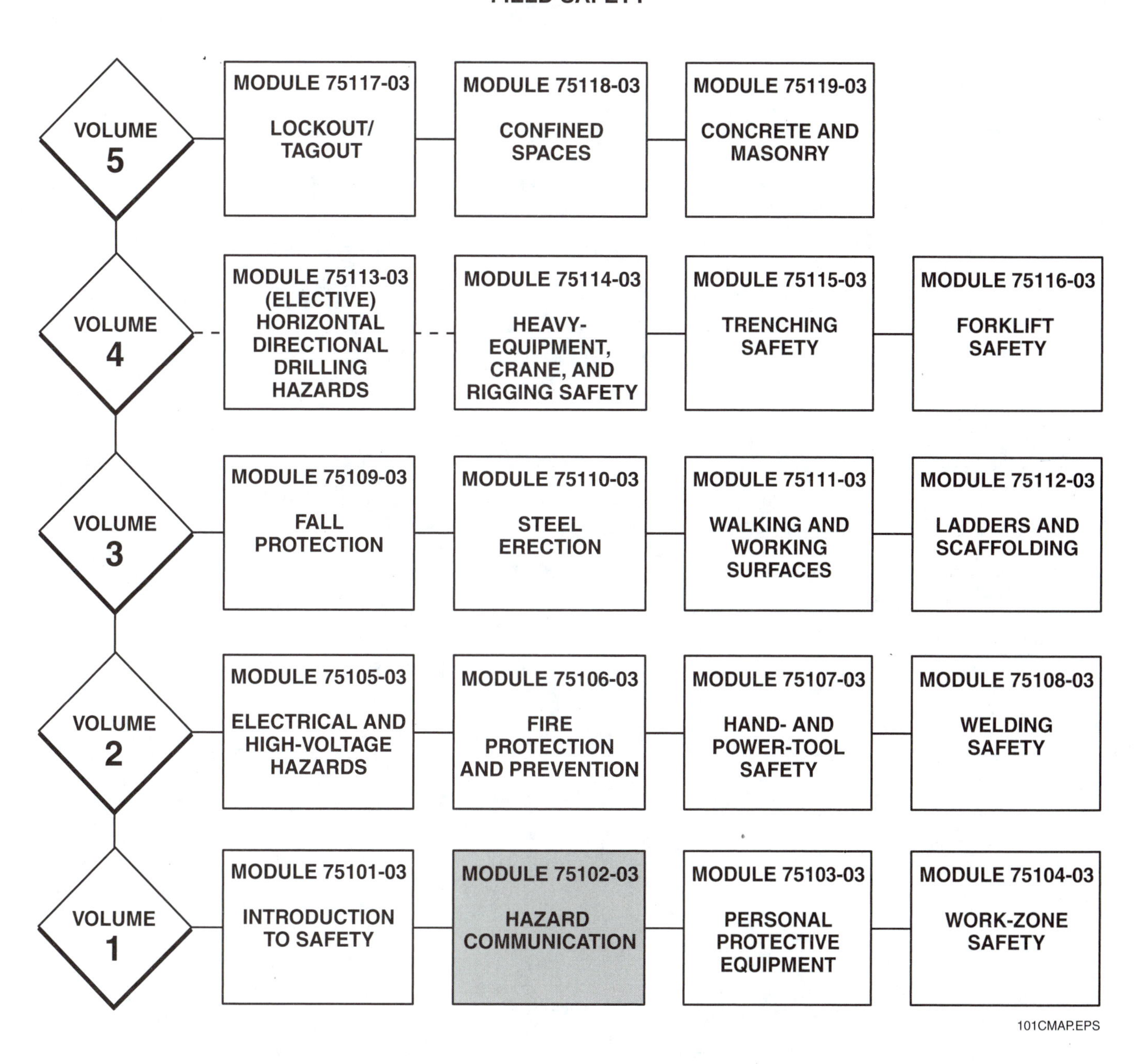

MODULE 75102-03 CONTENTS

Figures

Tables

MODULE 75102-03

Hazard Communication

Objectives

When you have completed this module, you will be able to do the following:

1. Identify different types of warning labels.
2. Explain how a material safety data sheet (MSDS) is used.
3. Identify and apply the safety information on an MSDS.
4. Demonstrate and explain proper on-site safety and emergency-response procedures.

Prerequisites

Before you begin this module, it is recommended that you successfully complete the following: Field Safety, Module 75101-03.

Required Materials

1. Pencil and paper
2. Appropriate personal protective equipment

1.0.0 ◆ INTRODUCTION

Using hazardous materials improperly can cause health problems. These can include minor problems such as skin or eye irritation, or serious conditions such as cancer. These materials also present physical hazards such as fire, corrosion, and reactivity. You have the right to know the hazards of all the chemicals you will be exposed to on the job. OSHA directs employers to tell workers about hazardous materials on the job site through the **Hazard Communication (HazCom) Standard**. You may have heard of it as the worker right-to-know program.

Employers must educate their workers about the hazardous materials they might use on the job. This is done through a written HazCom program and training. In addition, all materials must have proper labels and **material safety data sheets (MSDSs)**. Labels and MSDSs provide information about health hazards, safety precautions, and emergency responses. You need to understand this information in order to protect yourself.

The final responsibility for your safety rests with you. Workers have the following responsibilities when it comes to HazCom.

- Learn to recognize hazardous materials labels.
- Know where MSDSs are kept on your job site.
- Report any hazards you spot on the job site to your supervisor.
- Know the physical and health hazards of the materials you use.
- Know how to protect yourself from hazards.
- Know what to do in an emergency.
- Understand your employer's HazCom Program.

2.0.0 ◆ HAZARDOUS MATERIALS

There are many types of hazardous materials on a construction site. These materials and related hazards are shown in *Table 1*.

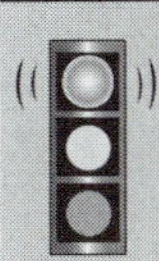

WARNING!

Radioactive materials require special handling, training, and personal protective equipment. Never enter an area where radioactive materials are being used unless you have the proper training and appropriate personal protective equipment.

Hazardous materials come in many different forms. Chemicals are not only liquids. They can also be solids, gases, fumes, and mists. Many common products contain several chemicals. For example, some paints contain cadmium and lead.

Many chemicals pose health hazards like disease or burns. Others pose physical hazards, including fire or explosion. Some have both health and physical hazards. The MSDS will help you identify the hazards and understand how to protect yourself.

2.1.0 Radioactive Hazards

Radioactive materials can be a serious hazard on a job site. They are used in construction during radiographic testing of welds in piping, vessels, or pumps. All radioactive materials must be properly labeled and warning signs must be posted in the work area. You must have special training and equipment. Only trained workers are allowed in these areas. If you see signs like those shown in *Figure 1*, stay away from that area.

Table 1 Typical Hazardous Materials

Substance*	Hazard
Glue and adhesives	Toxic fumes, explosion, fire
Gasoline	Toxic fumes, explosion, fire
Paints	Toxic fumes, explosion, fire
Concrete and mortar	Toxic fumes, lung disease, cement dermatitis, chemical burns
Fuels and oils	Toxic fumes, explosion, fire
Wood	Flying particles
Compressed gases	Toxic fumes, explosion, fire
Corrosive liquids	Burns, respiratory hazards
Radioactive materials	Burns, radiation sickness

* Each of these materials must have an MSDS. Do not use them unless you have read and understand the MSDS. The MSDS will help you determine the hazards that you may be exposed to and understand how to protect yourself.

102F01.EPS

Figure 1 ◆ Radioactive warning signs.

Always Follow the Rules

A worker was given the job of filling some drums with a corrosive liquid chemical. The company's policy required him to wear complete chemical personal protective equipment including a helmet with a visor, rubber gloves, and boots. Twenty minutes before quitting time, the worker decided to remove his safety equipment. While he was filling the drums, the hose broke and he suffered serious chemical burns on his hands and arms and had to be rushed to the hospital. He survived the accident but suffered greatly and missed a lot of work in the process.

The Bottom Line: Know the hazards of the chemicals you use. Always wear the proper personal protective equipment when working with any type of chemical.

Source: The Occupational Safety and Health Administration (OSHA)

DID YOU KNOW?

Pressure-Treated Lumber

Did you know that pressure-treated lumber is hazardous? The chemicals used to treat it, including arsenic, can cause illness and death. Refer to the applicable MSDS and follow these precautions when working with pressure-treated lumber.

- When cutting pressure-treated lumber, always wear eye protection and a dust mask.
- Wash any skin that is exposed while cutting or handling the lumber.
- Wash clothing that is exposed to sawdust separately from other clothing.
- Do not burn pressure-treated lumber because the ashes it creates are dangerous.
- If you have to dispose of pressure-treated lumber, bury it or put it in the trash.
- Be sure to read and follow the manufacturer's safety instructions.

Radioactive materials are a special type of physical hazard. You cannot see the radiation affecting your body. Excessive radiation exposure can cause skin burns, nausea, vomiting, infertility, and cancer. You can minimize radiation exposure by limiting the amount of time you are exposed and/or increasing your distance from the source. Use the proper shielding or personal protective equipment. Check with your supervisor to find out if there are any radioactive hazards and how to avoid them.

2.2.0 Physical Hazards of Chemicals

Chemicals can pose physical hazards including acid reactions or burns. Acids can create toxic vapors or react violently when mixed with other chemicals. Other chemicals are flammable, combustible, or explosive (*Figure 2*). For example, solvents and compressed gases are often flammable. These materials can catch fire in their liquid or gaseous state. Some, like blasting caps, are dangerous solids. These items should be kept away from open flames, sparks, intense heat, or other ignition sources to prevent fires or explosions. You need to know the physical hazards of the chemicals you use in order to prevent fires and chemical reactions.

Compressed gas is under very high pressure in the tank. Compressed gases must be handled carefully so that the tank is not damaged. A damaged tank may leak hazardous fumes. If the tank is severely damaged, it may explode.

Figure 2 ◆ Common labels for flammable, combustible, and explosive materials.

2.3.0 Health Hazards of Chemicals

Many chemicals are hazardous to your health. For example, adhesives and paints can be poisonous. Even mild exposure can cause skin irritation, breathing problems, or allergic reactions. Hazardous products are marked TOXIC or POISON (*Figure 3*). A skull and cross bones symbol is frequently used to note poisons. Exposure to poisons can result in death.

Figure 3 ◆ Common labels for toxic materials.

Top Ten Most Hazardous Substances

This is a list of the top 10 most hazardous substances used on job sites. These chemicals are ingredients in many products. The solvent you use may contain one of these chemicals. Always check labels. Review the MSDS for each product before using it.

1. Arsenic
2. Lead
3. Mercury
4. Vinyl chloride
5. Polychlorinated biphenyls (PCBs)
6. Benzene
7. Cadmium
8. Benzopyrene
9. Polycyclic aromatic hydrocarbons
10. Benzofluoranthene

Source: Agency for Toxic Substances and Disease Registry (ATSDR)

DID YOU KNOW?

Cement Hazards

Dry cement dust can enter open wounds and cause blood poisoning. If cement dust comes in contact with body fluids, it can cause chemical burns to the membranes of the eyes, nose, mouth, throat, or lungs. It can also cause a fatal lung disease known as silicosis. Wet cement or concrete can cause chemical burns to the eyes and skin. Repeated contact with cement or wet concrete can result in an allergic skin reaction known as cement dermatitis. Always use the proper personal protective equipment to prevent cement dust from getting into your eyes, nose, mouth, throat, or lungs.

2.3.1 Chemical Exposure

Chemicals can cause harm when they get into your body. This is known as chemical exposure. You can be exposed to toxic chemicals through your skin, mouth, eyes, and lungs. These are known as routes of exposure. Chemicals can be absorbed through direct contact with your skin. Open cuts increase exposure. Smoking or eating without washing your hands after handling chemicals may cause you to ingest toxins. You can get chemicals in your eyes by rubbing them. Splashes, mists, or vapors can also allow chemicals to get in your eyes. Breathing chemical vapors, mists, or dusts will allow chemicals to enter your lungs. The personal protective equipment you use should reduce or eliminate the routes of chemical exposure. For example, respirators should be worn to protect your lungs from airborne chemicals (*Figure 4*).

The health effect depends on the individual, the toxicity, and the dose. Each person reacts to chemicals differently. Some are more sensitive than others. Your chemical sensitivity can change over time. Toxicity is a measurement of how poisonous a chemical is. Some chemicals are very poisonous while others are only moderately irritating. The dose is how much of the chemical gets into your body over a certain time period. It is similar to a dose of medicine. It is often said that the dose makes the poison. A big dose of a mild toxin can be deadly. You need to understand the dangers and know how to protect yourself.

2.3.2 Effects of Chemical Exposure

If you swallow a whole bottle of aspirin, you will get very sick. However, taking one or two aspirins for a headache can make you feel better. Different exposures have different results. Health officials calculate safety limits for various routes of exposure (eyes, skin, and mouth) over different time periods (15 minutes, 1 hour, and 8 hours). Generally, we can think of the two extremes: acute and chronic. A short-term, intense contact with a chemical is known as **acute exposure**. Some chemical exposure can affect you right away. This is known as an **acute effect**. Dizziness, headaches, and skin rashes are common acute effects. For example, inhaling acetone vapors (exposure) will cause immediate dizziness (effect).

Some effects caused by chemicals do not appear for many months or years. Low-level exposure over a long period of time can also result in health problems. This is known as **chronic exposure**. Often people do not realize they are being affected. Breathing asbestos dust over many years will cause asbestosis. This is a **chronic effect**.

(A) SELF-CONTAINED BREATHING APPARATUS

(B) SUPPLIED-AIR RESPIRATOR

(C) FULL-FACE PIECE RESPIRATOR

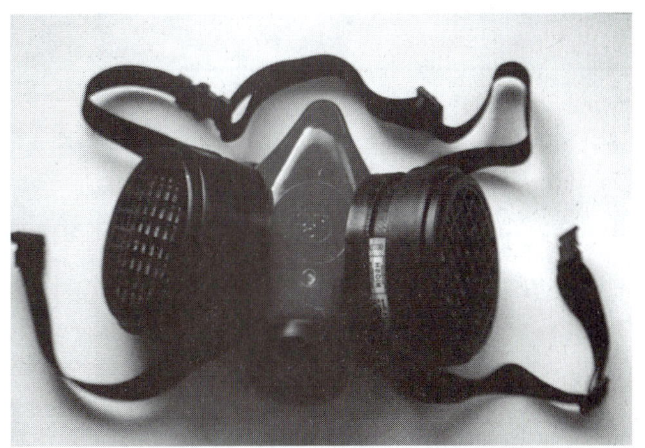

(D) HALF-MASK RESPIRATOR

102F04.EPS

Figure 4 ◆ Different types of respiratory equipment.

Personal protective equipment can prevent both chronic and acute chemical exposure. This can range from safety goggles and gloves to full body suits *(Figure 5)*.

It is important to know what types of chemicals you are working with and how to protect yourself. You also need to know what to do when you are exposed or chemicals are spilled. The MSDS and other safety guides like the *NIOSH Pocket Guide to Chemical Hazards* can help you (see the *Appendix* at the back of this module). Always follow all established safety guidelines, wear the proper personal protective equipment, and ask the safety person on site if you are unsure of how to protect yourself from chemical hazards.

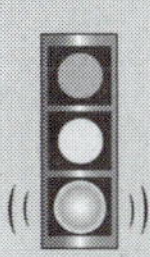

NOTE

For more information on personal protective equipment, please refer to the *Personal Protective Equipment* module in Volume One.

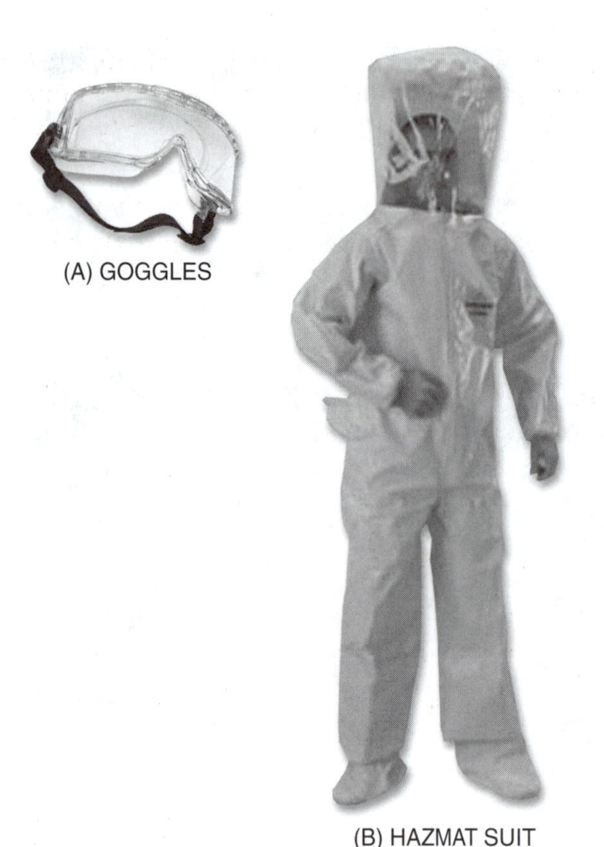

Figure 5 ◆ Personal protective equipment.

3.0.0 ◆ LABELING

On a construction site, all materials in containers must have a label. Labels describe what is in a container. They also warn you of chemical hazards.

The HazCom Standard states that hazardous material containers must be labeled, tagged, or marked. The label must include the name of the material, the appropriate hazard warnings, and the name and address of the manufacturer. OSHA does not require specific labels. Label information can be any type of message using words, pictures, or symbols. However, labels must describe the hazards present. Labels must be readable and easily seen.

Two common labeling systems are from the National Fire Protection Association (NFPA) and the U.S. Department of Transportation (DOT). The NFPA developed *NFPA 704*. You may have heard it called the NFPA Hazard Warning Diamond (*Figure 6*). The four-color diamond can be a container label. It is also used on doors to note the hazard in a room or building. Each section and color represents a hazard: health, flammability, instability, and specific hazards. Numbers from zero to four indicate increasing hazards. The DOT also requires labels for hazardous materials shipments (*Figure 7*). Either of these labels can be part of your company's HazCom labeling program. You may see these labels at your job site.

Hazardous materials at the site must be properly labeled. If the material is transferred from a labeled container, the new container must be labeled. The new label must have all of the information from the original label. Make sure that any materials you work with are labeled. Be sure that you understand your company's labels.

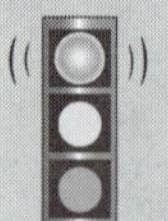

WARNING!

Never use chemicals from an unlabeled container.

FLAMMABILITY HAZARD (RED)

- **4** Extremely flammable – Rapidly vaporizes at normal pressure and temperature or is readily dispersed in air and will burn readily.
- **3** Flammable – Ignites at normal temperatures.
- **2** Ignites when moderately heated.
- **1** Ignites when preheated.
- **0** Will not burn.

HEALTH HAZARD (BLUE)

- **4** Extreme – Fatal with very short exposure. Wear special full protective suit and breathing apparatus.
- **3** Serious – Serious injury with short exposure. Wear full protective suit and breathing apparatus.
- **2** Moderate – Continued exposure can cause injury. Use breathing apparatus.
- **1** Slight – Exposure can cause irritation. Breathing apparatus may be worn.
- **0** Normal – No hazard.

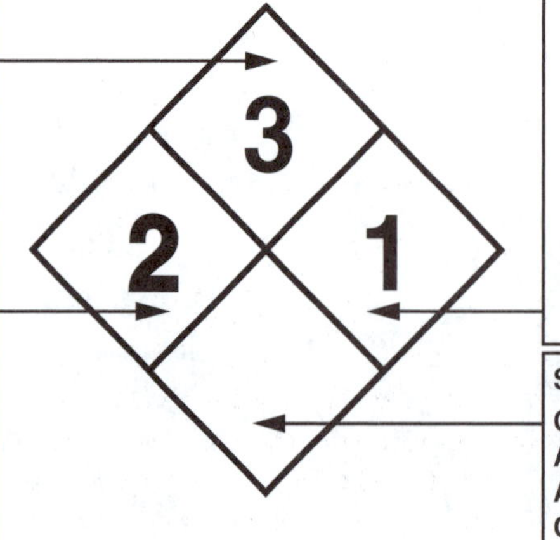

INSTABILITY HAZARD (YELLOW)

- **4** Readily capable of detonation or of explosive decomposition or reaction at normal temperatures and pressures.
- **3** Capable of detonation or explosive reaction if shocked, heated under confinement, or mixed with water.
- **2** Violent chemical change possible but does not detonate. May react violently or form explosive mixtures with water.
- **1** Normally stable. Can become unstable at high temperatures and pressures. May react with water but not violently.
- **0** Normally stable. Not reactive to water.

SPECIFIC HAZARD (WHITE)

OXY	Oxidizer
ACID	Acid
ALK	Alkali
COR	Corrosive
W̶	Use NO WATER
☢	Radioactive

102F06.EPS

Figure 6 ◆ NFPA Hazard Warning Diamond.

4.0.0 ◆ MATERIAL SAFETY DATA SHEETS (MSDSs)

Material safety data sheets (MSDSs) are fact sheets prepared by the chemical manufacturer or importer. Each product used on a construction site must have an MSDS. An MSDS describes the substance, hazards, safe handling, first aid, and emergency spill procedures. OSHA does not have a mandatory form, but they do require inclusion of specific information. The Chemical Manufacturers Association has developed a standard form that meets national and international standards. Most chemical manufacturers use this format. The sections of the form include:

- Chemical product and company information
- Composition/information on ingredients
- Hazard identification
- First-aid measures
- Fire-fighting information
- Accidental release measures
- Handling and storage
- Exposure controls/personal protection
- Physical and chemical properties
- Stability and reactivity
- Toxicological properties
- Ecological properties
- Disposal considerations
- Transportation information
- Regulatory information
- Other information

An MSDS can be difficult to read. The scientific information is fairly technical. *Figure 8* shows a sample MSDS for a common construction adhesive. The most important things to look for on an MSDS are the specific hazards, personal protection, handling procedures, and first aid information. Most MSDSs have a 24-hour emergency-response number.

Using *Figure 8*, try to find the information you would need to use the adhesive described on the sample MSDS. First locate the hazards. Section 3 of the MSDS in *Figure 8* shows that the adhesive is extremely flammable. It is also harmful or fatal if swallowed. Repeated exposure over many years could cause brain damage.

Next, find out how to minimize these hazards. Section 8 tells you that mechanical ventilation is needed to reduce hazardous vapors. This can be a fan in an open window. If ventilation is not enough, respiratory protection is needed. Section 8 also tells you how to protect your eyes and skin.

102F07.EPS

Figure 7 ◆ DOT hazardous materials labels.

Asbestos

Asbestos is a hazardous, fibrous substance that causes lung disease, including cancer. It was once used regularly in construction. Pipe insulation, shingles, wallboard, floor covering, and blown-in insulation are just a few products that may contain asbestos.

The federal government stopped production of most asbestos products in the early 1970s. However, installation of these products continued through the late 1970s and early 1980s. Today, asbestos fibers can be released during renovations of older buildings. Breathing asbestos dusts can have chronic effects. Smoking increases the risk of serious illness from asbestos. Never work around asbestos unless you have the proper training.

Source: Agency for Toxic Substances and Disease Registry (ATSDR)

M A T E R I A L   S A F E T Y   D A T A   S H E E T

SECTION 1 - CHEMICAL PRODUCT AND COMPANY IDENTIFICATION

PRODUCT NAME : 2000 CONSTRUCTION ADHESIVE
UPC NUMBER : 7079825010, 70709825013, 7079825014, 7079825016, 7079825107
PRODUCT USE/CLASS : Construction Adhesive

MANUFACTURER:
DAP INC.
2400 BOSTON ST.
BALTIMORE, MD 21224

24 HOUR EMERGENCY:
TRANSPORTATION: 1-800-535-5053 (352-323-3500)
MEDICAL : 1-800-327-3874 (513-558-5111)

PREPARE DATE : 02/03/1999
REVISION NO. : 6
REVISION DATE: 02/14/2000

GENERAL INFORMATION:
DAP INC.: 1-800-DAP-TIPS (1-888-327-8477)

SECTION 2 - COMPOSITION/INFORMATION ON INGREDIENTS

ITEM	CHEMICAL NAME	CAS NUMBER	WT/WT % RANGE
01	N-Hexane	110-54-3	5.0-10.0 %
02	Aliphatic Petroleum Distillate	64742-89-8	15.0-20.0 %
03	2-Methyl Pentane	107-83-5	5.0-10.0 %
04	3-Methyl Pentane	96-14-0	5.0-10.0 %

EXPOSURE LIMITS

ITEM	ACGIH TLV-TWA	ACGIH TLV-STEL	OSHA PEL-TWA	OSHA PEL-CEILING	COMPANY TLV-TWA	SKIN
01	50 ppm	N.E.	50 ppm	N.E.	N.E.	NO
02	400 ppm	N.E.	400 ppm	N.E.	N.E.	NO
03	N.E.	N.E.	N.E.	N.E.	N.E.	NO
04	N.E.	N.E.	N.E.	N.E.	N.E.	NO

(See Section 16 for abbreviation legend)

Remaining ingredients are not considered hazardous per the OSHA Hazard Communication Standard.

Listed Permissible Exposure Levels (PEL) are from the U.S. Dept. of Labor OSHA Final Rule Limits (CFR 29 1910.1000); limits may vary between states.

SECTION 3 - HAZARDS IDENTIFICATION

EMERGENCY OVERVIEW: DANGER! Extremely flammable liquid and vapor. Vapor harmful. Harmful or fatal if swallowed. Vapors may cause flash fire or explosion. Aspiration hazard if swallowed - can enter lungs and cause damage. Harmful if inhaled.

POTENTIAL HEALTH EFFECTS:

EFFECTS OF OVEREXPOSURE - EYE CONTACT: May cause eye irritation.

EFFECTS OF OVEREXPOSURE - SKIN CONTACT: May irritate skin. Prolonged or repeated contact can result in defatting and drying of the skin which may result in skin irritation and dermatitis (rash).

EFFECTS OF OVEREXPOSURE - INHALATION: Vapor harmful if inhaled. Vapor may irritate nose and upper respiratory tract. Vapor inhalation may affect the brain or nervous system causing dizziness, headache or nausea.

102F08A.EPS

Figure 8 ◆ DAP 2000 construction adhesive MSDS (1 of 5).

EFFECTS OF OVEREXPOSURE - INGESTION: This material may be harmful or fatal if swallowed. Aspiration of material into the lungs due to vomiting can cause chemical pneumonitis which can be fatal. If ingested, this product may cause vomiting, diarrhea, and depressed respiration.

EFFECTS OF OVEREXPOSURE - CHRONIC HAZARDS: Reports have associated permanent brain and nervous system damage with prolonged and repeated occupational overexposure to solvents. Hexane exposure may cause damage to the arms and legs which may be permanent. Symptoms include: loss of memory, loss of intellectual ability, and loss of coordination.

MEDICAL CONDITIONS WHICH MAY BE AGGRAVATED BY CONTACT: None known.

PRIMARY ROUTE(S) OF ENTRY: SKIN CONTACT INHALATION

SECTION 4 - FIRST AID MEASURES

EYE CONTACT: Flush with large quantities of water until irritation subsides. Contact a physician.

SKIN CONTACT: Wash with soap and water.

INHALATION: Remove to fresh air. If not breathing, give artificial respiration. If breathing is difficult, give oxygen. Contact a physician immediately.

Note: Only trained personnel should give artificial respiration or administer oxygen.

INGESTION: DO NOT INDUCE VOMITING. If irritation or complications arise, contact a physician or Regional Poison Control Center immediately.

COMMENTS: None.

SECTION 5 - FIRE FIGHTING MEASURES

FLASH POINT: -30 F
(SETAFLASH CLOSED CUP)

LOWER EXPLOSIVE LIMIT: N.A.
UPPER EXPLOSIVE LIMIT: N.A.

AUTOIGNITION TEMPERATURE: N.E.

EXTINGUISHING MEDIA: CO2 DRY CHEMICAL FOAM

UNUSUAL FIRE AND EXPLOSION HAZARDS: Extremely flammable. Material will readily ignite at room temperature. Vapors may form an explosive mixture with air. Vapors can travel long distances to a source of ignition and flashback. Containers may explode if exposed to extreme heat. Eliminate sources of ignition: heat, electrical equipment, sparks, and flames. Do not put in contact with oxidizing or caustic materials.

SPECIAL FIREFIGHTING PROCEDURES: Full protective equipment, including self-contained breathing apparatus, is recommended to protect from combustion products. Cool exposed containers with water.

SECTION 6 - ACCIDENTAL RELEASE MEASURES

SPILL OR LEAK PROCEDURES: Dike spill area. Immediately eliminate sources of ignition. Use absorbent material or scrape up dried material and place into containers.

102F08B.EPS

Figure 8 ◆ DAP 2000 construction adhesive MSDS (2 of 5).

SECTION 7 - HANDLING AND STORAGE

HANDLING INFORMATION: KEEP OUT OF REACH OF CHILDREN. Avoid skin and eye contact. Avoid breathing vapors. Use only in a well ventilated area.

STORAGE INFORMATION: Store away from caustics and oxidizers. Keep away from heat, spark, and flame. Keep containers tightly closed when not in use. Keep containers from excessive heat and freezing. Do not store at temperatures above 120 degrees F.

OTHER PRECAUTIONS: Intentional misuse by deliberately concentrating and inhaling vapors may be harmful or fatal. Do not take internally. Construction and repair activities can adversely affect indoor air quality. Consult with the occupants or a representative (i.e. maintenance, building manager, industrial hygienist, or safety officer) to determine ways to minimize any impact.

SECTION 8 - EXPOSURE CONTROLS/PERSONAL PROTECTION

ENGINEERING CONTROLS: Provide sufficient mechanical ventilation (local or general exhaust) to maintain exposure below PEL and TLV. Vapors are heavier than air and will collect in low areas. Check all low areas (basements, sumps, etc.) for vapors before entering.

RESPIRATORY PROTECTION: If 8 hour exposure limit or value is exceeded for any component, use an approved NIOSH/OSHA respirator. Consult your safety equipment supplier and the OSHA regulation, 29 CFR 1910.134 for respirator requirements. A respiratory protection program that meets OSHA 1910.134 and ANSI Z88.2 requirements must be followed whenever workplace conditions warrant a respirator's use.

EYE PROTECTION: Goggles or safety glasses with side shields.

SKIN PROTECTION: Solvent impervious gloves.

OTHER PROTECTIVE EQUIPMENT: Provide eyewash and solvent impervious apron if body contact may occur.

HYGIENIC PRACTICES: Remove contaminated clothing and wash before reuse.

SECTION 9 - PHYSICAL AND CHEMICAL PROPERTIES

BOILING RANGE	: 150 - 160 F	VAPOR DENSITY	: Is heavier than air
ODOR	: Gasoline-like		
APPEARANCE	: Gray color paste	EVAPORATION RATE:	Is faster than Butyl Acetate
SOLUBILITY IN H2O	: Negligible		
SPECIFIC GRAVITY	: 1.3058		
VAPOR PRESSURE	: 125 mm Hg @ 68F		
PHYSICAL STATE	: Paste		

(See Section 16 for abbreviation legend)

SECTION 10 - STABILITY AND REACTIVITY

CONDITIONS TO AVOID: Excessive heat and freezing.

INCOMPATIBILITY: Strong oxidizers and caustics.

HAZARDOUS DECOMPOSITION PRODUCTS: Normal decomposition products, i.e. COx, NOx

HAZARDOUS POLYMERIZATION: Will not occur under normal conditions.

STABILITY: This product is stable under normal storage conditions.

102F08C.EPS

Figure 8 ◆ DAP 2000 construction adhesive MSDS (3 of 5).

SECTION 11 - TOXICOLOGICAL PROPERTIES

No product or component toxicological information is available.

SECTION 12 - ECOLOGICAL INFORMATION

No Information.

SECTION 13 - DISPOSAL CONSIDERATIONS

WASTE MANAGEMENT/DISPOSAL: Dispose of according to Federal, State, and Local Standards. Discarded material should be incinerated at a permitted facility. Liquids cannot be disposed of in a landfill. Do not reuse empty container. State and Local regulations/restrictions are complex and may differ from Federal regulations. Responsibility for proper waste disposal is with the owner of the waste.

EPA WASTE CODE - If discarded (40 CFR 261): None, yields no liquid component when evaluated by EPA method 1311 (TCLP)

SECTION 14 - TRANSPORTATION INFORMATION

DOT PROPER SHIPPING NAME: Adhesive(Consumer Commodity*)

DOT HAZARD CLASS: 3(ORM-D*)

DOT UN/NA NUMBER: UN 1133(NONE*) PACKING GROUP: III(NONE*)

* For containers of 1 gallon or less.
Note: The shipping information provided is applicable for domestic ground transport only. Different categorization may apply if shipped via other modes of transportation and/or to non-domestic destinations.

SECTION 15 - REGULATORY INFORMATION

U.S. FEDERAL REGULATIONS: AS FOLLOWS -

OSHA: Hazardous by definition of Hazard Communication Standard (29 CFR 1910.1200)

SARA SECTION 313:
This product contains the following substances subject to the reporting requirements of Section 313 of Title III of the Superfund Amendments and Reauthorization Act of 1986 and 40 CFR Part 372:

----------- CHEMICAL NAME -----------	CAS NUMBER	WT/WT % RANGE
N-Hexane	110-54-3	5.0-10.0%

TOXIC SUBSTANCES CONTROL ACT:
This product contains the following chemical substances subject to the reporting requirements of TSCA 12(B) if exported from the United States:

----------- CHEMICAL NAME -----------	CAS NUMBER
No information is available.	

All ingredients are listed on the TSCA inventory

NEW JERSEY RIGHT-TO-KNOW:
The following materials are non-hazardous, but are among the top five components in this product:

102F08D.EPS

Figure 8 ◆ DAP 2000 construction adhesive MSDS (4 of 5).

---------- CHEMICAL NAME ----------	CAS NUMBER
Hydrocarbon resin	TSRN-1223370035031P
Magnesium Carbonate	546-93-0
SB Rubber	TSRN-618608-5085P
Calcium Carbonate	1317-65-3

PENNSYLVANIA RIGHT-TO-KNOW:
The following non-hazardous ingredients are present in the product at greater than 3%:

---------- CHEMICAL NAME ----------	CAS NUMBER
Hydrocarbon resin	proprietary
Magnesium Carbonate	546-93-0
SB Rubber	proprietary
Calcium Carbonate	1317-65-3

CALIFORNIA PROPOSITION 65:
WARNING: The chemical(s) noted below and contained in this product, are known to the state of California to cause cancer, birth defects or other reproductive harm:

---------- CHEMICAL NAME ---------- CAS NUMBER
No Proposition 65 chemicals are present in this product.

INTERNATIONAL REGULATIONS: AS FOLLOWS -

CANADIAN WHMIS: This MSDS has been prepared in compliance with Controlled Product Regulations except for use of the 16 headings.

CANADIAN WHMIS CLASS: No information available.

SECTION 16 - OTHER INFORMATION

HMIS RATINGS - HEALTH: 2 FLAMMABILITY: 3 REACTIVITY: 1

PREVIOUS MSDS REVISION DATE: 07/17/1999
REASON FOR REVISION:
SECTION 1: New Emergency Contact Phone Numbers

VOC less water, less exempt solvent: 330-340 gm/l(25-26%)
VOC material: 330-340 gm/l

LEGEND:
ACGIH - AMERICAN CONFERENCE OF GOVERNMENTAL INDUSTRIAL HYGIENISTS
N.A. - NOT APPLICABLE
N.E. - NOT ESTABLISHED
PEL - PERMISSIBLE EXPOSURE LIMIT
NTP - NATIONAL TOXICOLOGY PROGRAM
SARA - SUPERFUND AMENDMENTS AND REAUTHORIZATION ACT OF 1986
STEL - SHORT TERM EXPOSURE LIMIT
TLV - THRESHOLD LIMIT VALUE(8 HR. TIME WEIGHTED AVERAGE OR TWA)
VOC - VOLATILE ORGANIC COMPOUND
NJRTK - NEW JERSEY RIGHT TO KNOW LAW
N.D. - NOT DETERMINED

MSDS# 10302

This data is offered in good faith as typical values and not as a product specification. No warranty either expressed or implied, is hereby made. The recommended industrial hygiene and safe handling procedures are believed to be generally applicable. However, each user should review the recommendations in specific context of the intended use and determine if they are appropriate.

< End OF MSDS >

102F08E.EPS

Figure 8 ◆ DAP 2000 construction adhesive MSDS (5 of 5).

Section 7 gives general handling and storage information.

Section 4 lists the first aid measures for eye contact, skin contact, or inhalation. Section 5 explains fire hazards and fire-fighting measures. Now you have the information you need in case of an emergency.

All MSDSs must be kept on site. Ask your supervisor to tell you where the MSDSs are located. Have him or her point out the sections that relate to your job. The health and safety of you and your co-workers depends on it.

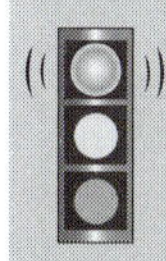

WARNING!

The MSDSs must be kept in the work area and be readily accessible to all workers. Material safety data sheets must be written in English. The company's safety officer or competent person should review the MSDS before the hazardous material is used.

5.0.0 ◆ RESPONDING TO EMERGENCIES

Every job site should have a plan for responding to emergencies. Planning is especially important if the accident happens in a remote area that does not have a telephone. Make sure you know what your employer's emergency plan involves. Find out who needs to be called in case of an emergency and what you need to do to protect yourself.

5.1.0 Emergency-Response Teams

Emergency-response teams are the first line of defense in emergencies. Only employees who are physically capable of performing the duties and who have been properly trained can be assigned to an emergency-response team. Depending on the size of the operation, there may be one or several teams trained in the following areas:

- Use of various types of fire extinguishers
- First aid, including cardiopulmonary resuscitation (CPR)
- Shutdown procedures
- Evacuation procedures
- Chemical spill-control procedures
- Use of self-contained breathing apparatus (SCBA)
- Search and emergency-rescue procedures

The type and extent of the emergency will depend on site operations. The response will vary according to the type of process, the material handled, the number of employees, and the availability of outside resources.

Emergency-response teams must be trained in the types of emergencies that could happen, and must know the emergency actions that need to be performed. They should be informed about special hazards, such as storage and use of flammable materials, toxic chemicals, and radioactive sources to which they may be exposed during fire and other emergencies.

It is important to determine when not to intervene. For example, team members must be able to determine if the fire is too large for them to handle or whether it is safe to perform an emergency rescue. If emergency-response team members could be injured or incapacitated, they should wait for professional firefighters and/or emergency-response groups.

5.2.0 Personal Protective Equipment (PPE)

Effective personal protective equipment is essential for any person who may be exposed to potentially hazardous substances. In emergency situations, employees may be exposed to a wide variety of hazardous circumstances, including:

- Chemical splashes or contact with toxic materials
- Falling objects and flying particles
- Unknown atmospheres that may contain toxic gases or inadequate oxygen to sustain life
- Fire and electrical hazards

It is extremely important that employees be adequately protected in these situations. Some of the personal protective equipment that may be used includes:

- Safety glasses, goggles, or face shields for eye protection
- Hard hats and safety shoes for head and foot protection
- Proper respirators for breathing protection
- Whole body coverings, gloves, hoods, and boots for full body protection from chemicals
- Protection from environmental conditions such as extreme temperatures

Summary

Working around hazardous materials is a serious matter. You have a right to know what hazards you may be exposed to. Your employer has the responsibility to tell you about these hazards. Reading and understanding labels and MSDSs is an important part of overall job safety. It is also important to wear the proper protective clothing and know your company's emergency response plan. Many accidents happen because workers are not properly protected from hazardous materials and do not know how to respond when they are exposed to them.

Review Questions

1. OSHA's Hazard Communication (HazCom) Standard requires all employers to educate employees about on-site hazardous chemicals.
   a. True
   b. False

2. Ordinary concrete can cause lung disease.
   a. True
   b. False

102E01.EPS

Figure 1

Refer to *Figure 1* to answer Question 3.

3. This symbol indicates a _____ hazard.
   a. biological
   b. combustible
   c. fire
   d. radiation

4. Washing your hands can help to prevent chemical exposure.
   a. True
   b. False

5. Breathing asbestos for many years is known as an acute exposure.
   a. True
   b. False

6. Nausea from breathing solvent vapors is _____.
   a. grounds for an OSHA fine
   b. an acute effect of chemical exposure
   c. not mentioned on the MSDS
   d. no cause for concern

7. A NFPA label with a flammability rating of 3 would indicate that the product _____.
   a. will not burn
   b. must be preheated to burn
   c. ignites when moderately heated
   d. ignites at normal temperatures

Refer to *Figure 2* to answer Question 8.

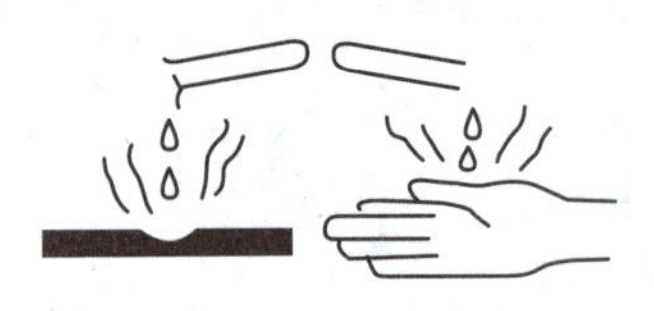

102E02.EPS

Figure 2

8. This symbol is found on containers that hold _____ materials.
   a. poisonous
   b. explosive
   c. radioactive
   d. corrosive

9. The information on an MSDS includes _____.
   a. cost and availability
   b. ecological properties
   c. local fire codes
   d. warranty limitations

10. The information on an MSDS is for _____.
    a. the safety officer
    b. anyone who uses the product
    c. chemical manufacturers
    d. the OSHA inspector

GLOSSARY

Trade Terms Introduced in This Module

Acute effect: An immediate reaction to chemical exposure, such as nausea or death.

Acute exposure: A short term, usually intense, exposure to a chemical.

Chronic effect: A delayed or long-term reaction to chemical exposure. Chronic effects can also be deadly, such as cancer.

Chronic exposure: A long-term, usually low-dose, exposure to a chemical.

Hazard Communication (HazCom) Standard: A federal OSHA regulation requiring employers to educate and inform workers about chemical hazards on the job site *(29 CFR 1910.1200)*.

Material safety data sheet (MSDS): A document that must accompany any hazardous material. The MSDS identifies the substance and gives the exposure limits, the physical and chemical characteristics, the kind of hazard it presents, precautions for safe handling and use, and specific control measures.

NIOSH Pocket Guide to Chemical Hazards
First Aid Procedures

Code	Definition
Eye: Irrigate immediately	If this chemical contacts the eyes, immediately wash the eyes with large amounts of water, occasionally lifting the lower and upper lids. Get medical attention immediately. Contact lenses should not be worn when working with this chemical.
Eye: Irrigate promptly	If this chemical contacts the eyes, promptly wash the eyes with large amounts of water, occasionally lifting the lower and upper lids. Get medical attention if any discomfort continues. Contact lenses should not be worn when working with this chemical.
Eye: Frostbite	If eye tissue is frozen, seek medical attention immediately; if tissue is not frozen, immediately and thoroughly flush the eyes with large amounts of water for at least 15 minutes, occasionally lifting the lower and upper eyelids. If irritation, pain, swelling, lacrimation, or photophobia persist, get medical attention as soon as possible.
Eye: Medical attention	Self-explanatory
Skin: Blot/brush away	If irritation occurs, gently blot or brush away excess.
Skin: Dust off solid; water flush	If this solid chemical contacts the skin, dust it off immediately and then flush the contaminated skin with water. If this chemical or liquids containing this chemical penetrate the clothing, promptly remove the clothing and flush the skin with water. Get medical attention immediately.
Skin: Frostbite	If frostbite has occurred, seek medical attention immediately; do NOT rub the affected areas or flush them with water. In order to prevent further tissue damage, do NOT attempt to remove frozen clothing from frostbitten areas. If frostbite has NOT occurred, immediately and thoroughly wash contaminated skin with soap and water.
Skin: Molten flush immediately/solid-liquid soap wash immediately	If this molten chemical contacts the skin, immediately flush the skin with large amounts of water. Get medical attention immediately. If this chemical (or liquids containing this chemical) contacts the skin, promptly wash the contaminated skin with soap and water. If this chemical or liquids containing this chemical penetrate the clothing, immediately remove the clothing and wash the skin with soap and water. If irritation persists after washing, get medical attention.
Skin: Soap flush immediately	If this chemical contacts the skin, immediately flush the contaminated skin with soap and water. If this chemical penetrates the clothing, immediately remove the clothing and flush the skin with water. If irritation persists after washing, get medical attention.
Skin: Soap flush promptly	If this chemical contacts the skin, promptly flush the contaminated skin with soap and water. If this chemical penetrates the clothing, promptly remove the clothing and flush the skin with water. If irritation persists after washing, get medical attention.
Skin: Soap promptly/molten flush immediately	If this solid chemical or a liquid containing this chemical contacts the skin, promptly wash the contaminated skin with soap and water. If irritation persists after washing, get medical attention. If this molten chemical contacts the skin or nonimpervious clothing, immediately flush the affected area with large amounts of water to remove heat. Get medical attention immediately.

102A01.EPS

Skin: Soap wash	If this chemical contacts the skin, wash the contaminated skin with soap and water.
Skin: Soap wash immediately	If this chemical contacts the skin, immediately wash the contaminated skin with soap and water. If this chemical penetrates the clothing, immediately remove the clothing, wash the skin with soap and water, and get medical attention promptly.
Skin: Soap wash promptly	If this chemical contacts the skin, promptly wash the contaminated skin with soap and water. If this chemical penetrates the clothing, promptly remove the clothing and wash the skin with soap and water. Get medical attention promptly.
Skin: Water flush	If this chemical contacts the skin, flush the contaminated skin with water. Where there is evidence of skin irritation, get medical attention.
Skin: Water flush immediately	If this chemical contacts the skin, immediately flush the contaminated skin with water. If this chemical penetrates the clothing, immediately remove the clothing and flush the skin with water. Get medical attention promptly.
Skin: Water flush promptly	If this chemical contacts the skin, flush the contaminated skin with water promptly. If this chemical penetrates the clothing, immediately remove the clothing and flush the skin with water promptly. If irritation persists after washing, get medical attention.
Skin: Water wash	If this chemical contacts the skin, wash the contaminated skin with water.
Skin: Water wash immediately	If this chemical contacts the skin, immediately wash the contaminated skin with water. If this chemical penetrates the clothing, immediately remove the clothing and wash the skin with water. If symptoms occur after washing, get medical attention immediately.
Skin: Water wash promptly	If this chemical contacts the skin, promptly wash the contaminated skin with water. If this chemical penetrates the clothing, promptly remove the clothing and wash the skin with water. If irritation persists after washing, get medical attention.
Breath: Respiratory support	If a person breathes large amounts of this chemical, move the exposed person to fresh air at once. If breathing has stopped, perform mouth-to-mouth resuscitation. Keep the affected person warm and at rest. Get medical attention as soon as possible.
Breath: Fresh air	If a person breathes large amounts of this chemical, move the exposed person to fresh air at once. Other measures are usually unnecessary.
Breath: Fresh air, 100% O_2	If a person breathes large amounts of this chemical, move the exposed person to fresh air at once. If breathing has stopped, perform artificial respiration. When breathing is difficult, properly trained personnel may assist the affected person by administering 100% oxygen. Keep the affected person warm and at rest. Get medical attention as soon as possible.
Swallow: Medical attention immediately	If this chemical has been swallowed, get medical attention immediately.

102A02.EPS

Figure Credits

North Safety Products	102F04 (A) through (C)
Becki Swinehart	102F04 (D)
Bacou-Dalloz	102F05 (A)
TNT Safety-Quip	102F05 (B)
DAP, Inc.	102F08
The National Institute for Occupational Safety and Health (NIOSH)	Appendix

NCCER CURRICULA — USER UPDATE

NCCER makes every effort to keep its textbooks up-to-date and free of technical errors. We appreciate your help in this process. If you find an error, a typographical mistake, or an inaccuracy in NCCER's curricula, please fill out this form (or a photocopy), or complete the online form at **www.nccer.org/olf**. Be sure to include the exact module ID number, page number, a detailed description, and your recommended correction. Your input will be brought to the attention of the Authoring Team. Thank you for your assistance.

Instructors – If you have an idea for improving this textbook, or have found that additional materials were necessary to teach this module effectively, please let us know so that we may present your suggestions to the Authoring Team.

NCCER Product Development and Revision

13614 Progress Blvd., Alachua, FL 32615

Email: curriculum@nccer.org
Online: www.nccer.org/olf

❑ Trainee Guide ❑ AIG ❑ Exam ❑ PowerPoints Other ____________________

Craft / Level: Copyright Date:

Module ID Number / Title:

Section Number(s):

Description:

Recommended Correction:

Your Name:

Address:

Email: Phone:

Module 75103-03

Personal Protective Equipment

COURSE MAP

This course map shows all of the modules in Field Safety. The suggested training order begins at the bottom and proceeds up. The local Training Program Sponsor may adjust the training order.

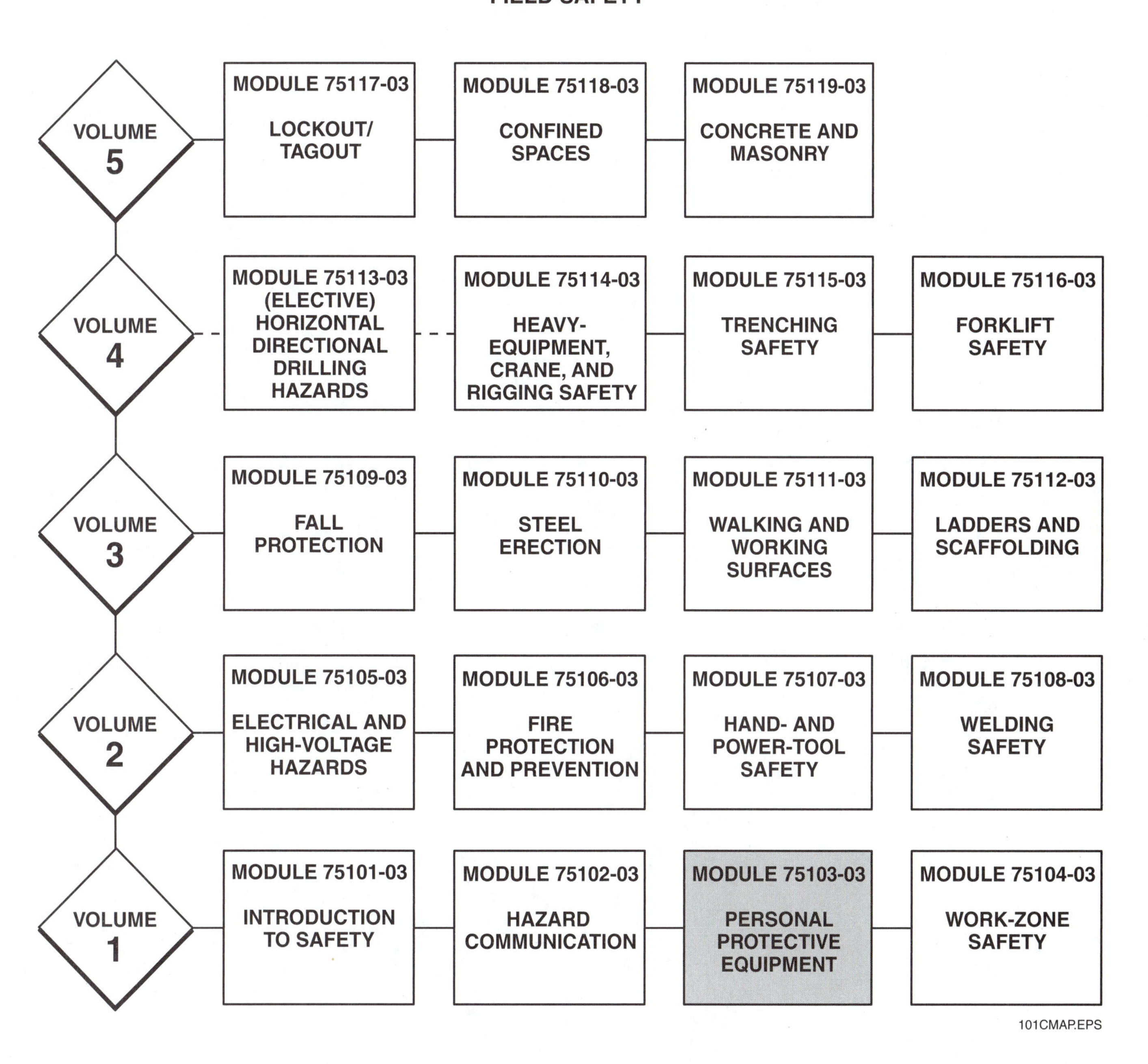

MODULE 75103-03 CONTENTS

Figures

Tables

MODULE 75103-03

Personal Protective Equipment

Objectives

When you have completed this module, you will be able to do the following:

1. Describe the common types of personal protective equipment (PPE) needed for working on a construction site.
2. Describe how to properly use and care for personal protective equipment (PPE).
3. Identify and describe the three main types of respirators used on construction sites.

Prerequisites

Before you begin this module, it is recommended that you successfully complete the following: Field Safety, Modules 75101-03 and 75102-03.

Required Materials

1. Pencil and paper
2. Appropriate personal protective equipment

1.0.0 ◆ INTRODUCTION

Personal protective equipment (PPE) is designed to protect you from injury. Many workers are injured on the job because they are not using personal protective equipment. In one incident, a worker suffered a serious eye injury after debris from the work site blew into his eyes. This worker had been told several times by his supervisor to wear safety glasses. Had he worn the proper safety glasses, this worker would not have been injured, nor would he have had to take time off from work to recover.

You won't see all the potentially dangerous conditions just by looking around a job site. It's important to stop and consider what type of accidents could happen before performing any job. Using common sense and knowing how to use personal protective equipment will greatly reduce your chances of getting hurt.

2.0.0 ◆ PERSONAL PROTECTIVE EQUIPMENT USE AND CARE

The best personal protective equipment is of no use to you unless you do four things:

- Regularly inspect it.
- Properly care for it.
- Use it properly when it is needed.
- Never alter or modify it in any way.

Protective equipment commonly used on construction sites includes the following:

- Hard hats
- Eye protection
- Safety harnesses
- Gloves
- Safety shoes
- Hearing protection
- Respiratory protection

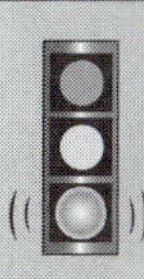

NOTE

Don't take chances with your safety. Use your personal protective equipment every time the job requires it.

2.1.0 Hard Hats

Figure 1 shows a typical hard hat. The outer shell of the hat can protect your head from a hard blow. The webbing inside the hat maintains a space between the shell and your head. Adjust the headband so that the webbing fits your head and there is at least 1" of space between your head and the shell. Do not alter your hard hat in any way.

Figure 1 ◆ Typical hard hat.

Hard hats used to be made of metal. However, metal conducts electricity, so most hard hats are now made of reinforced plastic or fiberglass.

Figure 2 ◆ Typical safety goggles and glasses.

2.2.0 Safety Glasses, Goggles, and Face Shields

Wear eye protection (*Figure 2*) whenever there is even the slightest chance of an eye injury. Areas where there are potential eye hazards from falling or flying objects are usually identified, but you should always be on the lookout for possible hazards.

Regular safety glasses will protect you from objects flying at you from the front such as large chips, particles, sand, or dust. You can add side shields for protection from the sides. In some cases, you may need a face shield. Safety goggles give your eyes the best protection from all directions.

Welders must use tinted goggles or welding hoods. Tinted lenses protect the eyes from the bright welding **arc** or flame.

Table 1 shows an eye and face protection selection chart.

No Training + No PPE = Death

A carpenter apprentice was killed when he was struck in the head by a nail that was fired from a powder-actuated tool in another room. The tool operator was attempting to anchor a plywood form in preparation for pouring a concrete wall. When he fired the gun, the nail passed through the hollow wall and traveled 27' before striking the victim. The tool operator had never received training in the proper use of the tool, and none of the employees in the area were wearing personal protective equipment.

The Bottom Line: You can be injured by the actions of others. Wear your PPE as a first line of defense against accidents.

Source: The Occupational Safety and Health Administration (OSHA)

Table 1 Eye and Face Protection Selection Chart

Eye and Face Protection Selection Chart		
Source	**Assessment of Hazard**	**Protection**
IMPACT - Chipping, grinding, machining, drilling, chiseling, riveting, sanding, etc.	Flying fragments, objects, large chips, particles, sand, dirt, etc.	Glasses with side protection, goggles, face shields. For severe exposure, use face shield over primary eye protection.
CHEMICALS - Acid and chemical handling	Splash	Goggles, eyecup and cover types. For severe exposure, use face shield over primary eye protection.
	Irritating mists	Special-purpose goggles.
DUST - Woodworking, buffing, general dusty conditions	Nuisance dust	Goggles, eyecup and cover types.
LIGHT and/or RADIATION		
Welding - electric arc	Optical radiation	Welding helmets or welding shields. Typical shades: 10-14.
Welding - gas	Optical radiation	Welding goggles or welding face shield. Typical shades: gas welding 4-8, cutting 3-6, brazing 3-4.
Cutting, torch brazing, torch soldering	Optical radiation	Spectacles or welding face shield. Typical shades: 1.5-3.
Glare	Poor vision	Spectacles with shaded or special-purpose lenses, as suitable.

103T01.TIF

2.3.0 Safety Harnesses

Safety harnesses (*Figure 3*), are heavy-duty harnesses that buckle around your body. They have leg, shoulder, chest, and pelvic straps.

Safety harnesses have a D-ring attached to one end of a short section of rope called a lanyard (*Figure 4*). The other end of the lanyard must be attached to a strong anchor point located above the work area. A qualified person will select the anchor point. The lanyard must be long enough to let you work but short enough to keep you from falling more than 6'.

Use a safety harness and lanyard when you are working in the following situations:

- More than 6' above ground or according to company policy
- Near a large opening in a floor
- Near a deep hole
- When there is a danger of falling on protruding rebar

Your life can depend on a secure safety harness. Carefully inspect the harness each time you use it. Check that the buckles and D-ring are not bent or deeply scratched. Check the harness for any cuts or rough spots. If you find any damage, turn in the harness for testing or replacement.

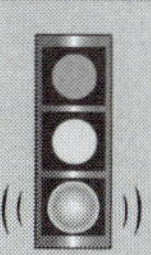

NOTE

For more information on fall protection, refer to the *Fall Protection* module in Volume Three.

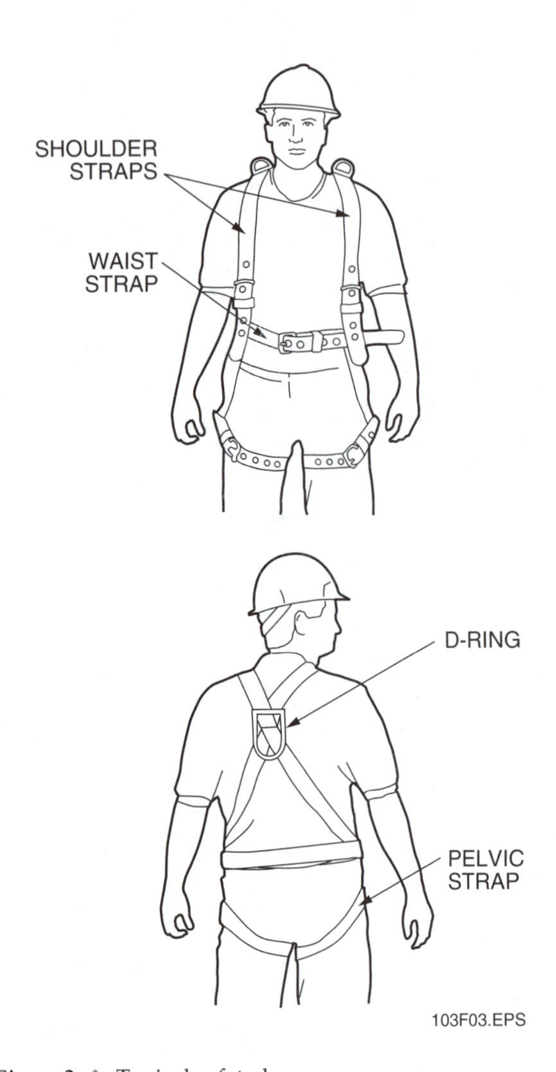

Figure 3 ◆ Typical safety harness.

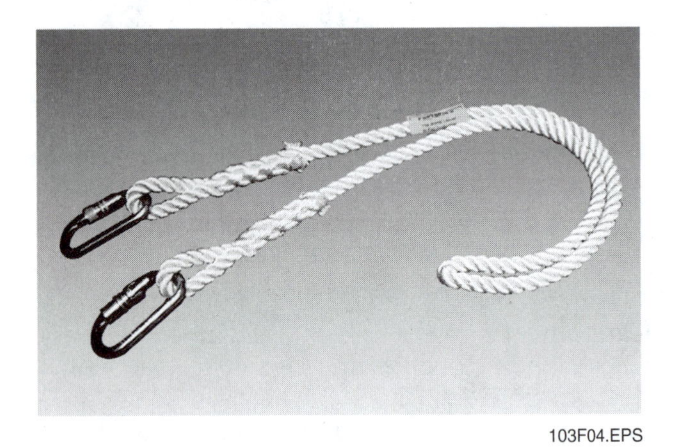

Figure 4 ◆ Lanyard.

2.4.0 Gloves

On many construction jobs, you must wear heavy-duty gloves to protect your hands (*Figure 5*). Construction work gloves are usually made of cloth, canvas, or leather. Never wear cloth gloves around rotating or moving equipment.

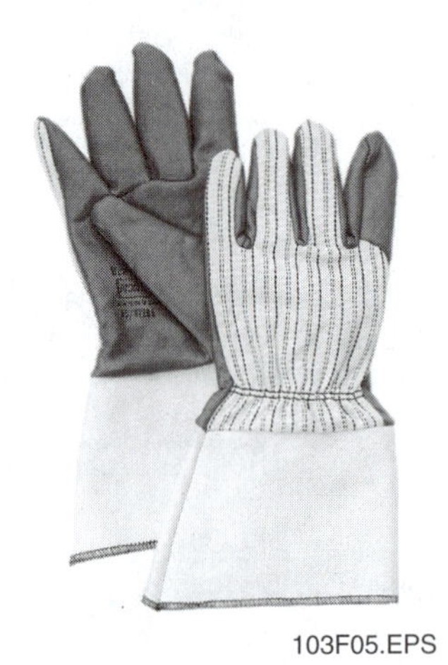

Figure 5 ◆ Work gloves.

Gloves help prevent cuts and scrapes when you handle sharp or rough materials. Heat-resistant gloves are sometimes used for handling hot materials. Electricians use special rubber-insulated gloves when they work on or around live circuits. Latex or other nonporous gloves protect your hands from chemicals, paints, or adhesives. Choose the right gloves for the job.

Inspect your gloves before putting them on. Replace gloves when they become worn, torn, or soaked with oil or chemicals.

Electrician's rubber-insulated gloves must be tested regularly to ensure that they will protect the wearer. After visually inspecting rubber-insulated gloves, other defects may be observed by applying the air test shown in *Figure 6*:

Step 1 Stretch the glove and look for any defects.

Step 2 Twirl the glove around quickly or roll it down from the glove gauntlet to trap air inside.

Step 3 Trap the air by squeezing the gauntlet with one hand. Use the other hand to squeeze the palm, fingers, and thumb to check for weaknesses and defects.

Step 4 Hold the glove up to your ear to try to detect any escaping air.

Step 5 If the glove does not pass this inspection, it must be turned in for disposal.

2.5.0 Safety Shoes

The best shoes to wear on a construction site are steel-toed, steel-soled safety shoes (*Figure 7*). The steel toe protects your toes from falling objects. The steel sole keeps nails and other sharp objects from puncturing your foot. The next best footwear material is heavy leather. Never wear canvas shoes or

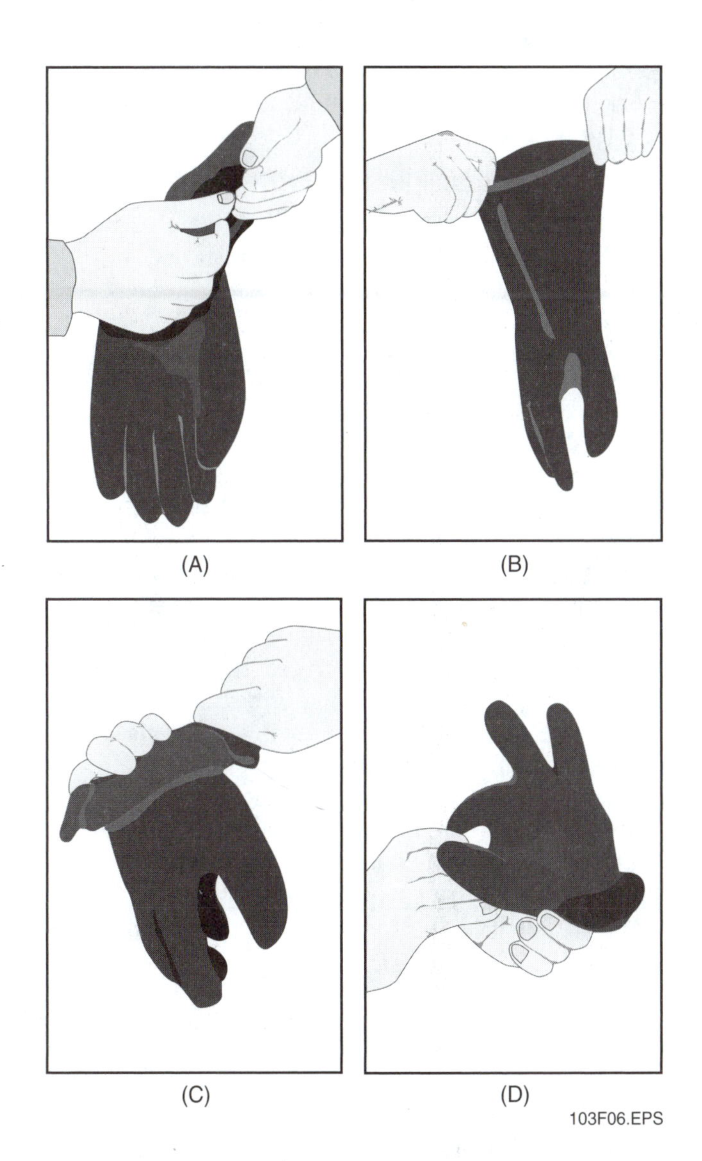

Figure 6 ◆ Glove inspection.

Figure 7 ◆ Safety shoes.

sandals on a construction site. Canvas shoes can catch fire easily and they do not protect you from sharp objects. Always replace boots or shoes when the sole tread becomes worn or the shoes have holes, even if the holes are on top. Don't wear oil-soaked shoes when you are welding due to the risk of fire.

Use Your Head and Your Hard Hat

Employees were dismantling grain spouts at a grain elevator. Sections of the spout were connected by collars. A 10' section of spout weighing 600 pounds was being pulled through a vent hole by a 5-ton winch. As the spout was being pulled through the opening, it became wedged at the point where the collar was to pass through. When the employees used pry bars to free the collar, the spout popped out of the vent. An employee standing beside the spout was struck in the head and killed. He was not wearing a hard hat.

The Bottom Line: Always wear your hard hat to protect your head from injuries. It could save your life.

Source: The Occupational Safety and Health Administration (OSHA)

2.6.0 Hearing Protection

Damage to most parts of the body causes pain. Ear damage, however, does not always cause pain. Exposure to loud noise over a long period of time can cause hearing loss, even if the noise is not loud enough to cause pain.

Most construction companies follow the Occupational Safety and Health Administration (OSHA) rules in deciding when hearing protection must be used. Specially designed earplugs that fit into your ears and filter out noise (*Figure 8*) are one type of hearing protection. You need to clean earplugs regularly with soap and water to prevent ear infections.

Another type of hearing protection is earmuffs, which are large, padded covers for the entire ear (*Figure 9*). You must adjust the headband on earmuffs for a snug fit. If the noise level is very high, you may need to wear both earplugs and earmuffs.

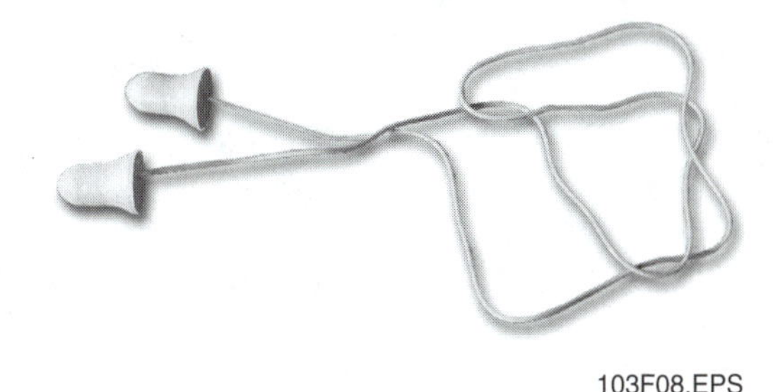

Figure 8 ◆ Ear plugs for hearing protection.

103F09.EPS

Figure 9 ◆ Ear muffs for hearing protection.

Noise-induced hearing loss can be prevented by limiting exposure and using appropriate personal protective equipment. *Table 2* shows the recommended duration of exposure to sound levels rated 90 decibels and higher.

Table 2 Maximum Noise Levels

Sound Level (decibels)	Maximum Hours of of Continuous Exposure per Day	Examples
90	8	Power lawn mower
92	6	Belt sander
95	4	Tractor
97	3	Hand drill
100	2	Chain saw
102	1.5	Impact wrench
105	1	Spray painter
110	.5	Power shovel
115	.25 or less	Hammer drill

2.7.0 Respiratory Protection

Wherever there is danger of suffocation or other breathing hazards, you must use a respirator. Federal law specifies which type of respirator to use for various breathing hazards. Respirators are grouped into three main types based on how they work to protect the wearer from contaminants. The types are:

- Air-purifying respirators
- Supplied-air respirators (SAR)
- Self-contained breathing **apparatus** (SCBA)

2.7.1 Air-Purifying Respirators

Air-purifying respirators provide the lowest level of protection. They are made for use only in atmospheres that have enough oxygen to sustain life (at least 19.5%). Air-purifying respirators use special filters and cartridges to remove specific gases, vapors, and particles from the air. The respirator cartridges contain charcoal, which absorbs certain toxic vapors and gases. When the wearer detects any taste or smell, it indicates the charcoal's absorption capacity has been reached. This means the cartridge can no longer remove the contaminants. The respirator filters remove particles such as dust, mists, and metal fumes by trapping them within the filter material. Filters should be changed when breathing becomes difficult.

Depending on the contaminants, cartridges can be used alone or in combination with a filter/prefilter and filter cover. Air-purifying respirators should be used for protection against only the types of contaminants listed on the filters and cartridges. Also refer to The National Institute for Occupational Safety and Health (NIOSH) approval label affixed to each respirator carton and replacement filter/cartridge carton. Respirator manufacturers typically classify air-purifying respirators into four groups:

- No maintenance
- Low maintenance
- Reusable
- Powered air-purifying respirators (PAPRs)

No-maintenance and low-maintenance respirators are typically used for residential or light commercial work that does not call for constant and

Be Prepared

An employee sitting in a looped chain was lowered approximately 17' into a 21'-deep manhole. Twenty seconds later he started gasping for air and fell from the chain seat face down into the accumulated water at the bottom of the manhole. An autopsy determined that oxygen deficiency was the cause of death. Air testing and a respirator could have saved his life.

The Bottom Line: You can't see a breathing hazard. Make sure the air is tested and you have the right respirator for the job.

Source: The Occupational Safety and Health Administration (OSHA)

heavy respirator use. No-maintenance respirators are typically half-mask respirators with permanently attached cartridges or filters. The entire respirator is discarded when the cartridges or filters are spent. Low-maintenance respirators are generally half-mask respirators that use replaceable cartridges and filters. However, they are not designed for constant use.

Reusable respirators (*Figure 10*) are made in half-mask and full-face piece styles. These respirators require the replacement of cartridges, filters, and respirator parts. Their use also requires a complete respirator maintenance program.

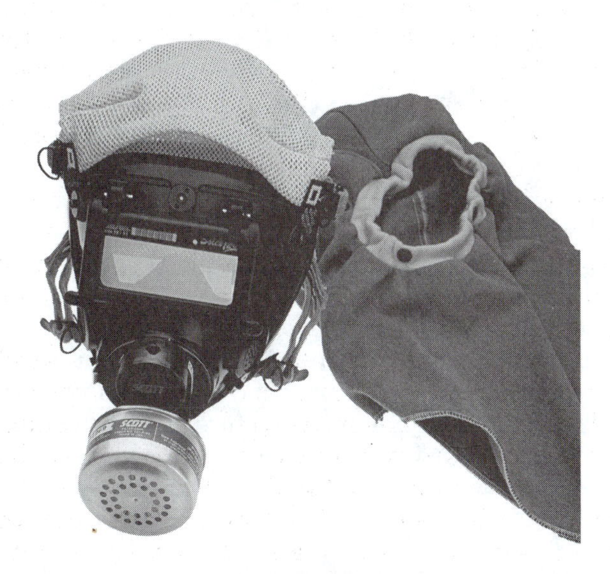

103F10.EPS

Figure 10 ◆ Reusable half-mask air-purifying respirator.

Powered air-purifying respirators (PAPRs) are made in half-mask, full-face piece, and hood styles. They use battery-operated blowers to pull outside air through the cartridges and filters attached to the respirator. The blower motors can be either mask- or belt-mounted. Depending on the cartridges used, they can filter particles, dust, fumes, and mists along with certain gases and vapors. PAPRs like the one shown in Figure 11 have a belt-mounted, powered air-purifier unit connected to the mask by a breathing tube. Many models also have an audible and visual alarm that is activated when the airflow falls below the required minimum level. This feature gives an immediate indication of a loaded filter or low battery charge condition. Units with the blower mounted in the mask do not use a belt-mounted powered air purifier connected to a breathing tube.

103F11.EPS

Figure 11 ◆ Powered air-purifying respirator.

2.7.2 Supplied-Air Respirators

Supplied-air respirators (*Figure 12*) provide a supply of air for extended periods of time through a high-pressure hose connected to an external source of air, such as a compressor, compressed-air cylinder, or pump. They provide a higher level of protection in atmospheres where air-purifying respirators are not adequate. Supplied-air respirators are typically used in toxic atmospheres. Some can be used in atmospheres that are immediately dangerous to life and health (IDLH) as long as they are equipped with an air cylinder for emergency escape. An atmosphere is considered IDLH if it poses an immediate hazard to life or produces immediate, irreversible, and debilitating effects on health. There are two types of supplied-air respirators: continuous flow and pressure demand.

A continuous-flow supplied-air respirator provides air to the user in a constant stream. One or two hoses are used to deliver the air from the air source to the face piece. Unless the compressor or pump is specially designed to filter the air, or a portable air-filtering system is used, the unit must be located where there is breathable air. Continuous-flow respirators are made with tight-fitting half-masks or full-face pieces. They are also made with hoods.

103F12.EPS

Figure 12 ◆ Supplied-air respirator.

The flow of air to the user may be adjusted either at the air source (fixed flow) or on the unit's regulator (adjustable flow). The pressure-demand supplied-air respirator is similar to the continuous-flow type except that it supplies air to the user's face piece via a pressure-demand valve as the user inhales and fresh air is required. It typically has a two-position exhalation valve that allows the worker to switch between the pressure-demand and negative-pressure modes to facilitate entry into, movement within, and exit from a work area.

2.7.3 Self-Contained Breathing Apparatus (SCBA)

SCBAs provide the highest level of respiratory protection. They can be used in oxygen-deficient atmospheres (below 19.5% oxygen), in poorly ventilated or **confined spaces**, and in IDLH atmospheres. These respirators provide a supply of air for about 30 to 60 minutes from a compressed-air cylinder worn on the user's back. An emergency escape breathing apparatus (EEBA) is a smaller version of a SCBA cylinder (*Figure 13*). EEBAs are used for escape from hazardous environments and generally provide a five- to ten-minute supply of air.

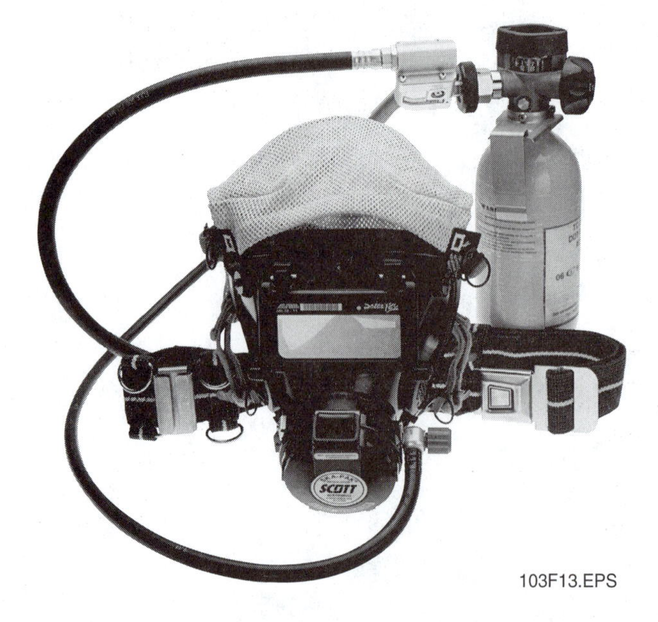

103F13.EPS

Figure 13 ◆ Emergency escape breathing apparatus.

2.8.0 Respiratory Program

A respirator must be properly selected for the contaminant present and its concentration level (*Figure 14*). It must be properly fitted and used in accordance with the manufacturer's instructions. It must be worn during all times of exposure.

BEFORE USING A RESPIRATOR YOU MUST DETERMINE THE FOLLOWING:

1. THE TYPE OF CONTAMINANT(S) FOR WHICH THE RESPIRATOR IS BEING SELECTED
2. THE CONCENTRATION LEVEL OF THE CONTAMINANT(S)
3. WHETHER THE RESPIRATOR CAN BE PROPERLY FITTED ON THE WEARER'S FACE

ALL RESPIRATOR INSTRUCTIONS, WARNINGS, AND USE LIMITATIONS CONTAINED ON EACH PACKAGE MUST ALSO BE READ AND UNDERSTOOD BY THE WEARER BEFORE USE.

103F14.EPS

Figure 14 ◆ Use the correct respirator.

Employers must have a respiratory protection program with the following components:

- Standard operating procedures for selection and use
- Employee training
- Regular cleaning and disinfecting
- Sanitary storage
- Regular inspection
- Annual fit testing
- Pulmonary function testing

As an employee, you are responsible for wearing respiratory protection when needed. In certain concentrations, vapors or fumes can be eliminated by the use of air-purifying devices as long as the oxygen levels are acceptable. Smoke billowing from a fire and the fumes generated when welding are examples of fumes. Always check the cartridge on your respirator to ensure it is the correct type for the air conditions and contaminants found on your job site.

When selecting a respirator to wear while working with specific materials, you must first determine the hazardous ingredients contained in the material and their exposure levels, and then choose the proper respirator to protect yourself at those levels. Always read the product's material safety data sheet (MSDS). It identifies the hazardous ingredients and should list the type of respirator and cartridge recommended for use with the product.

Limitations that apply to all half-mask (air-purifying) respirators are as follows:

- These respirators do not completely eliminate exposure to contaminants, but they will reduce the exposure to an acceptable level.
- These respirators do not supply oxygen and must not be used in areas where the oxygen level is below 19.5%.
- These respirators must not be used in areas where chemicals have poor warning signs, such as no taste or odor.

If breathing becomes difficult, if you become dizzy or nauseated, if you smell or taste the chemical, or if you have other noticeable effects of exposure, leave the area immediately, return to a fresh air area, and seek any necessary assistance.

2.8.1 Positive and Negative Fit Checks

Respirators are useless unless properly fit-tested to each individual. To obtain the best protection from your respirator, you must perform positive and negative fit checks each time you wear it. These fit checks must be repeated until you have obtained a good face seal.

To perform a positive fit check, do the following:

Step 1 Adjust the face piece for the best fit, then adjust the head and neck straps to ensure good fit and comfort.

> **WARNING!**
> Do not overtighten the head and neck straps. Tighten them only enough to stop leakage. Overtightening can cause face-piece distortion and dangerous leaks.

Step 2 Block the exhalation valve with your hand or other material.

Step 3 Breathe out into the mask.

Step 4 Check for air leakage around the edges of the face piece.

Step 5 If the face piece puffs out slightly for a few seconds, a good face seal has been obtained.

To perform a negative fit check, do the following:

Step 1 Block the inhalation valve with your hand or other material.

Step 2 Attempt to inhale.

Step 3 Check for air leakage around the edges of the face piece.

Step 4 If the face piece caves in slightly for a few seconds, a good face seal has been obtained.

A respirator must be clean, in good condition, and all of its parts must be in place for it to give you proper protection. Respirators must be cleaned every day. Failure to do so will limit their effectiveness and offer little or no protection. For example, suppose you wore the respirator for two weeks and did not clean it. The bacteria from breathing into the respirator, plus the airborne contaminants that managed to enter the face piece, have made the inside of your respirator very unsanitary. Continued use may cause you more harm than good. Remember, only a clean and complete respirator will provide you with the necessary protection. Follow these general guidelines:

- Inspect the condition of your respirator before and after each use.
- Do not wear a respirator if the face piece is distorted or if it is worn and cracked. You will not be able to get a proper face seal.

- Do not wear respirators if any part is missing. Replace worn straps or missing parts before use.
- Do not expose respirators to excessive heat or cold, chemicals, or sunlight.
- Clean and wash your respirator each day. Remove the cartridge and filter, hand wash the respirator using mild soap and a soft brush, and let it air dry overnight.
- Sanitize your respirator each week. Remove the cartridge and filter, and then soak the respirator in a sanitizing solution for at least two minutes. Thoroughly rinse with warm water and let it air dry overnight. Store the clean and sanitized respirator in its resealable plastic bag. Do not store the respirator face down. This will cause distortion of the face piece.

Summary

It's important to be aware of the hazards of your job and know how to protect yourself from them. Using personal protective equipment is an important part of overall job safety. Many accidents happen because workers are not properly protected from hazards. Hard hats, eye and face protection, body harnesses, safety shoes and gloves, hearing protection, and respiratory protection must be used to help avoid accidents and injuries on a work site.

Respiratory protection is a particularly important type of personal protective equipment. It protects you from suffocation and other breathing hazards. It's important to know how to use a respirator and to use it properly every time.

Review Questions

1. When using a hard hat, there should be _____ of space between your head and the shell of the hard hat.

a. ¼"
b. ½"
c. 1"
d. 2"

2. Safety glasses give your eyes protection from all directions.

a. True
b. False

3. Safety harnesses have all of the following types of straps *except* _____.

a. chest
b. leg
c. neck
d. pelvic

4. A lanyard should keep you from falling more than _____.

a. 6'
b. 8'
c. 10'
d. 12'

5. It is safe to wear cloth gloves around rotating and moving equipment.

a. True
b. False

6. The best shoes to wear on a construction site are _____.

a. canvas shoes
b. heavy leather shoes
c. rubber-soled sneakers
d. steel-toed shoes

7. If the noise level on a construction site is very high, workers should wear both ear plugs and ear muffs.

a. True
b. False

8. Continuous exposure to sound levels of 102 decibels, such as the noise of an impact wrench, should last no more than _____ hour(s).

a. 0.25
b. 1.5
c. 2.5
d. 4

9. The highest level of respiratory protection is provided by _____.

a. air-purifying respirators
b. supplied-air respirators (SAR)
c. self-contained breathing apparatus (SCBA)
d. exhaust hoods

10. It is only necessary to perform a positive and negative fit check the first time you wear a respirator.

a. True
b. False

GLOSSARY

Trade Terms Introduced in This Module

Apparatus: An assembly of machines used together to perform a particular job.

Arc: The flow of electrical current from one point to another through a gas such as air.

Confined space: An area that is large enough for a person to work in, but with limited ways of entering and exiting, and that is not meant for continuous human occupancy.

Figure Credits

Bullard Classic Head Protection	103F01
Bacou-Dalloz	103F02
Protecta International, Inc.	103F04
North Safety Products	103F05
Milwaukee Electric Tool Company	103F07
Charles Rogers	103F08, 103F09
Scott Health and Safety	103F10, 103F11, 103F13
Hornell, Inc.	103F12

NCCER CURRICULA — USER UPDATE

NCCER makes every effort to keep its textbooks up-to-date and free of technical errors. We appreciate your help in this process. If you find an error, a typographical mistake, or an inaccuracy in NCCER's curricula, please fill out this form (or a photocopy), or complete the on-line form at **www.nccer.org/olf**. Be sure to include the exact module ID number, page number, a detailed description, and your recommended correction. Your input will be brought to the attention of the Authoring Team. Thank you for your assistance.

Instructors – If you have an idea for improving this textbook, or have found that additional materials were necessary to teach this module effectively, please let us know so that we may present your suggestions to the Authoring Team.

NCCER Product Development and Revision
13614 Progress Blvd., Alachua, FL 32615

Email: curriculum@nccer.org
Online: www.nccer.org/olf

❑ Trainee Guide ❑ AIG ❑ Exam ❑ PowerPoints Other ______________________________

Craft / Level: Copyright Date:

Module ID Number / Title:

Section Number(s):

Description:

Recommended Correction:

Your Name:

Address:

Email: Phone:

Module 75104-03

Work-Zone Safety

COURSE MAP

This course map shows all of the modules in Field Safety. The suggested training order begins at the bottom and proceeds up. The local Training Program Sponsor may adjust the training order.

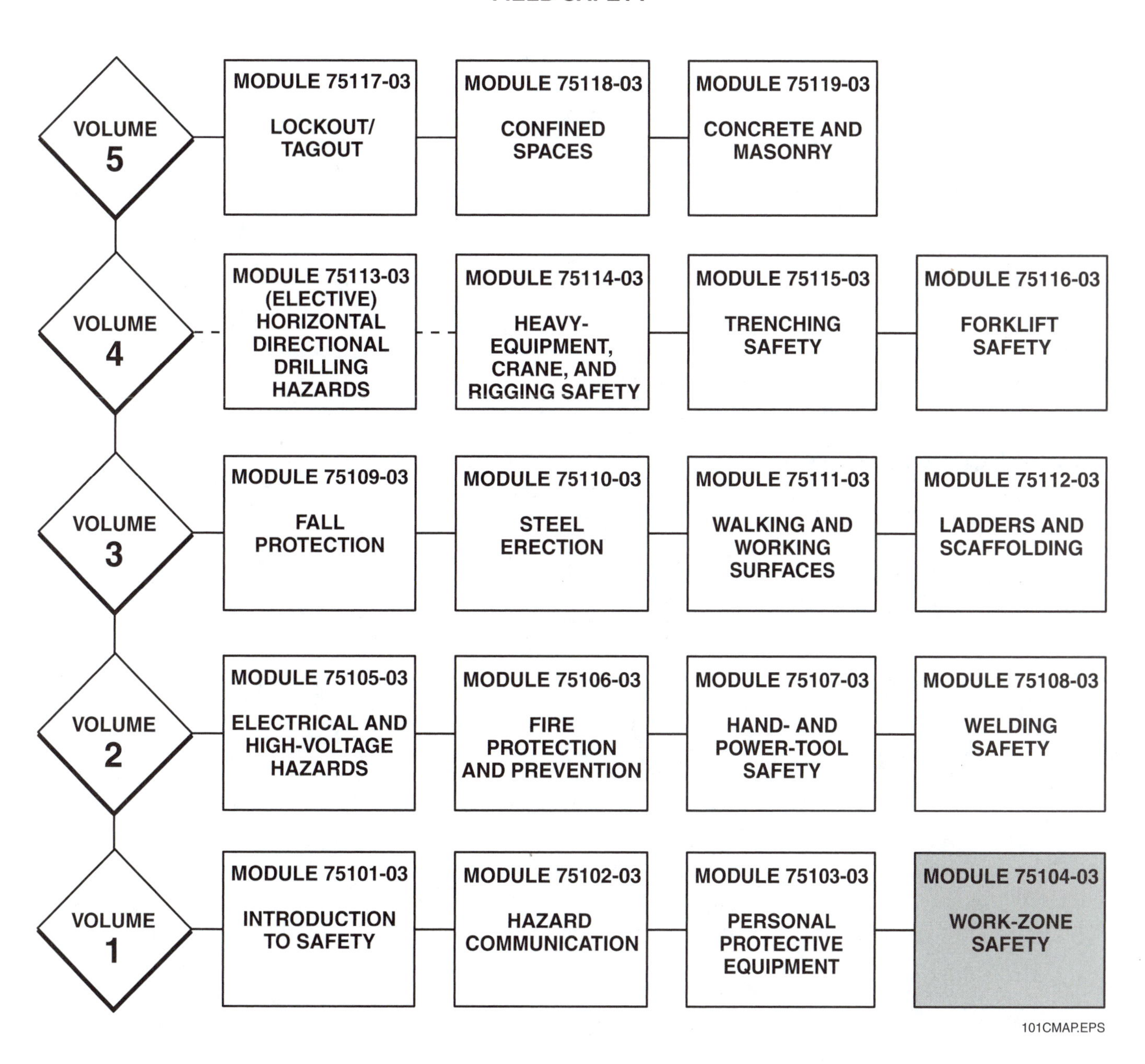

MODULE 75104-03 CONTENTS

Figures

MODULE 75104-03

Work-Zone Safety

Objectives

When you have completed this module, you will be able to do the following:

1. Identify signs, signals, and barricades that will help you perform your job safely.
2. Identify the hazards and safeguards of working in a highway work zone.

Prerequisites

Before you begin this module, it is recommended that you successfully complete the following: Field Safety, Modules 75101-03 through 75103-03.

Required Materials

1. Pencil and paper
2. Appropriate personal protective equipment

1.0.0 ◆ INTRODUCTION

Construction work is often done in or near public areas. Creating a clear work zone is an important part of working safely. Barricades, fencing, caution tape, signs, and cones are used to mark a construction work zone. To ensure everyone's safety, you must keep people and vehicles away from your work area.

Signs, tags, and color codes are used in the workplace to protect employees from hazardous conditions, and to assist them in responding to emergencies (*Figure 1*). For signs, tags, and color codes to be effective, all workers must understand what they mean and know what action they are required to take. This helps to avoid confusion and ensures their effectiveness.

104F01.EPS

Figure 1 ◆ Communication tags and signs.

Signals such as alarms, bells, buzzers, whistles, and horns can also be used to communicate hazards to workers. For example, back-up alarms are used on forklifts, construction equipment, and trucks. Fire alarms are used to clear work areas. Conveyer belt lines have buzzers, bells, and/or whistles to let workers know they are about to be started.

Barricades are another way to warn of potential danger. They are used on construction sites to keep out unauthorized personnel and control traffic (*Figure 2*). Barricades are also used to control pedestrian traffic outside or walking traffic in rooms and hallways that have been recently washed or waxed.

104F02.EPS

Figure 2 ◆ Typical uses of barricades.

It's important to recognize all of the signs and signals related to your job site and make sure they are properly placed and working correctly. Doing this can save your life.

2.0.0 ◆ SIGNS

All work sites have specific markings and signs to identify hazards and provide emergency information (*Figure 3*). Learn to recognize these types of signs:

- Danger signs
- Caution signs
- Informational signs
- Safety signs
- Safety tags

104F03.EPS

Figure 3 ◆ Examples of signs.

2.1.0 Danger Signs

Danger signs are usually red, black, and white. They are used to inform workers that an immediate hazard exists and specific precautions must be observed to avoid an accident (*Figure 4*). Examples include:

- DANGER – HIGH VOLTAGE
- DANGER – NO SMOKING, MATCHES, OR OPEN LIGHTS
- DANGER – KEEP AWAY

104F04.EPS

Figure 4 ◆ Typical danger sign.

Some of the dangers you may experience on a site are radiation and biological hazards. Be on the lookout for these types of signs as well.

2.1.1 Radiation Signs

Radiation signs are used in the workplace to alert workers to radiation hazards (*Figure 5*). The radiation hazard sign contains the word RADIATION as well as the radiation symbol. The sign's background is yellow and the panel is a reddish-purple color with yellow letters. Any additional lettering used against the yellow background is black.

104F05.EPS

Figure 5 ◆ Example of a radiation sign.

2.1.2 Biological Hazard Signs

Biological hazard signs are used to warn workers of the actual or potential presence of a biological hazard (*Figure 6*). Biological hazards, or biohazards, can be any infectious agents that create a real or potential health risk. Biological hazard signs are commonly used to identify contaminated equipment, containers, rooms, materials, and areas housing experimental animals.

The biohazard symbol used on the sign is fluorescent orange or orange-red in color. Background colors may vary, but must contrast enough for the symbol to be easily identified. Wording is used on the sign to indicate the nature of the hazard or to identify the specific hazard.

104F06.EPS

Figure 6 ◆ Biological hazard sign.

2.2.0 Caution Signs

Caution signs are used to inform workers about potential hazards or unsafe practices (*Figure 7*). When you see a caution sign, take action to protect yourself. Caution signs are yellow with black letters. Examples include:

- CAUTION – DO NOT OPERATE
- CAUTION – KEEP AISLES CLEAR
- CAUTION – ELECTRIC FENCE

104F07.EPS

Figure 7 ◆ Typical caution sign.

Yellow is the basic color used for caution. It is used to identify places where physical hazards may be caused by striking against objects, stumbling, falling, tripping, and being caught between obstacles. Solid yellow, yellow and black stripes, or yellow and black checkers caution workers against these hazards.

Caution signs and piping systems that contain dangerous materials are also yellow. Yellow is used to warn workers against starting machinery under repair. Painted barriers and flags should be located at the starting point or power source. They should be displayed so that workers will notice them easily on such things as electrical controls, ladders, scaffolds, vaults, valves, dryers, boilers, elevators, and tanks.

2.3.0 Informational Signs

Informational signs are used where it is necessary to provide general information that is not related to safety (*Figure 8*). The standard color used is blue. The background, the entire sign, or just a panel may be blue. Examples include:

- NO ADMITTANCE
- NO TRESPASSING
- FOR EMPLOYEES ONLY

104F08.EPS

Figure 8 ◆ Common informational sign.

2.3.1 Traffic and Housekeeping Signs

Black and white informational signs are used as traffic and housekeeping markings. These signs identify things such as:

- Dead ends of aisles or passageways
- Location of trash cans
- Location and width of aisle-ways
- Rooms or passageways
- Stairways (risers, direction, borders)
- Drinking fountains and food-dispensing machines

2.3.2 Directional Signs

Directional signs help workers and visitors find locations such as restrooms, stairways, and locker rooms. Standard colors for directional signs are black and white.

Exit and fire extinguisher signs are specialized directional signs that provide vital safety information and must be distinctive in color. They are red and white, as shown in *Figure 9*.

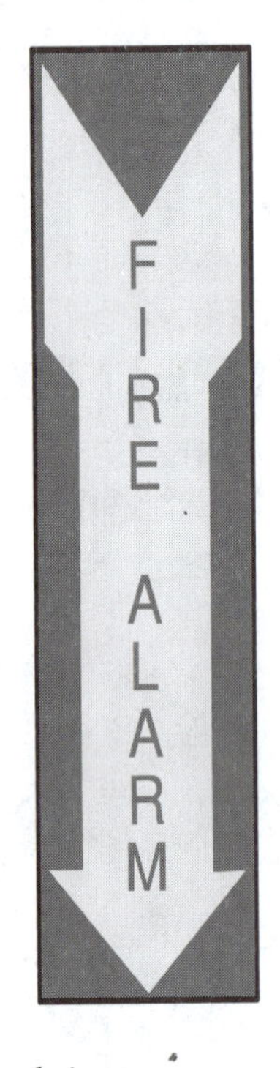

104F09.EPS

Figure 9 ◆ Directional signs.

2.3.3 On-Site Traffic Signs

On-site traffic signs help in the safe movement of vehicles and pedestrians. Just as traffic signs must be obeyed on public highways and streets, they must also be observed in the workplace.

A slow-moving vehicle emblem is used on vehicles that move at speeds of 25 miles per hour (mph) or less. The emblem is a fluorescent yellow-orange triangle with a dark red reflective border (*Figure 10*). It is frequently seen on construction and farm equipment.

104F10.EPS

Figure 10 ◆ Slow-moving vehicle sign.

2.4.0 Safety Signs

Safety instruction signs are used when there is a need for general instructions and suggestions related to safety measures (*Figure 11*). The background and lettering on these signs are white and green, but can vary depending on the message and the location of the sign. Any letters used against the white background are black. Examples include:

- REPORT ALL UNSAFE CONDITIONS TO YOUR SUPERVISOR
- WALK, DON'T RUN
- HELP KEEP THIS PLANT SAFE AND CLEAN

2.5.0 Safety Tags

Accident-prevention tags are used as a temporary way of warning workers about immediate and potential hazards (*Figure 12*). They are similar to signs; however, they are not designed to be used in place of signs or as a permanent means of protection. For example, an Out of Order tag may be used on damaged equipment until it can be disposed of or repaired. A Do Not Start tag may be placed on machinery during lockout procedures. Tags can be an effective means of protecting workers and property. Tags and the devices used to attach them must meet specific physical requirements to ensure their durability and effectiveness.

104F11.EPS

Figure 11 ◆ Common safety sign.

2.5.1 Do Not Start Tags

Do Not Start tags are used to prevent the unexpected energizing of equipment that could result in injury, equipment damage, or both. The tags must be placed in a **conspicuous location** that effectively blocks the starting mechanism. The tag is white. The square panel is red and the letters may be white, grey, or visibly etched on the tag.

2.5.2 Danger Tags

Danger tags are used where an immediate hazard exists. They tell workers that specific precautions must be observed. Danger tags are white. White letters appear on a red oval in a black square. A danger tag may read DANGER – UNSAFE – DO NOT USE.

2.5.3 Caution Tags

Caution tags are used to warn workers of potential hazards and to caution workers against unsafe practices. These tags inform workers that proper precautions should be taken. For example, a tag may read CAUTION – STOP MACHINERY TO CLEAN, OIL, OR REPAIR. Caution tags are yellow. They contain a black square panel with yellow letters.

2.5.4 Out of Order Tags

Out of Order tags should only be used to notify workers that a particular machine or tool is out of order and that using it may present a hazard. Out of Order tags are white. They contain a black square with white letters.

2.5.5 Radiation Tags

Radiation tags are yellow with a reddish-purple panel. Any letters used against the yellow background are black. The radiation symbol should also appear on the tag. Radiation tags notify workers of actual or potential radiation hazards.

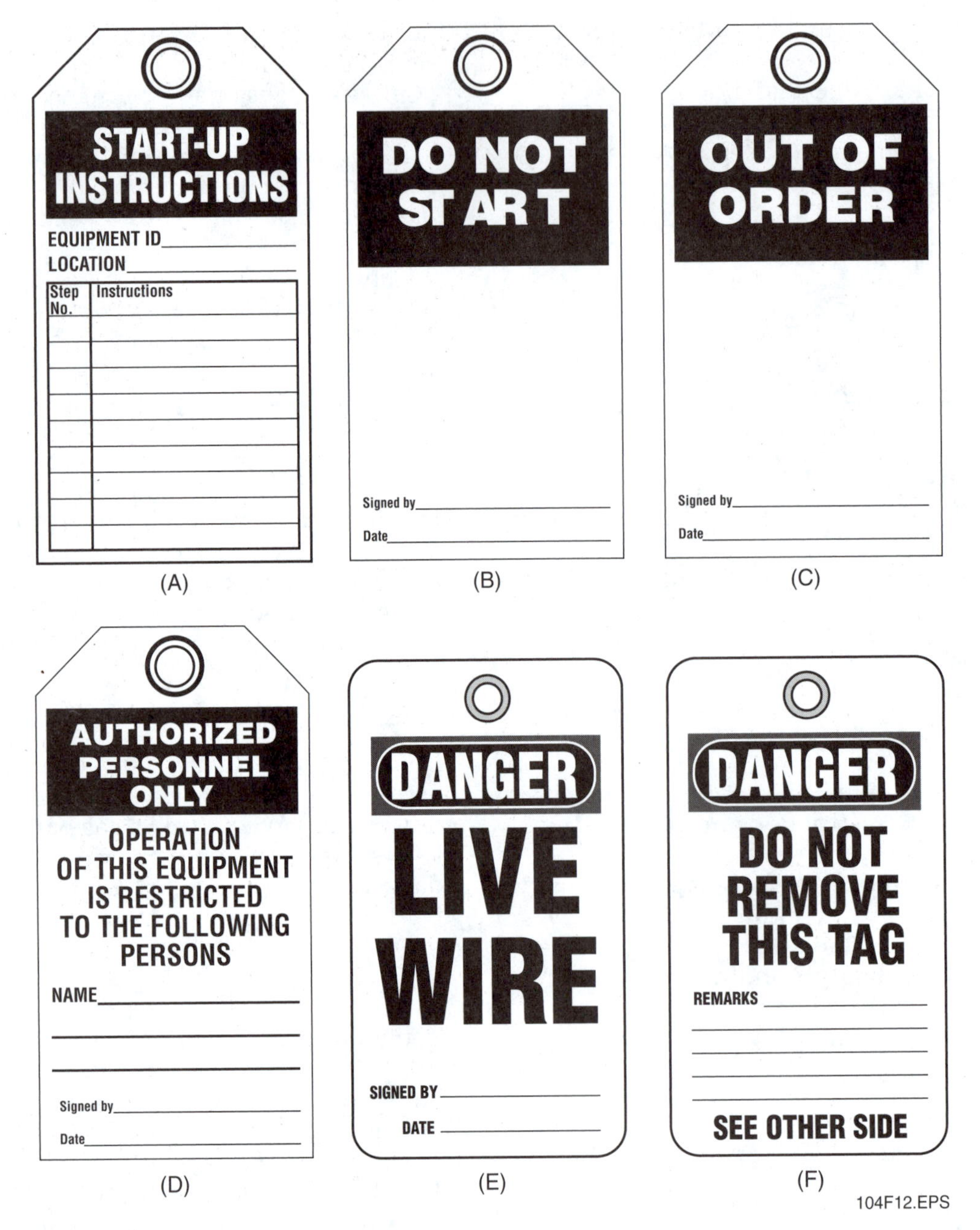

Figure 12 ◆ Examples of safety tags.

2.5.6 Biological Hazard Tags

Biological Hazard tags signify the presence of an actual or potential biological hazard. Biological hazards can include any infectious agents that present a risk or potential risk to workers. Biological hazard tags are usually white or fluorescent orange with black letters. The tag also shows the biological hazard or biohazard symbol (refer to *Figure 6*).

3.0.0 ◆ SIGNALS

Signals are used to inform workers of potential dangers. There are several types of signals that can be used. They are:

- Alarms
- Bells
- Buzzers
- Whistles
- Horns
- Hand signals

Hand signals (*Figure 13*) control vehicle traffic, handling of materials, and assisting equipment operators. It is essential that all affected workers know what each hand signal means before it is used. The meaning should be confirmed between the equipment operator and spotter or person giving the operator the signals before a task is started.

Figure 13 ◆ Hand signals used during the operation of cranes and other vehicles used to lift loads.

Are You Getting the Signal?

A 45-year-old supervisor for an excavation construction company was killed when a bulldozer slid sideways as he drove it onto a flatbed trailer. This happened because after one unsuccessful attempt to drive the bulldozer onto the platform, he decided to drive up the ramps instead. As he did this, the tracks of the machine began to slide sideways towards the ditch, and it fell off the trailer. He tried to jump out of the cab, but was pinned beneath the cab structure when it landed on top of him. Air bags and blocks were used to elevate the bulldozer and remove the victim's body. A spotter was not used to direct the bulldozer during this incident.

The Bottom Line: Always arrange for a qualified observer (spotter) to provide verbal or visual directions when loading track equipment onto a trailer.

Source: The National Institute for Occupational Safety and Health (NIOSH)

4.0.0 ◆ BARRICADES AND BARRIERS

Any opening in a wall, floor, or ground is a safety hazard. There are two types of protection for these openings: they can be guarded or they can be covered. Cover any hole whenever possible. When it is not practical to cover a hole, use barricades. If the bottom edge of a wall opening is less than 3' above the floor and would allow someone to fall 4' or more, place guards around the opening. There are several different guard methods:

- Railings are used across wall openings or as barriers around floor openings to prevent falls. See *Figure 14(A)*.
- Warning barricades alert workers to hazards but provide no real protection. See *Figure 14(B)*. Typical warning barricades are made of plastic tape or rope strung from wire or between posts.

The tape or rope is color-coded:

- Red means danger. No one can enter an area with a red warning barricade. A red barricade is used when there is danger from falling objects or when a load is suspended over an area.
- Yellow means caution. You can enter an area with a yellow barricade, but you must know what the hazard is, and be careful. Yellow barricades are used around wet areas or areas containing loose dust. Yellow with black lettering warns of physical hazards such as bumping into something, stumbling, or falling.
- Yellow and purple indicates a radiation warning. No one may pass a yellow and purple barricade. These barricades are often used where piping welds are being X-rayed.

Be Alert

A 23-year-old apprentice lineman died of injuries he received after being run over by the tandem dual rear tires of a digger derrick truck. He was part of a five-person crew that had been working on the job for three days. At the time of the incident, the victim and two other crew members had just finished setting and back-filling around a utility pole. They proceeded to the next pole requiring framing and setting, walking 30' ahead of the digger derrick in one lane of the road. The digger derrick moved slowly in reverse to the same pole. At a point approximately midway between the two poles, the victim knelt with his back to the truck to apparently inscribe a word or initials into some seal coating on the roadway. He was hit and run over by the backing truck's passenger-side tandem dual rear tires. The digger derrick truck did not have a back-up alarm system.

The Bottom Line: Keep your mind on the job. Communication with other workers can save your life. Check equipment for warning signals before you use it. If it does not have warning signals, set hand or radio signals.

Source: The National Institute for Occupational Safety and Health (NIOSH)

(A)

(B)

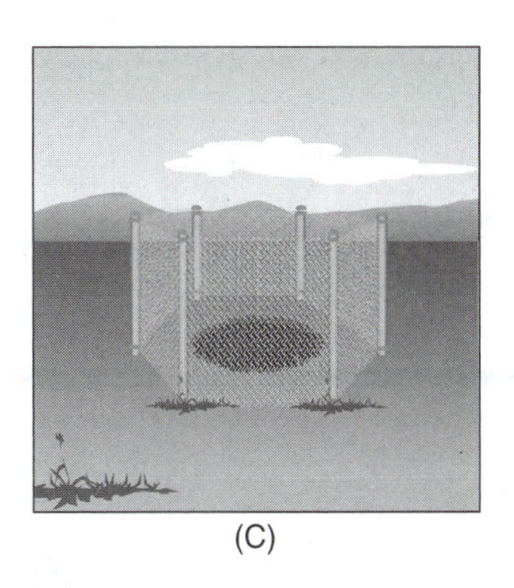

(C)

(D)

(E)

104F14.EPS

Figure 14 ◆ Common types of barricades and barriers.

- Protective barricades provide both a visual warning and protection from injury. See *Figure 14(C)*. They can be wooden posts and rails, posts and chain, or steel cable. People cannot get past protective barricades.
- Blinking lights are placed on barricades so they can be seen at night. See *Figure 14(D)*.
- Hole covers are used to cover open holes in the floor or ground. See *Figure 14(E)*. They must be secured and labeled. They must be strong enough to support twice the weight of anything that may be placed on top of them.

The types of barriers and barricades you see will vary between job sites. There may also be different procedures for when and how barricades are put up. Learn and follow the policies at your job site.

5.0.0 ◆ HIGHWAY WORK-ZONE SAFETY

Construction contractors, contracting agencies, and others responsible for work-zone safety face the challenge of providing a safe workplace while ensuring the safe movement of the public through

Stay Alert

An 18-year-old flagger, outfitted in full reflective vest, pants, and hard hat, was directing traffic at one end of a bridge approach during a night milling operation. The work zone was correctly marked with cones and signs, and the entire bridge was lit with street lights. The flagger was standing under portable flood lights in the opposing traffic lane close to the center line, facing oncoming traffic. A pickup truck traveling in the wrong lane at an estimated speed of 55 to 60 miles per hour struck the flagger head on and carried him approximately 200'. He died of multiple traumatic injuries at the scene.

The Bottom Line: Highway construction is very dangerous. Even if you are following all of the safety rules, you are still at risk. Stay alert.

Source: The National Institute for Occupational Safety and Health (NIOSH)

Work Inside the Zone

A 41-year-old electrician died after he fell from the basket of an aerial lift truck. He and two co-workers were installing electrical wiring for a sign on the side of a highway bridge. The two left-most traffic lanes were closed to traffic. The workers arrived at the scene with a pickup truck and a truck equipped with a basket-type work platform attached to a hydraulic boom. At this time the basket was positioned above the closed traffic lanes. While the co-workers cut the pipe, the victim moved the basket to a location above the nearest open traffic lane. A tractor trailer approached the site and observed the warning cones and signs directing traffic to merge to the right. The driver proceeded toward the work site in the open traffic lane nearest the closed lanes. When the truck neared the work area, the driver noticed the basket over the traffic lane in which he was driving. Before the driver could slow down or safely move to the next lane, the tractor trailer struck the basket. The collision caused the victim to be thrown from the basket to the surface of the roadway. This worker died because he chose to work outside of the established work safety zone.

The Bottom Line: Working outside the safety zone can be fatal.

Source: The National Institute for Occupational Safety and Health (NIOSH)

the work zone. Highway and street construction present complex work situations in which workers face multiple injury risks under conditions that may change without warning.

Highway and street construction workers are at risk of fatal and serious nonfatal injury when working in the vicinity of passing motorists, construction vehicles, and equipment. Each year, more than 100 workers are killed and over 20,000 are injured in the highway and street construction industry. Vehicles and equipment operating in and around the work zone are involved in over half of the worker fatalities in this industry.

The most common sources of injury in highway work zones are trucks, road grading and surfacing machinery, and cars. In one case, a 27-year-old male worker died from injuries received when a speeding car entered the highway work zone and smashed the rear of the truck he was working on.

It is important to know how to stay safe in a consistently unsafe environment. The following guidelines can help save your life:

- Ensure that all of the appropriate signs (*Figure 15*), warning devices, barriers, and barricades (*Figure 16*) are in place.
- If you are a flagger, be especially aware of vehicle traffic.
- Always wear a high-visibility, reflective vest (*Figure 17*) and protective clothing including hats, armbands, and boots.
- Know and use the appropriate hand signals.
- Make sure you have visual contact with anyone using heavy machinery.

104F15.EPS

Figure 15 ◆ Typical highway construction sign.

5.1.0 Traffic Control

Highway construction workers are responsible for knowing the local traffic regulations, as well as the safety practices of their employer. Traffic control devices such as signs, signals, and markings are used to regulate or guide traffic near highway construction sites.

104F16.EPS

Figure 16 ◆ Barricades used in road construction.

104F17.EPS

Figure 17 ◆ Typical high-visibility, reflective vest.

5.1.1 Signs

Traffic control zone signs provide both general and specific messages. They can be as simple as a stop sign or as complicated as an electronic sign that provides detour directions. They are especially important in a highway work zone where there is danger from both the road traffic and the heavy equipment being used.

5.1.2 Flaggers

Workers who provide traffic control are called flaggers. Flaggers are responsible for the safety of both the workers on the site and the public. Flaggers wear high-visibility clothing and use hand-signaling devices, such as STOP/SLOW paddles, lights, and red flags to control road users through traffic control zones. Flaggers must maintain a safe distance from both road traffic and the heavy equipment used on the site.

5.2.0 Moving Heavy Highway Equipment

Highway construction workers operate many different types and sizes of heavy equipment such as dump trucks, backhoes, and graders that include very large bulldozers and excavators. Highway construction workers are needed for projects involving highway construction, bridge construction, and highway maintenance.

Safety Gear for Flaggers

For daytime work, the flagger's vest, shirt, or jacket shall be either orange, yellow, yellow-green, or a fluorescent version of these colors.

For nighttime work, the vest, shirt, or jacket must be reflective. The reflective material must be either orange, yellow, white, silver, yellow-green, or a fluorescent version of these colors.

Source: Department of Transportation (DOT), Manual on Uniform Traffic Control Devices 2001

Wear Your Seatbelt, and Watch Where You're Going

A 30-year-old construction laborer died at a highway construction site when the vehicle he was driving collided with an earth-moving vehicle called a scraper. No seatbelt restraint system was installed in the truck. At approximately 2:30 p.m., the victim was driving the truck south at an estimated 45 miles per hour and watering the roadbed. At the same time, another worker was driving a fully loaded scraper in the opposite direction on the same section of roadbed. As the vehicles approached one another, the water truck veered sharply to the left into the path of the oncoming scraper. The front of the water truck struck the left front tire, wheel assembly, and fender of the scraper and the truck overturned. The victim died instantly as a result of massive head injuries sustained in the collision and was pronounced dead at the scene by the county coroner.

The Bottom Line: Wear your seatbelt and be alert for oncoming vehicles.

Source: The National Institute for Occupational Safety and Health (NIOSH)

Loading and moving equipment can present many unexpected hazards. When loading heavy equipment be sure to:

- Always signal before moving.
- Use the proper crawler blocks and safety chains.
- Use a signal person when clearances are close.
- Never overload a trailer.
- Always know the height, width, and weight of your load.
- Use substantial ramps or blocks when loading.
- Always set the swing brake and/or lock when the loading is completed.

Follow these guidelines when moving heavy equipment:

- Be sure all loose gear, outrigger beams, and pads are secure.
- Follow the manufacturer's instructions when towing.
- Always have your lights on and use proper signs and warning flags.
- Do not exceed bridge load limits.
- Watch for narrow spots and low clearances.
- Allow for boom overhang and structure clearances when turning corners.
- Slow down when crossing obstacles, such as bridge expansion joints, cattle guards, or potholes.
- Be sure to watch height clearances when moving boom equipment.

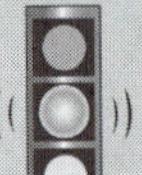

CAUTION

Uneven ground may cause the boom to bob and weave enough to contact obstructions.

Summary

Work-zone safety is an important part of overall job safety. Many accidents happen because of a lack of communication. Signs, signals, and barricades make communication clear on a work site and in public areas where work is being done. It's important to know the hazards of your job and be able to recognize the signs, signals, and barricades that could save your life.

You must set up a clear work zone with tape, fencing, cones, or barriers. This is particularly important when working on a busy street or highway. Make sure that your work zone includes all work areas. Stay alert for traffic.

Review Questions

1. The sign used to inform workers that an immediate hazard exists is a(n) _____ sign.
   a. caution
   b. danger
   c. informational
   d. warning

2. Red is the standard color used for caution.
   a. True
   b. False

3. Hand signals are used for controlling vehicular traffic, handling materials, and assisting equipment operators.
   a. True
   b. False

4. A barricade that alerts workers to hazards but provides no real protection is called a(n) _____.
   a. hole cover
   b. temporary barricade
   c. railing
   d. warning barricade

5. Vehicles and equipment operating in and around highway work zones are responsible for over half of the worker deaths in the highway and street construction industry.
   a. True
   b. False

For Questions 6 through 10, match the type o… or tag with the corresponding description.

6. _____ Danger
7. _____ Caution
8. _____ Informational
9. _____ Safety
10. _____ Accident prevention

a. This sign or tag is used where it is necessary to tell workers about general information not related to safety.
b. This sign or tag is used when there is a need for general instructions and suggestions related to safety measures.
c. This sign or tag is used to inform workers that an immediate hazard exists and specific precautions must be observed to avoid an accident.
d. This sign or tag is used as a temporary way of warning workers about immediate and potential hazards.
e. This sign or tag is used to inform workers about potential hazards or unsafe practices.

GLOSSARY

Trade Term Introduced in This Module

Conspicuous location: A particularly noticeable spot, as would be appropriate for posting an important sign or tag to ensure it is seen.

Figure Credits

Marzetta Signs	104F01, 104F12 (E) and (F)
Brigid McKenna	104F02, 104F15, 104F16
IDESCO	104F12 (A) through (D)
Champion America	104F17

NCCER CURRICULA — USER UPDATE

NCCER makes every effort to keep its textbooks up-to-date and free of technical errors. We appreciate your help in this process. If you find an error, a typographical mistake, or an inaccuracy in NCCER's curricula, please fill out this form (or a photocopy), or complete the online form at **www.nccer.org/olf**. Be sure to include the exact module ID number, page number, a detailed description, and your recommended correction. Your input will be brought to the attention of the Authoring Team. Thank you for your assistance.

Instructors – If you have an idea for improving this textbook, or have found that additional materials were necessary to teach this module effectively, please let us know so that we may present your suggestions to the Authoring Team.

NCCER Product Development and Revision

13614 Progress Blvd., Alachua, FL 32615

Email: curriculum@nccer.org

Online: www.nccer.org/olf

❑ Trainee Guide ❑ AIG ❑ Exam ❑ PowerPoints Other ______________________________

Craft / Level: Copyright Date:

Module ID Number / Title:

Section Number(s):

Description:

Recommended Correction:

Your Name:

Address:

Email: Phone:

Module 75105-03

Electrical and High-Voltage Hazards

COURSE MAP

This course map shows all of the modules in Field Safety. The suggested training order begins at the bottom and proceeds up. The local Training Program Sponsor may adjust the training order.

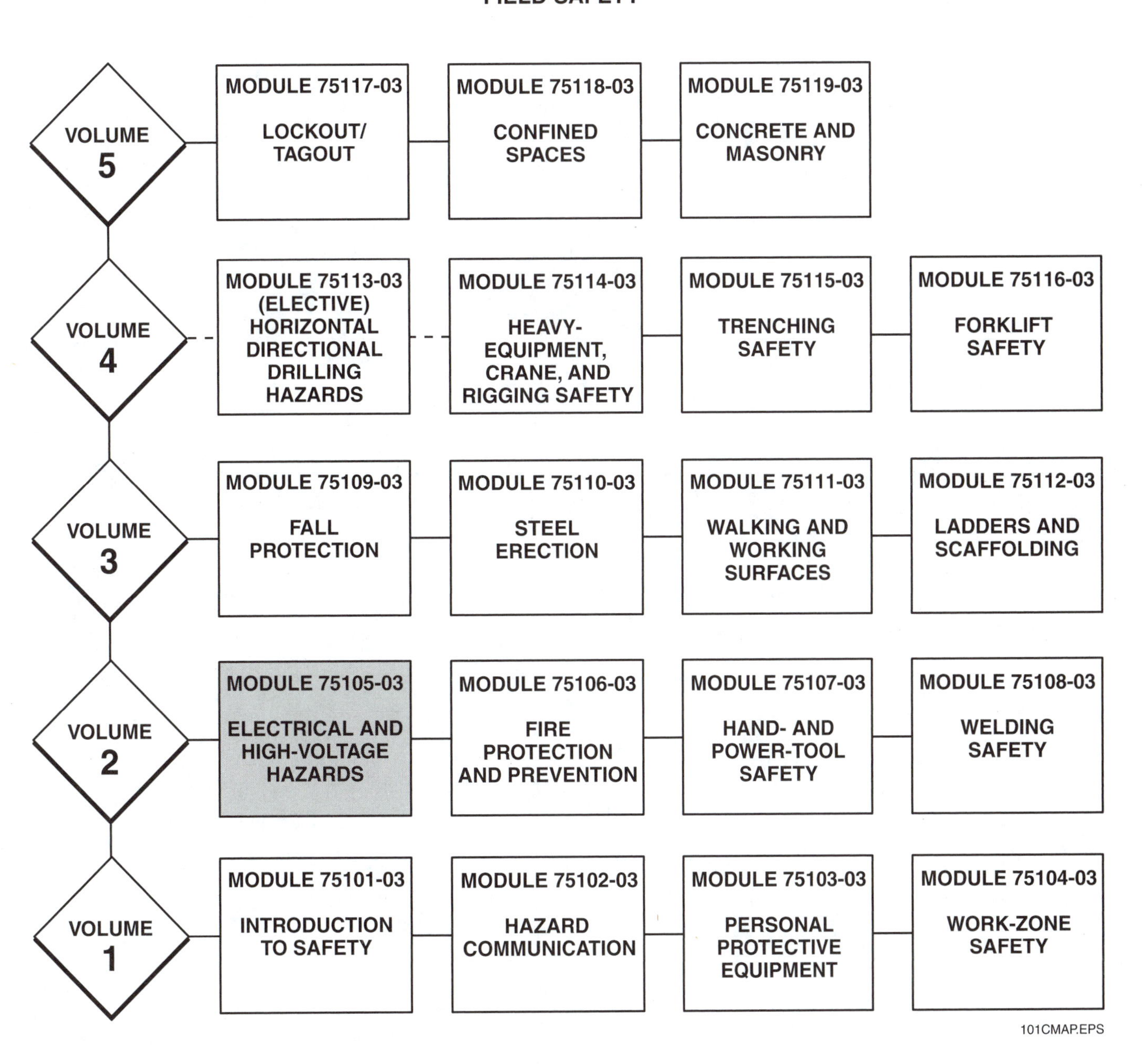

MODULE 75105-03 CONTENTS

Figures

Electrical and High-Voltage Hazards

Objectives

When you have completed this module, you will be able to do the following:

1. Describe the risks associated with working around electricity and high voltage.
2. Describe the effects of electrical shock on the human body.
3. Define *insulation* and *grounding*.
4. Describe how a ground fault circuit interrupter (GFCI) works.
5. Explain where a ground fault circuit interrupter is required.
6. Discuss the purpose of an assured equipment grounding conductor program.
7. Define *lockout/tagout* and describe how it protects workers.

Prerequisites

Before you begin this module, it is recommended that you successfully complete the following: Field Safety, Modules 75101-03 through 75104-03.

Required Materials

1. Pencil and paper
2. Appropriate personal protective equipment

1.0.0 ◆ INTRODUCTION

On the job site, you will be exposed to many potentially hazardous conditions. No training manual, set of rules and regulations, or listing of hazards can make working conditions completely safe. This is especially true when working with electrical tools or working near high voltage. Electrical accidents are the third leading cause of death in the workplace. In fact, each year in the U.S. there are approximately 700 electricity-related deaths in the workplace. It is possible, however, to work around electricity without serious accident or injury. To be safe, you need to be aware of potential hazards and stay alert to them. You must take the proper precautions and practice the basic rules of safety. You must be safety-conscious at all times. Safety should become a habit. Keeping a safe attitude on the job will go a long way toward reducing the number and severity of accidents. Remember, your safety is up to you.

2.0.0 ◆ ELECTRICAL SHOCK

The major risk when working around electricity, especially high voltage, is electrical shock. A primary cause of death from electrical shock is electrical current changing the heart's rhythm. Normally, the heart's operation uses a very low-level electrical signal to cause the heart to contract and pump blood. When an abnormal electrical signal, such as current from an electrical shock, reaches the heart, the low-level heartbeat signals are overcome. The heart begins twitching in an irregular manner and goes out of rhythm with the pulse. This twitching is called **fibrillation**. Unless the normal heartbeat rhythm is restored using special defibrillation equipment (*Figure 1*), the individual will die. No known case of heart fibrillation has ever been corrected without the use of a defibrillator operated by a qualified medical practitioner. Other effects of electrical shock may include immediate heart stoppage and burns. In addition, the body's reaction to the shock can cause a fall or other accident. Delayed internal problems can also result.

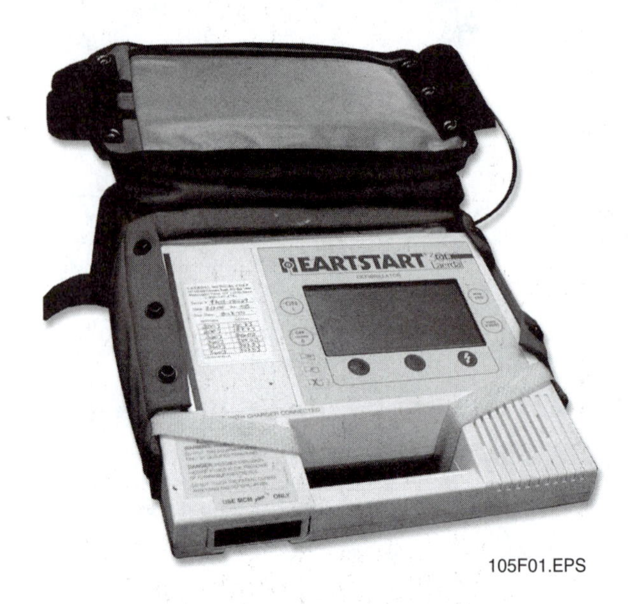

105F01.EPS

Figure 1 ◆ Defibrillation equipment.

2.1.0 The Effect of Current

The amount of current that passes through a body determines the outcome of an electrical shock. The higher the voltage, the greater the chance for a fatal shock. In a recent study in California, the following results were observed by the State Division of Industry Safety:

- In one ten-year study, investigators reviewed 9,765 electrical injuries. Two-thirds of these accidents involved low-voltage conductors, those carrying 600 volts (V) or less.
- Portable, electrically operated hand tools made up the second largest number of injuries. Most of those injuries happened when the frame or case of the tool became energized. These injuries could have been prevented by following proper safety practices, using grounded or double-insulated tools, and using **ground fault circuit interrupter (GFCI)** protection.
- Over 18% of these injuries involved contact with voltage levels over 600V. When tools or equipment touch high-voltage overhead lines, the chance that a resulting injury will be fatal climbs to 28%.

High-voltage currents are primarily found in power lines. Electrical equipment, building wiring, and portable generators generally use low voltages (under 600V). Due to the frequency of contact, most electrocution deaths actually occur at low voltages. A lax attitude about low voltages can kill you.

Understanding the risks involved in working with electrical equipment is the first step in learning to work safely.

DID YOU KNOW?

Current Kills

Current, measured in amps, increases if the resistance, measured in ohms, decreases and the voltage, measured in volts, remains the same. For example, a shock from a 110V plug across dry skin will be uncomfortable; however, the same 110V shock across sweaty skin can be fatal because sweaty skin offers less resistance and increases the amperage of the shock.

2.2.0 Body Resistances

Electricity travels in closed circuits, and its normal route is through a conductor. Shock occurs when the body becomes part of the electric circuit. This is shown in *Figure 2*.

The current must enter the body at one point and leave at another. Shock normally occurs in one of three ways: the person must come in contact with both wires of the electric circuit, one wire of the electric circuit and the ground, or a metallic part that is in contact with an energized wire while the person is in contact with the ground. To fully understand the harm done by electrical shock, we need to understand something about the human body and its parts: the skin, the heart, and muscles.

Skin covers the body and is made up of three layers. The most important layer, as far as electric shock is concerned, is the outer layer of dead cells referred to as the horny layer. This layer is composed mostly of a protein called keratin, and it is the keratin that provides the largest percentage of the body's electrical resistance. When it is dry, the outer layer of skin may have a resistance of several thousand ohms, but when it is moist, there is a radical drop in resistance. This is also the case if there is a cut or abrasion that pierces the horny layer. The amount of resistance provided by the skin varies widely from individual to individual. A worker with a thick horny layer will have a much higher resistance than someone with a thin horny layer, such as a child. The resistance also varies at different parts of the body. For instance, a worker with high-resistance hands may have low-resistance skin on the back of his calf.

The skin, like any insulator, has a **breakdown voltage** at which it ceases to act as a resistor and is simply punctured, leaving only the lower-resistance body tissue to impede the flow of current. The breakdown voltage varies with the individual, but is in the area of 600V. Since most industrial power-distribution systems operate at 480V or higher, technicians working at these levels need to be aware of the shock potential.

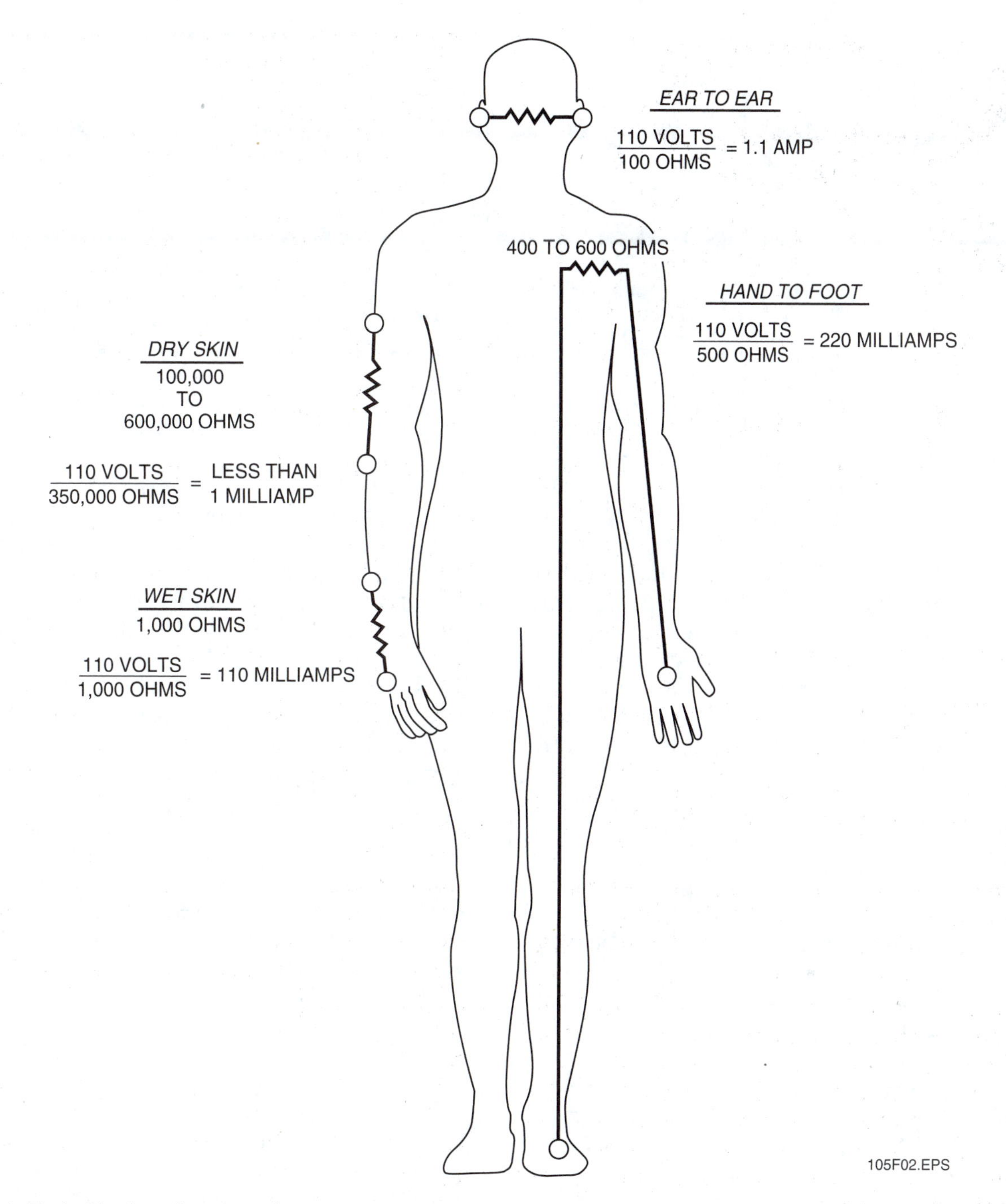

Figure 2 ◆ Typical body resistance and currents.

The heart is the pump that generates blood flow throughout the body. The blood flow is caused by the contractions of the heart muscle, which is controlled by electrical impulses. The electrical impulses are delivered by an intricate system of nerve tissue with built-in timing mechanisms, which make the chambers of the heart contract at exactly the right time. An outside electric current of as little as 75 milliamperes can upset the rhythmic, coordinated beating of the heart by disturbing the nerve impulses. When this happens, the heart is said to be in fibrillation, and the pumping action stops. Death will occur quickly if the normal beat is not restored. Remarkable as it may seem, what is needed to defibrillate the heart is a shock of an even higher intensity.

The other muscles of the body are also controlled by electrical impulses delivered by nerves. Electric shock can cause loss of muscular control, resulting in the inability to let go of an electrical conductor. Electric shock can also cause injuries of an indirect nature in which involuntary muscle reaction from the electric shock can cause bruises, fractures, and even deaths resulting from collisions or falls.

The Added Risk of Wearing Jewelry

People who work around electricity must not wear any jewelry, including wedding rings, bracelets, and necklaces. One unfortunate worker lost his ring finger when his gold wedding band came in contact with an electrical capacitor.

The Bottom Line: Jewelry is usually made of materials that are excellent conductors, such as gold, silver, or platinum. Wearing these items greatly increases the risk of severe injury or death from electrical shock.

Source: Topaz Publications

The severity of shock received when a person becomes a part of an electric circuit is affected by three primary factors: the amount of current flowing through the body (measured in amperes), the path of the current through the body, and the length of time the body is in the circuit. Other factors that may affect the severity of the shock are the frequency of the current, the phase of the heart cycle when shock occurs, and the general health of the person prior to the shock. Effects can range from a barely perceptible tingle to immediate cardiac arrest. A difference of only 100 milliamperes exists between a current that is barely perceptible and one that can be fatal.

A severe shock can cause considerably more damage to the body than is visible. For example, a person may suffer internal hemorrhages and destruction of tissues, nerves, and muscle. In addition, shock is often only the first injury in a chain of events. The final injury may well be from a fall, cuts, burns, or broken bones.

2.3.0 Burns

The most common electrical injury is a burn. Burns suffered in electrical accidents may be of three types: electrical burns, arc burns, and thermal contact burns.

Electrical burns are the result of electric current flowing through the tissues or bones. Tissue damage is caused by the heat generated by the current flow through the body. An electrical burn is one of the most serious injuries you can receive, and should be given immediate attention. Since the most severe burning is likely to be internal, what may appear to be a small surface wound could, in fact, be an indication of severe internal burns.

Arc burns make up a large portion of the injuries from electrical malfunctions. The electric arc between metals can be up to 35,000°F, which is about four times hotter than the surface of the sun. Workers several feet from the source of the arc can receive severe or fatal burns. Since most electrical safety guidelines recommend safe working distances based on shock considerations, workers following these guidelines can still be at risk for arc burns. Electric arcs can occur due to poor electrical contact or failed **insulation**. Electrical arcing is caused by the passage of substantial amounts of current through vaporized terminal material, usually metal or carbon.

A thermal contact burn is caused by contact with objects thrown during the blast associated with an electric arc. This blast comes from the pressure developed by the near-instantaneous heating of the air surrounding the arc, and from the expansion of the metal as it is vaporized. (Copper expands by a factor in excess of 65,000 times when vaporized.) These pressures can be great enough to propel people, switchgear, and cabinets considerable distances. Another hazard associated with the blast is the explosion of molten metal droplets, which can also cause thermal contact burns and associated damage. A possible beneficial side effect of the blast is that it could throw a nearby person away from the arc, thereby reducing the effect of arc burns.

3.0.0 ◆ POWER CORD HAZARDS

Flexible power cords (extension cords) used to supply power to tools and equipment during construction are very common on a job site. Inside a power cord are conductors that carry potentially lethal voltages. The insulation on the outside of the power cord protects against shock. However, power cord insulation is subject to damage because the cords are often given rough use and are exposed to foot and vehicle traffic, sharp edges, and strain from being pulled. Over time, the protective insulation can wear away or be cut. Once the insulation becomes damaged, you can receive a shock by touching the exposed conductor. You can also be shocked if you and the exposed conductor are touching the same conductive surface, such as metal or water. For example, if you are standing on a metal roof, and an exposed conductor touches the roof, you can receive a shock.

The following are important tips that can protect you from electrical shock:

- Never use tape to repair a damaged power cord. It is no substitute for the original insulation, and may not prevent a shock.
- Never use a cord from which the grounding conductor (*Figure 3*) has been removed.
- Always connect power tools to a circuit protected by a ground fault circuit interrupter (described later in this module).
- Be extremely careful when using a power cord in a wet or damp area. Keep the cord out of any water. Electricity from an exposed conductor can travel along the outside of the cord, or along a wet surface, and give you a shock.

4.0.0 ◆ HIGH-VOLTAGE HAZARDS

High-voltage electricity, defined as 600V or more, is 10 times more likely to kill you than normal household voltages, which are around 120V or 240V.

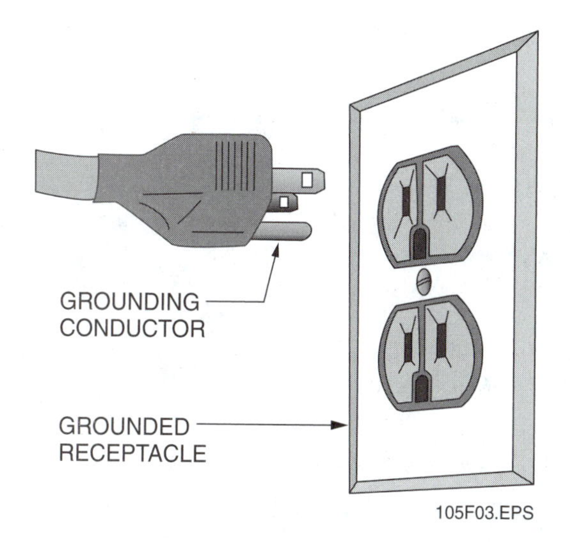

Figure 3 ◆ Electrical grounding conductor.

DID YOU KNOW?

Have you ever wondered why a bird can perch safely on an electrical wire, but other animals, like squirrels, commonly get electrocuted on power lines or at substations? It is because birds never complete the circuit by touching the power lines and ground at the same time. Squirrels may touch the charged conductors and ground (the power pole) simultaneously and be electrocuted.

High-voltage hazards come from several sources; however, the most common are:

- Contacting overhead power lines with ladders, scaffolds, or equipment with booms, such as cranes.
- Contacting underground power lines with backhoes and other digging equipment. Every community has laws that require contractors to contact the local power company or independent agency before digging. The agency will mark the locations of power lines and other utilities so they can be avoided.

Main power lines carry thousands of volts. Touching them will result in severe burns or death. Even getting too close is dangerous, because high voltages can arc to a grounded conducting source, such as a person.

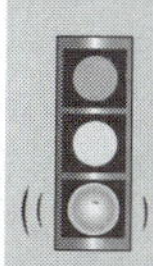

NOTE

Anyone who cuts power, gas, or communications lines is subject to the cost of repairing the lines and may also be charged for the equipment downtime. Imagine what the loss-of-service cost would be if you cut a cable carrying thousands of phone lines.

Know What's Over Your Head

Five employees were constructing a chain-link fence in front of a house and directly below a 7,200V energized power line. They were installing 21' sections of metal top rail on the fence. One employee picked up a 21' section of top rail and held it up vertically. The top rail contacted the 7,200V line, and the employee was electrocuted.

The Bottom Line: Following its inspection, OSHA determined that the employee who was killed had never received any safety training from his employer nor any specific instructions for avoiding the hazards posed by overhead power lines.

Source: The Occupational Safety and Health Administration (OSHA)

Before beginning any job, survey the site for overhead power lines. Contact the utility company to shut down or insulate the lines if you must work near them. Treat all power lines as energized until the utility company has certified that it is safe to proceed. The Occupational Safety and Health Administration (OSHA) recommends a minimum safe working distance of 10' from energized overhead lines. Equipment should be used at slower-than-normal speeds to maintain safe distances. Use warning signs. A spotter may be needed to prevent accidental contact with power lines.

5.0.0 ◆ INSULATION AND GROUNDING

Insulation and **grounding** are two recognized means of preventing injury from electrical equipment. Conductor insulation may be provided by placing nonconductive material such as plastic around the conductor. Grounding may be achieved through the use of a direct connection to a known ground, such as a metal cold water pipe.

Consider, for example, the metal housing or enclosure around a motor or on a portable tool or generator. Such enclosures protect the equipment from dirt and moisture and prevent accidental contact with exposed wiring. However, there is a hazard associated with housings and enclosures. A malfunction within the equipment, such as damaged insulation, may increase the risk of electrical shock. Many metal enclosures are connected to a ground to eliminate this hazard. If a hot wire contacts a grounded enclosure, it results in a ground fault, which will normally trip a circuit breaker or blow a fuse. Metal enclosures and housings are usually grounded by a wire connecting them to ground. This is called the **equipment grounding conductor**. Most portable electric tools and appliances are grounded this way. There is one disadvantage to grounding: a break in the grounding system may occur without the user's knowledge.

Insulation may be damaged by dropping the tool, hard use on the job, or simply by aging. If this damage causes a break in the grounding system, a shock hazard exists and the tool must be taken out of service. Double insulation may be used as additional protection on the live parts of a tool, but double insulation does not protect

Working Safely Around High Voltage

If you are working near high-voltage equipment, DANGER signs like the one shown here should be posted on the job site. Overhead power lines don't have any signs to remind you of the voltage hazard. Whenever you are working with ladders, scaffolds, or equipment with booms, make sure there are no power lines nearby.

There may not be any warning about buried power lines either. To prevent accidental contact with underground lines, you are required by law to have the designated authority locate and mark any buried utilities before digging in the area.

105SA01.EPS

Grounding and Insulation Saves Lives

One worker was climbing a metal ladder to hand an electric drill to the journeyman installer on a scaffold about 5' above him. When the worker climbing the ladder reached the third rung from the bottom of the ladder, he received a fatal shock. He died because the cord of the drill he was carrying had an exposed wire that made contact with the conductive ladder.

The Bottom Line: Inspect all equipment before using it. Never use electrical equipment with a damaged power cord. Use fiberglass ladders when working with electrical equipment.

Source: The Occupational Safety and Health Administration (OSHA)

against defective cords and plugs, or against heavy-moisture conditions. *Figure 4* shows an example of a double-insulated ungrounded tool.

The use of a ground fault circuit interrupter (GFCI) is one method used to overcome grounding and insulation problems.

5.1.0 Ground Fault Circuit Interrupters

A ground fault circuit interrupter (GFCI) is a fast-acting circuit breaker that senses small imbalances in the circuit caused by current leakage to ground. A GFCI continually matches the amount of current going to an electrical device against the amount of current returning from the device. Whenever the two values differ by more than 5 milliamps, the GFCI interrupts the electric power within $\frac{1}{40}$ of a second. *Figure 5* shows an extension cord with a GFCI.

A GFCI will not protect you from line-to-line contact hazards such as holding either two hot wires or a hot and a neutral wire in each hand. It does provide protection against the most common form of electrical shock, which is a ground fault. It also provides protection from fires, overheating, and wiring insulation deterioration.

GFCIs can be used successfully to reduce electrical hazards on construction sites. Tripping of GFCIs—interruption of circuit flow—is sometimes caused by wet connectors and tools. Limit the amount of water that tools and connectors come into contact with by using watertight or sealable connectors. Having more GFCIs or shorter circuits can prevent tripping caused by cumulative leakage from several tools or from extremely long circuits.

5.2.0 GFCI Requirements

OSHA has determined that to minimize the risk of electrical shock at a construction site, it is the employer's responsibility to provide either:

- Ground fault circuit interrupters are needed for all 120V, single-phase, 15A and 20A receptacle outlets being used if they are not part of the permanent wiring of the building or structure. That might include receptacles at the end of extension cords, even if the extension cord is plugged into the building's permanent wiring.

NOTE: COLORED AREAS SHOW INSULATING MATERIAL.

105F04.EPS

Figure 4 ◆ Double-insulated electric drill.

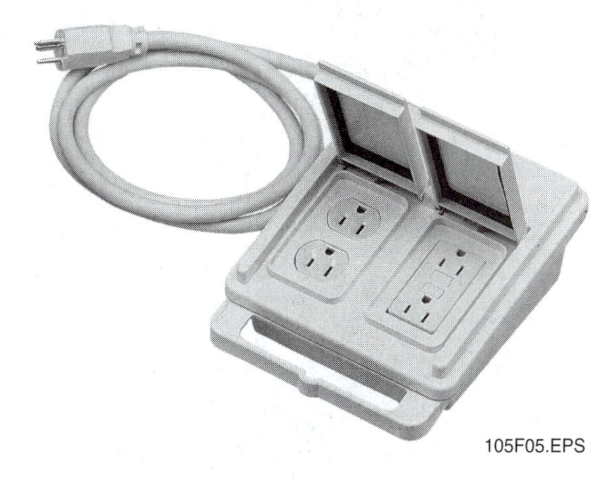

105F05.EPS

Figure 5 ◆ Extension cord with a GFCI.

GFCIs Can Save Your Life

A self-employed builder was using a metal cutting tool on a metal carport roof and was not using GFCI protection. The male and female plugs of his extension cord partially separated, and the active pin touched the metal roofing. When the builder grounded himself on the gutter of an adjacent roof, he received a fatal shock.

The Bottom Line: Always use GFCI protection and be aware of potential hazards.

Source: The Occupational Safety and Health Administration (OSHA)

- A scheduled and recorded **assured equipment grounding conductor program** is needed to cover cord sets, receptacles that are not part of the permanent wiring of the building, and equipment connected by cords and plugs that employees may use.

6.0.0 ◆ ASSURED EQUIPMENT GROUNDING CONDUCTOR PROGRAM

The assured equipment grounding conductor program covers all cord sets, receptacles that are not part of the permanent wiring, and equipment connected by a cord or plug that is available for employee use. OSHA requires that a written description of the employer's assured equipment grounding conductor program, including the specific procedures used, be kept at the job site. This program should outline the employer's specific procedures for the required equipment inspections, tests, and test schedule.

The required tests must be recorded, and the record maintained at the job site. The written program description and the test results must be made available to any employee upon request. The employer must designate one or more competent people to implement the program.

Electrical equipment noted in the assured equipment grounding conductor program must be visually inspected for damage or defects before each day's use. Any damaged or defective equipment must not be used until it is repaired. If you come across a questionable cord, outlet, or piece of equipment, do not use it. Show it to your supervisor immediately.

Two tests are required for the assured equipment grounding conductor program. One is a continuity test to ensure that the equipment grounding conductor is electrically continuous. It must be performed on all cord sets, nonpermanent receptacles, and cord- and plug-connected equipment that is required to be grounded. Many simple continuity testers are available, including:

- Lamp and battery
- Bell and battery
- Multimeter
- Receptacle tester

Figure 6 shows a continuity tester and a multimeter. This equipment can also be used to conduct the second test. In the second test, plugs and

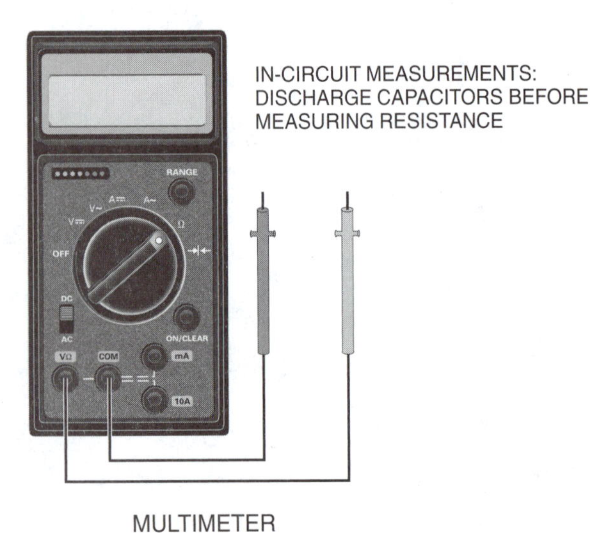

Figure 6 ◆ Test equipment.

receptacles are checked to make sure that the equipment grounding conductor is connected to its proper terminal. Any equipment that fails these tests cannot be used until it is repaired and passes these tests. If you discover an unsafe tool, machine, or piece of equipment, tell your supervisor. Nicked or broken extension cords, power tools with broken housings, or equipment with modified wiring can all pose electrical hazards. They must be tagged and replaced or repaired.

7.0.0 ◆ LOCKOUT/TAGOUT

Lockout/tagout procedures keep you safe from shock or accidental startup when you are working on an electrical circuit or near certain electrical equipment. The power to the system is shut off and then isolated with a tag and lock. The person who placed the lock is the only one who can remove it. This prevents others from starting the machine while you are working on it. Always notify your supervisor when performing a lockout/tagout procedure.

If more than one person is working on the same circuit, each person places a lock and tag. *Figure 7* shows a multiple lockout/tagout device.

NOTE
Lockout/tagout procedures are covered in more detail in the *Lockout/Tagout* module in Volume Five.

Figure 7 ◆ Multiple lockout/tagout device.

8.0.0 ◆ EMERGENCY RESPONSE

If someone near you is receiving an electrical shock, do not touch that person. Instead, immediately turn off the power source. If that is not possible, try using nonconductive material such as a blanket, rope, coat, or piece of dry wood to separate the person from the electrical source. Never use anything wet or damp. If you touch the person with your body, use a metal object, or use wet or damp material, you could also become a victim.

Once the person has been separated from the shock source, immediately call for medical help. If the victim is not breathing, a trained person should immediately begin artificial respiration.

Summary

Safety must be your primary concern at all times so that you do not become either the victim of an accident or the cause of one. Safety requirements and safe work practices are provided by OSHA and your employer. You must adhere to all safety requirements and follow your employer's safe work practices and procedures. Also, you must be able to identify the potential safety hazards at your job site. The consequences of unsafe job-site conduct can be expensive, painful, and even deadly.

Here is a summary of the important things to remember about electrical safety:

- Use three-wire extension cords and protect them from damage. Never use damaged cords.
- Make sure that panels, switches, outlets, and plugs are grounded.
- Never use metal ladders near any source of electricity.
- Always inspect electrical power tools before using them.
- Never operate any piece of electrical equipment that has a lockout device attached to it.
- Use a GFCI.

Review Questions

1. Some injuries from electric shock may be due to a fall caused by the body's reaction to the shock.
   a. True
   b. False

2. In an electrical accident, the higher the voltage passing through the body, the _____.
   a. greater the resistance of the body's horny layer
   b. greater the chance for a fatal shock
   c. greater the chance the worker will be thrown out of harm's way
   d. lower the amperage required to kill you

3. Contact with low-voltage conductors causes the majority of electrocution deaths.
   a. True
   b. False

4. Fibrillation occurs when an outside electrical current as small as 75 milliamperes disturbs the heart's normal pumping rhythm.
   a. True
   b. False

5. Shock can occur in any of the following ways, *except* coming into contact with _____.
   a. water and a ground wire
   b. both wires of the electrical circuit
   c. one wire of a circuit and the ground
   d. both the ground and a metallic object energized through contact with the circuit

6. A(n) _____ burn is one of the most serious injuries you can receive, and should be given immediate medical attention.
   a. arc
   b. electrical
   c. powder
   d. thermal contact

7. A ground fault circuit interrupter is a fast-acting circuit breaker that senses small imbalances in the circuit and quickly shuts off the electricity.
   a. True
   b. False

8. A GFCI will provide protection from all of the following *except* _____.
   a. ground faults
   b. line-to-line contact hazards
   c. overheating
   d. wiring insulation deterioration

9. All of the following pieces of test equipment may be used to test the continuity of the equipment grounding conductor *except* a(n) _____.
   a. ammeter
   b. lamp and battery
   c. receptacle tester
   d. multimeter

10. All of the following should be part of the lockout/tagout procedure *except* _____.
    a. testing the equipment to find out why it is faulty
    b. notifying your supervisor or safety technician
    c. de-energizing the equipment
    d. tagging and locking out the power source to the equipment

PROFILE IN SUCCESS

Paul Foley, Foster Wheeler Corp.
Safety Manager

How did you choose a career in the Safety Industry?
Early in my career, I noticed that there was no one on site to provide immediate medical care. I went to night school to become an Emergency Medical Technician (EMT). During the course of a scheduled power plant outage, an employee broke his arm and I splinted it. The next day, the owner asked me why I interfered with the superintendent to provide this care. I explained that I was trained to handle emergency medical situations, and instead of firing me, he offered me a job as a safety specialist.

What types of training have you been through?
I have been through the NCCER Safety Technician Course, OSHA 500, Georgia Institute of Technology, Scaffold Trainer Competency Course, Safeway Scaffold Systems outreach program, Emergency Medical Services, Daytona Beach Community College, 3M Respirator Trainer, U.S. Army leadership and development training.

What kinds of work have you done in your career?
I have been a site safety technician, site safety supervisor, site safety manager, medical case manager, supervisor safety awareness instructor, and orientation specialist.

Tell us about your present job.
I am currently on a power plant project in Pennsylvania. My duties include medical case management, daily site audits, new-hire orientation, accident investigation (not many to investigate), confined-space hazard assessment, and first-aid administration.

What factors have contributed most to your success?
The military gave me the confidence and knowledge to lead. The Safety Tech course and the OTI 500 gave me credibility to approach prospective employers.

What advice would you give to those new to Safety industry?
No matter how much authority you are assigned or training you possess, do not forget that the only reason you have a job is because of the people around you. Some companies instill a lot of power in their safety professionals; do not abuse that power. Co-workers and subordinates will respect you if you are honest and humble; do not flex your authority for ego's sake.

GLOSSARY

Trade Terms Introduced in This Module

Assured equipment grounding conductor program: A detailed plan specifying an employer's required equipment inspections and tests and a schedule for conducting those inspections and tests.

Breakdown voltage: The voltage at which an insulator has a breakdown and ceases to act as a resistor.

Equipment grounding conductor: A wire that connects metal enclosures and containers to ground.

Fibrillation: Very rapid, irregular contractions of the muscle fibers of the heart that result in the heartbeat and pulse going out of rhythm with each other.

Ground fault circuit interrupter (GFCI): A fast-acting circuit breaker that senses small imbalances in the circuit caused by current leakage to ground and, in a fraction of a second, shuts off the electricity.

Grounding: The process of directly connecting an electrical circuit to a known ground to provide a zero-voltage reference level for the equipment or system.

Insulation: The practice of placing nonconductive material such as plastic around the conductor to prevent current from passing through it.

Figure Credits

Charles Burke	105F01
Coleman Cable Inc.	105F05
Veronica Westfall	105F06

NCCER CURRICULA — USER UPDATE

NCCER makes every effort to keep its textbooks up-to-date and free of technical errors. We appreciate your help in this process. If you find an error, a typographical mistake, or an inaccuracy in NCCER's curricula, please fill out this form (or a photocopy), or complete the online form at **www.nccer.org/olf**. Be sure to include the exact module ID number, page number, a detailed description, and your recommended correction. Your input will be brought to the attention of the Authoring Team. Thank you for your assistance.

Instructors – If you have an idea for improving this textbook, or have found that additional materials were necessary to teach this module effectively, please let us know so that we may present your suggestions to the Authoring Team.

NCCER Product Development and Revision
13614 Progress Blvd., Alachua, FL 32615

Email: curriculum@nccer.org
Online: www.nccer.org/olf

❑ Trainee Guide ❑ AIG ❑ Exam ❑ PowerPoints Other ______________________

Craft / Level: Copyright Date:

Module ID Number / Title:

Section Number(s):

Description:

Recommended Correction:

Your Name:

Address:

Email: Phone:

Module 75106-03

Fire Protection and Prevention

COURSE MAP

This course map shows all of the modules in Field Safety. The suggested training order begins at the bottom and proceeds up. The local Training Program Sponsor may adjust the training order.

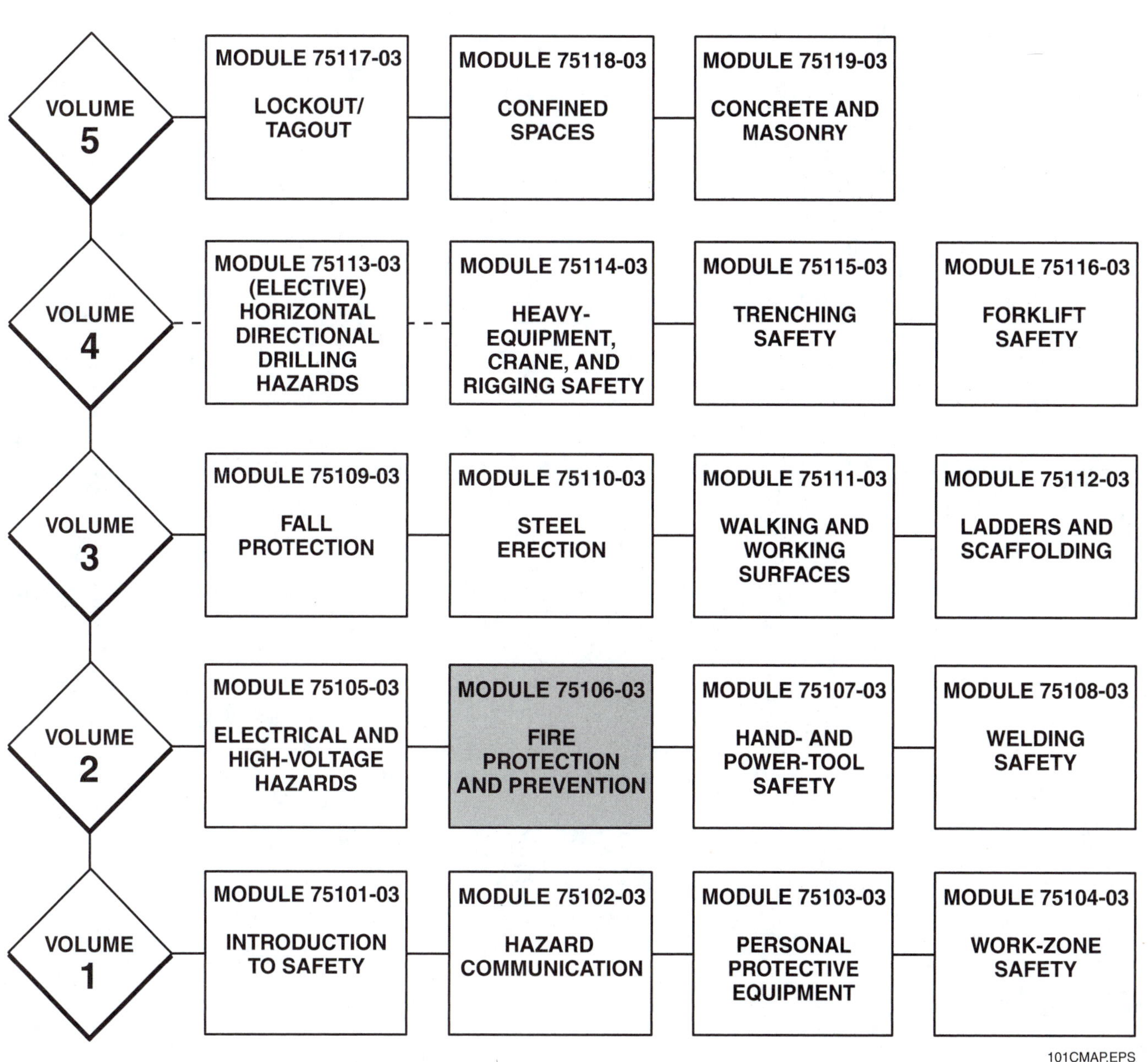

MODULE 75106-03 CONTENTS

Figures

MODULE 75106-03

Fire Protection and Prevention

Objectives

When you have completed this module, you will be able to do the following:

1. Identify the typical fuel sources found on a construction site.
2. Identify the typical sources of ignition found on a construction site.
3. Explain the procedures for proper handling and storage of flammable materials.
4. Explain the classes of fire extinguishers and name the type of fire for which each is most effective.
5. Identify the type and use of a fire extinguisher from its label.
6. Use a fire extinguisher to put out a fire.

Prerequisites

Before you begin this module, it is recommended that you successfully complete the following: Field Safety, Modules 75101-03 through 75105-03.

Required Materials

1. Pencil and paper
2. Appropriate personal protective equipment

1.0.0 ◆ INTRODUCTION

Fire is a hazard that is both extremely dangerous and highly preventable. Fire prevention can be as simple as keeping **fuel sources** away from heat sources. If you work carelessly around fuel sources, like propane, lumber, paper, or paint, you can easily start a fire which could result in serious injury, death, and extensive damage. Fire- related injuries and deaths are typically caused by inhaling smoke or other toxic gases, burns, falls due to low visibility or collapsing structures, and being trapped by falling debris.

This module discusses the typical fuel and ignition sources present on a construction work site, ways to prevent fires from starting, and how to protect yourself, your co-workers, and the site from fires.

2.0.0 ◆ FUEL SOURCES

If you look around a work site, you will see several items that could be considered fuel for a fire. Typical fuel sources include:

- **Combustible** building material and debris such as lumber and scrap wood
- Trash piles, bins, and dumpsters
- Bulk fuel, paint, solvents, and welding gases
- Gasoline, propane, or diesel fuel found at refueling areas
- Liquefied petroleum (LP) gas
- Solvents such as kerosene or paint thinner (*Figure 1*)

Some dangerously **flammable** liquids such as styrene, methyl acrylate, and acrylonitrile, may be found around the work site. They can give off enough vapor at normal temperatures to form flammable mixtures with air. Liquid propane (LP) gas, an alternate fuel source commonly found on work sites, is highly flammable and heavier than air. LP gas will settle down at floor level like a blanket so it isn't always easy to smell. If someone lights a match or creates a spark around an LP gas leak, it

Construction Fire Facts

- Each year there are an estimated 4,800 construction-site fires.
- Construction-site fires cause $35 million in property damage each year.
- Construction-site fires are responsible for 30 serious injuries and 10 fatalities each year. That means that every week and a half, a construction worker is killed or injured by fire in the United States.
- Construction-site fires are harder to bring under control and extinguish than residential or forest fires because of the construction materials and chemicals used on a work site.

Source: U.S. Fire Administration

106F01.EPS

Figure 1 ◆ Fuel source.

106F02.EPS

Figure 2 ◆ Fire triangle.

will cause a fire or explosion. Therefore, all workplace chemicals should be treated as serious fire hazards and all warnings and directions on storage and use should be carefully read and followed.

3.0.0 ◆ SOURCES OF IGNITION

What is needed for a fire to start can be shown as a **fire triangle** (*Figure 2*). **Sources of ignition** are those things that introduce heat or flame to an environment. Sources of ignition, or heat sources, are sometimes obvious and other times not obvious at all. Some obvious sources of ignition include welding equipment, grinding (*Figure 3*), roofing tar pots, space heaters, temporary lighting systems, and lighters and cigarettes. Each of these things produces enough heat and/or flame to start a fire. That is why they are considered sources of ignition.

Less obvious, but perhaps more dangerous, are fire hazards presented by the various chemicals present on a job site. Highly reactive chemicals such as polymers, acids, and petrochemicals, may undergo vigorous, uncontrolled reactions that can cause an explosion or enough heat to ignite a fire.

106F03.EPS

Figure 3 ◆ Ignition sources.

Even slow reactions can be hazardous if they involve large amounts of material or if the heat and gases are confined, such as in a sealed storage drum. Drums that are swollen and distorted from over-pressurization are very dangerous. They may rupture at any time without warning, release their contents, and ignite a fire. Fires involving dangerously reactive materials can be more hazardous than other types of fires. The heat from these fires can lead to violent, uncontrolled chemical reactions and potentially explosive ruptures of sealed containers.

It's important to know and recognized potential sources for ignition in your working environment. Always ask your supervisor if you are unsure about a potential ignition source.

Follow the Rules

Two employees were welding brackets onto a 55,000-gallon oil storage tank. The tank contained explosive vapors from a waste chemical, and oil materials from automobile and truck service stations. One worker was killed and another injured when the tank exploded and the top was blown off. Neither worker had permission to work in the tank nor were they wearing the proper personal protective equipment.

The Bottom Line: Hot work, like welding, can cause fires and explosions. Never work in a dangerous area without authorization and always wear appropriate personal protective equipment.

Source: Occupational Safety and Health Administration (OSHA)

4.0.0 ◆ EMERGENCY ACTION PLAN

Your company-appointed competent person, sometimes called the fire brigade chief or emergency action plan coordinator, is responsible for creating an emergency action plan. An emergency action plan describes the actions employees should take to ensure their safety if a fire or other emergency situation occurs. It consists of the following items:

- Conditions under which an evacuation would be necessary
- A clear chain of command
- Person authorized to order an evacuation
- Evacuation procedures and emergency escape route assignments (*Figure 4*)
- Designated meeting area outside of the building
- Procedures for assisting visitors to evacuate
- Procedures for assisting those with disabilities or difficulty understanding English
- Procedures for workers who remain to operate critical plant operations before evacuating
- Procedures for accounting for workers after the evacuation has been completed
- Designated rescue and first-aid personnel

If you are unsure about the action plan at your job site, ask your supervisor to explain it to you. It may mean the difference between life and death.

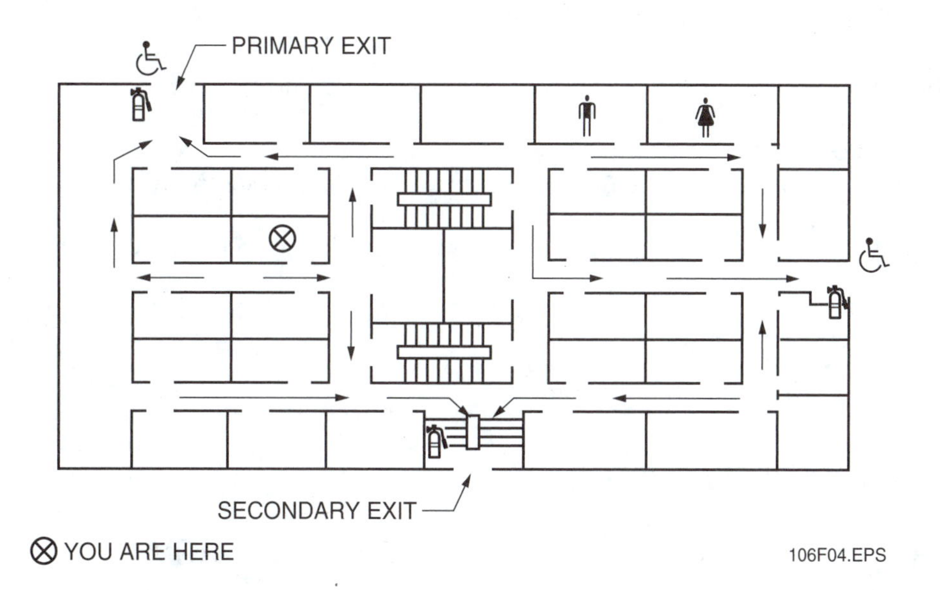

Figure 4 ◆ Typical emergency escape plan.

Don't Get Burned

Here are some facts about gasoline.

- One gallon of gasoline contains the same explosive force as 14 sticks of dynamite.
- Gasoline vapors are heavier than air, can travel a number of feet to an ignition source, and can ignite at temperatures of 45°F. To be safe, keep open gasoline containers well removed from all potential ignition sources.
- Gasoline has a low electrical conductivity. As a result, a charge of static electricity builds up on gasoline as it flows through a pipe or hose. Getting into and out of your vehicle during refueling can build up a static charge, especially during dry weather. That charge can cause a spark that can ignite gasoline vapors if it occurs near the fuel nozzle.

Source: U.S. Department of Energy

5.0.0 ◆ SITE LAYOUT

The risk of fire can be greatly reduced by reducing fire hazards. Consider all of the different things that go on at a construction site every day. One person may be grinding tools in one area while someone else is working with paint thinner in another area. The combination of sparks and combustible fumes would be disastrous. One way these disasters are avoided is by laying out the work site so that fuel sources are kept far enough away from sources of ignition so that there is very little chance of a fire. A **safety technician** is responsible for laying out a site to keep fire risks at acceptable levels, but you can do your part by keeping an eye out for problems and alerting your supervisor or the safety technician if:

- Fire exits and escape routes are not clear and passable.
- Any areas are not accessible for fire fighting.
- Fuel storage and dispensing areas are not set away from the rest of the site.
- People are smoking in areas where No Smoking signs are posted.
- Electrical systems are not properly bonded and grounded.

If you see any of these problems, or anything else that concerns you, tell your supervisor or safety technician immediately. It could save lives.

6.0.0 ◆ ADDITIONAL PRECAUTIONS

Using common sense is the best way to prevent fires. Here are some additional tips to help reduce the risk of fire at a work site.

- Make sure the tarps and covers you use are fire resistant. That way if they come into contact with a source of ignition, like a heater or lit cigarette, they won't start burning.
- Use temporary heaters only when and where they are absolutely needed. Heaters are obvious sources of ignition. Leaving them unattended or using too many of them increases the possibility of a fire.
- Have no more than a one-day supply of flammable material inside the building and keep the material in UL-listed **safety cans**. Keeping the amount of fuel on site as low as possible will lower the risk of a fire and decrease the severity if a fire does occur.
- Look around before you set an ignition source somewhere. Before setting down a tar kettle, using a grinder or welder, parking a piece of equipment with an internal combustion engine, or lighting a cigarette, check the area for flammable material like paper, scrap wood, or chemicals.
- Make sure you know how to report a fire, how to select and use a fire extinguisher, and how to evacuate the site if there is a fire. Saving lives depends on a quick and appropriate response to an emergency situation. Knowing when to tackle a fire with an extinguisher and when to call the fire department is critical to saving lives and preventing damage.
- Always be alert to potential fire hazards. Construction work sites can be large, busy places. The best way to stay safe is for everyone to be a part of the safety team. That means looking out for risks and hazards and either correcting them or telling your supervisor about them.

7.0.0 ◆ MATERIALS HANDLING AND STORAGE

When it comes to fire safety and prevention, materials handling and storage is very important. There are ways of handling and storing flammable and combustible materials that decrease explosion and fire risk. Only approved containers and portable tanks may be used for flammable and combustible liquid storage and handling. Approved storage and handling containers are UL-Listed items such as safety cans and **safety cabinets**. It is your responsibility to know what the handling and storage procedures are on your job site.

7.1.0 Safety Cabinets

Safety cabinets (*Figure 5*) are important for the proper storage of flammables and combustibles. These cabinets are common on a variety of job sites, including construction, maintenance, and renovation locations. Just like with safety cans, it is important that you use these cabinets properly. Appropriate safety cabinets are designed specifically to reduce the risk of explosion and fire. Some features include vents with flame arresters, special lips on the shelves to catch spills, grounding connectors to prevent buildup of static electricity. They are made of steel designed to withstand intense heat and are tailored to meet OSHA and NFPA guidelines. Some manufacturers produce self-closing models that automatically close in response to fire-level heat. Because characteristics of these cabinets vary by model and manufacturer, you must familiarize yourself with the specific cabinets and procedures used on your site.

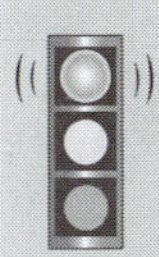

WARNING!
If you do not know how to use safety cabinets properly, ask your supervisor for instruction. Improper use is dangerous!

7.2.0 Safety Cans

Safety cans are special vessels used for safe transportation and storage of flammable materials (*Figure 6*). These essential pieces of equipment are available in a variety of styles and colors. Many job sites use color-coded safety cans to designate different types of liquids such as gasoline, kerosene, diesel fuel, and oils. Color coding not only helps segregate liquids that should not be used together; it also helps ensure that the right liquids are used for each job. Safety cans have many built-in safety features such as vapor control, flame arresters, and pressure relief valves, making them a very important piece of safety equipment. The features and usage requirements vary by manufacturer, so make sure you are familiar with the specifics for the safety cans on your site.

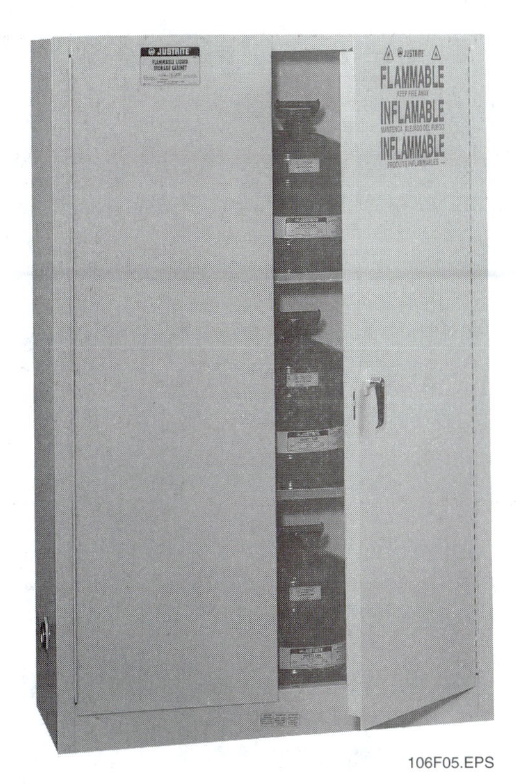

106F05.EPS

Figure 5 ◆ Safety cabinet.

106F06.EPS

Figure 6 ◆ Safety can.

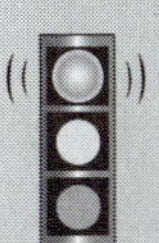

WARNING!
If you do not know how to use safety cans properly, ask your supervisor for instruction. Improper use is dangerous!

8.0.0 ◆ FIRE EXTINGUISHERS

If a fire does start, you must know how to act. According to the Occupational Safety and Health Administration (OSHA), workers expected to use fire extinguishers should have training and experience. If you haven't received the proper training and a fire occurs, do not attempt to extinguish it. Tell everyone immediately that there is a fire, then evacuate the area as quickly as possible. If you have experience and training, then you must quickly determine whether you should put out the fire with an extinguisher or evacuate the work site. Here are some general guidelines for fighting a fire:

- Call the fire department first.
- Make sure the building is being evacuated.
- Determine whether the fire is small enough to safely attempt putting it out with an extinguisher.
- Confirm that you have a safe path to an exit that is not threatened by the fire.
- Know how to properly and safely use a fire extinguisher.

Never fight a fire if any of the following are true:

- The fire is large or it is spreading beyond the immediate area in which it started.
- The fire could block your escape route.
- You are unsure of the proper operation of the extinguisher.
- You doubt that the extinguisher you have is designed for the type of fire at hand or is large enough to fight the fire.

8.1.0 Classes of Fires

All fire extinguishers are labeled, using standard symbols, for the classes of fires on which they can be used. There are four classes of fires (*Figure 7*):

- ***Class A fire*** – Ordinary combustibles such as wood, cloth, and paper
- ***Class B fire*** – Flammable liquids such as gasoline, oil, and oil-based paint
- ***Class C fire*** – Energized electrical equipment, including wiring, fuse boxes, circuit breakers, machinery, and appliances
- ***Class D fire*** – Combustible metals such as magnesium or sodium

A red slash through any of the symbols tells you the extinguisher cannot be used on that class of fire. A missing symbol tells you only that the extinguisher has not been tested for a given class of fire, but may be used if an extinguisher labeled for that class of fire is not available.

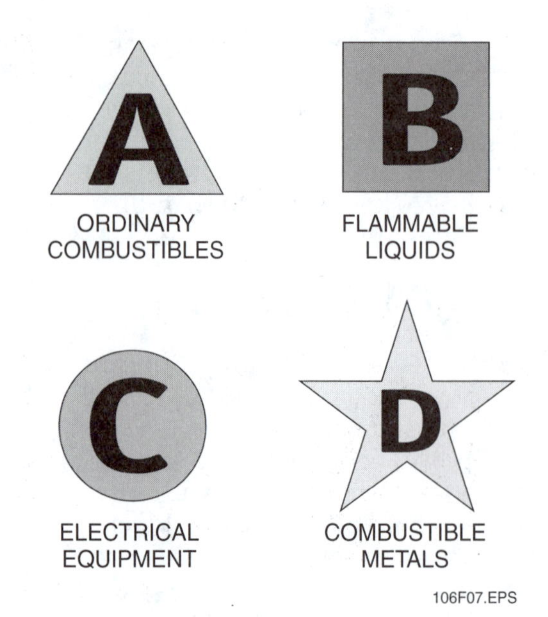

Figure 7 ◆ Classes of fire symbols.

Remember that the extinguisher must be appropriate for the type of fire being fought. Multipurpose fire extinguishers, labeled ABC, may be used on all three classes of fire. If you use the wrong type of extinguisher, you can endanger yourself and make the fire worse. It is also very dangerous to use water or an extinguisher labeled only for Class A fires on a gasoline or electrical fire.

8.2.0 Fire Extinguisher Sizes

Portable extinguishers (*Figure 8*) are rated for the size of fire they can handle (*Figure 9*). This rating is expressed as a number from 1 to 40 for Class A fires and from 1 to 640 for Class B fires. This rating will appear on the label. The larger the number, the

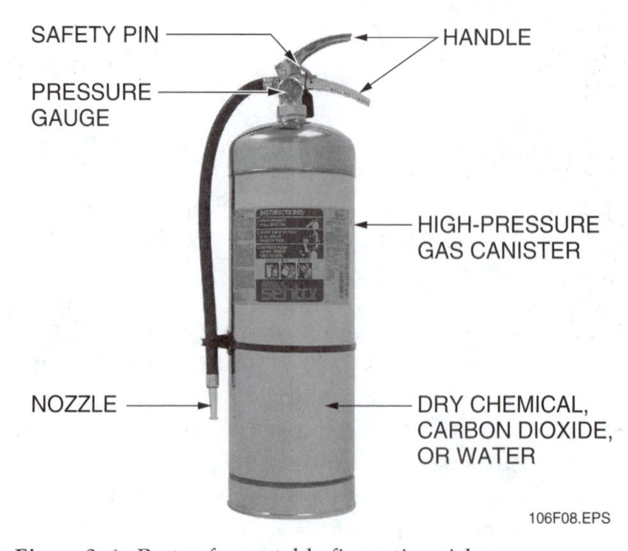

Figure 8 ◆ Parts of a portable fire extinguisher.

Figure 9 ◆ Fire extinguisher label.

larger the fire of a specific class on which the extinguisher can be used. Keep in mind that higher-rated models are often heavier. Make sure you can hold and operate an extinguisher before you need to use it. No number accompanies an extinguisher's Class C rating. The C on the label indicates only that the extinguisher is safe to use on electrical fires.

Extinguishers for Class D fires must match the type of metal that is burning. These extinguishers do not use numerical ratings. Extinguishers for Class D fires are labeled with a list detailing the metals that match the unit's extinguishing agent.

8.3.0 Types of Fire Extinguishers

Depending on their intended use, portable extinguishers (*Figure 10*) store specific extinguishing agents, which are expelled onto the fire.

- Pressurized water models are appropriate for use on Class A fires only. These must never be used on electrical or flammable-liquid fires.
- Carbon dioxide extinguishers contain pressurized liquid carbon dioxide that turns into a gas when expelled. These models are rated for use on Class B and C fires, but can also be used on Class A fires. Carbon dioxide does not leave a residue.

Static Electricity Can Kill

In recent incidents reported to the National Institute for Occupational Safety and Health (NIOSH), fires spontaneously ignited when workers or others attempted to fill portable gasoline containers (gas cans) in the backs of pickup trucks equipped with plastic bed liners or in cars with carpeted surfaces. Serious skin burns and other injuries resulted. Similar incidents in the last few years have resulted in warning bulletins from several private and government organizations.

These fires result from the buildup of static electricity. The insulating effect of the bed liner or carpet prevents the static charge generated by gasoline flowing into the container or other sources from grounding. The discharge of this buildup to the grounded gasoline dispenser nozzle may cause a spark and ignite the gasoline. Both ungrounded metal (most hazardous) and plastic gas containers have been involved in these incidents.

Source: The National Institute for Occupational Safety and Health (NIOSH)

EXTINGUISHER TYPE	TYPE OF FIRE

Water

Ordinary Combustibles

Fires in paper, cloth, wood, rubber, and many plastics require a water-type extinguisher labeled A.

CO_2

OR

Dry Chemical

Flammable Liquids

Fires in oils, gasoline, some paints, lacquers, grease, solvents, and other flammable liquids require an extinguisher labeled B.

Electrical Equipment

Fires in liquids, fuse boxes, energized electrical equipment, computers, and other electrical sources require an extinguisher labeled C.

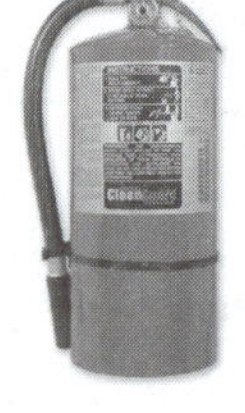

Multi-Purpose

Ordinary Combustibles, Flammable Liquids, or Electrical Equipment

Multi-purpose dry chemical extinguishers are suitable for use on Class A, B, and C fires.

Metals

Combustible metals such as magnesium and sodium require special extinguishers labeled D.

106F10.EPS

Figure 10 ◆ Fire extinguishers.

- Foam (or AFFF and FFFP) extinguishers coat the surface of a burning flammable liquid with a chemical foam. When using a foam extinguisher, blanket the entire surface of the liquid to exclude the air.
- Dry chemical extinguishers are either stored-pressure models or cartridge-operated models. The stored-pressure models have a lever above the handle for operation. The cartridge-operated models require two steps: first, depress the cartridge lever, and then squeeze the nozzle at the end of the hose. The dry chemicals leave a residue that must be cleaned up after use.
- Ammonium phosphate dry chemical extinguishers can be used on Class A, B, and C fires. However, they should never be used on commercial grease fryer fires because of the possibility of reflash and because it will make the fryer's automatic fire-protection system less effective.
- Sodium bicarbonate dry chemical extinguishers, used for fighting Class B and C fires, are preferred over other dry chemical extinguishers for fighting grease fires. Where provided, always use the extinguishing system first. This also shuts off the heat to the appliance.
- Potassium bicarbonate, urea-base potassium bicarbonate, and potassium chloride dry chemical extinguishers are more effective and use less chemical agent than sodium bicarbonate extinguishers.

8.4.0 Using Fire Extinguishers

Portable fire extinguishers apply an extinguishing agent that either cools burning fuel, displaces or removes oxygen, or stops the chemical reaction so a fire cannot continue to burn. When the handle of an extinguisher is compressed, it opens an inner canister of high-pressure gas that forces the extinguishing agent from the main cylinder through a siphon tube and out the nozzle. A fire extinguisher works much like a can of hair spray.

If fire extinguishers are available to you, it is your company's responsibility to teach you the principles and practices of using a fire extinguisher as well as the hazards associated with fighting small or developing fires. If you have been properly trained and are authorized to use a fire extinguisher, the following steps can help you fight a small fire.

Step 1 Sound the fire alarm and call the fire department, if appropriate.

Step 2 Identify a safe evacuation path before approaching the fire. Do not allow the fire, heat, or smoke to come between you and your evacuation path.

Step 3 Select the appropriate type of fire extinguisher.

Step 4 Discharge the extinguisher within its effective range using the **P.A.S.S.** (Pull pin, Aim, Squeeze, and Sweep) technique, aiming at the base of the flames (*Figure 11*).

Step 5 Back away from an extinguished fire in case it flames up again.

Step 6 Evacuate immediately if the extinguisher is empty and the fire is not out.

Step 7 Evacuate immediately if the fire gets too big.

Figure 11 ◆ P.A.S.S. technique.

Fire Extinguishers Can be Dangerous

A worker was killed when a gas-operated dry chemical fire extinguisher exploded when the worker tried to use it to put out a fire. An investigation determined that the fire extinguisher was severely corroded under the plastic end cap that provides a flat bottom for the unit to stand upright.

The Bottom Line: Fire extinguishers require periodic maintenance to ensure their readiness for emergency use. Never use a fire extinguisher that has not been inspected and maintained regularly.

Source: U.S. Department of Energy

Summary

Fires occur when something that burns, such as a fuel source, comes into contact with an ignition source. If this happens in an environment where oxygen is present, a fire will start. Making sure that fuel sources and ignition sources are always kept away from each other prevents fires. On a typical construction site where wood, paper, gasoline, paint, propane, kerosene, and other fuel sources are commonly used, and where heaters, welding tools, grinders, and tar kettles may all be in use, keeping sparks away from fuel is a tall order. To prevent fires, everyone must be knowledgeable of what can cause fires and aware of what's going on around them that might start a fire. Fire prevention isn't a job for just one person; it is everyone's job.

If a fire does start, be sure to call the fire department first and make sure the building is being evacuated. Then either put the fire out with a fire extinguisher or, if the fire is too large, if you have the wrong type of extinguisher, or if you are not trained on how to use the extinguisher, you should also evacuate.

Fires can be devastating and life-threatening. Preventing fires is the best way to keep yourself and your co-workers safe. Knowing what to do in case a fire does start will save lives and minimize damage. Make sure that you know what to do if a fire starts at your work site. It may save your life.

Review Questions

1. All of the following are considered fuel sources *except* _____.
   a. paint
   b. solvents
   c. trash
   d. metal

2. All of the following make up the fire triangle *except* _____.
   a. pressure
   b. oxygen
   c. heat
   d. fuel

3. Sparks caused by grinding are considered a fuel source.
   a. True
   b. False

4. Sources of ignition introduce _____ into the environment.
   a. flammable vapor
   b. fuel
   c. heat
   d. oxygen

5. _____ is responsible for creating an emergency action plan.
   a. Everyone on site
   b. OSHA
   c. The local fire department
   d. The company-appointed competent person

6. In case of a fire, knowing how to do all of the following is important *except* _____.
   a. how to report the fire
   b. when to bypass the chain of command, if needed
   c. how to use a fire extinguisher
   d. when to evacuate the work site

7. It is safe to fight a fire yourself as long as you know what type of material is burning.
   a. True
   b. False

8. A Class C fire is one in which _____.
   a. energized electrical equipment, including wiring, fuse boxes, circuit breakers, machinery, and/or appliances are burning
   b. ordinary combustibles such as wood, cloth, and/or paper are burning
   c. combustible metals such as magnesium or sodium are burning
   d. flammable liquids such as gasoline, oil, and oil-based paint are burning

9. A vessel used to safely transport and store flammable liquids is called a _____.
   a. safety cabinet
   b. fire cabinet
   c. safety can
   d. hazmat can

10. When using a fire extinguisher, you should aim the nozzle _____.
    a. in front of the flames
    b. at the top of the fire
    c. behind the flames
    d. at the base of the flames

GLOSSARY

Trade Terms Introduced in This Module

Class A fires: Fires in which ordinary combustibles such as wood, cloth, and/or paper are burning.

Class B fires: Fires in which flammable liquids such as gasoline, oil, and/or oil-based paint are burning.

Class C fires: Fires in which energized electrical equipment, including wiring, fuse boxes, circuit breakers, machinery, and/or appliances are burning.

Class D fires: Fires in which combustible metals such as magnesium or sodium are burning.

Combustible: Air or materials that can explode and cause a fire.

Fire triangle: Fire will only start if there is a source of ignition, a fuel source, and oxygen. These three things make up the sides of the fire triangle.

Flammable: Able to burn easily.

Fuel sources: Materials or gases that burn when exposed to heat, flames, or sparks. Wood, paper, and gas fumes are examples of fuel sources.

P.A.S.S.: The four steps to using a fire extinguisher are to Pull the pin, Aim at the fire, Squeeze the handle, and Sweep from side to side.

Safety cabinet: A special storage cabinet designed to reduce fire and explosion hazards from flammable materials.

Safety can: A special vessel designed to enable safe storage and handling of flammable liquids.

Safety technician: A professional trained to help keep the work environment safe.

Sources of ignition: Anything that creates excessive heat, flames, or sparks. Space heaters, matches, and grinders are examples of sources of ignition.

Figure Credits

Gary Wilson	106F02, 106F03
Justrite Mfg. Co.	106F05, 106F06
Ansul Incorporated	106F08, 106F10, 106F11
Kidde Safety	106F09

NCCER CURRICULA — USER UPDATE

NCCER makes every effort to keep its textbooks up-to-date and free of technical errors. We appreciate your help in this process. If you find an error, a typographical mistake, or an inaccuracy in NCCER's curricula, please fill out this form (or a photocopy), or complete the online form at **www.nccer.org/olf**. Be sure to include the exact module ID number, page number, a detailed description, and your recommended correction. Your input will be brought to the attention of the Authoring Team. Thank you for your assistance.

Instructors – If you have an idea for improving this textbook, or have found that additional materials were necessary to teach this module effectively, please let us know so that we may present your suggestions to the Authoring Team.

NCCER Product Development and Revision

13614 Progress Blvd., Alachua, FL 32615

Email: curriculum@nccer.org
Online: www.nccer.org/olf

❑ Trainee Guide ❑ AIG ❑ Exam ❑ PowerPoints Other ______________________

Craft / Level: Copyright Date:

Module ID Number / Title:

Section Number(s):

Description:

Recommended Correction:

Your Name:

Address:

Email: Phone:

Module 75107-03

Hand- and Power-Tool Safety

COURSE MAP

This course map shows all of the modules in Field Safety. The suggested training order begins at the bottom and proceeds up. The local Training Program Sponsor may adjust the training order.

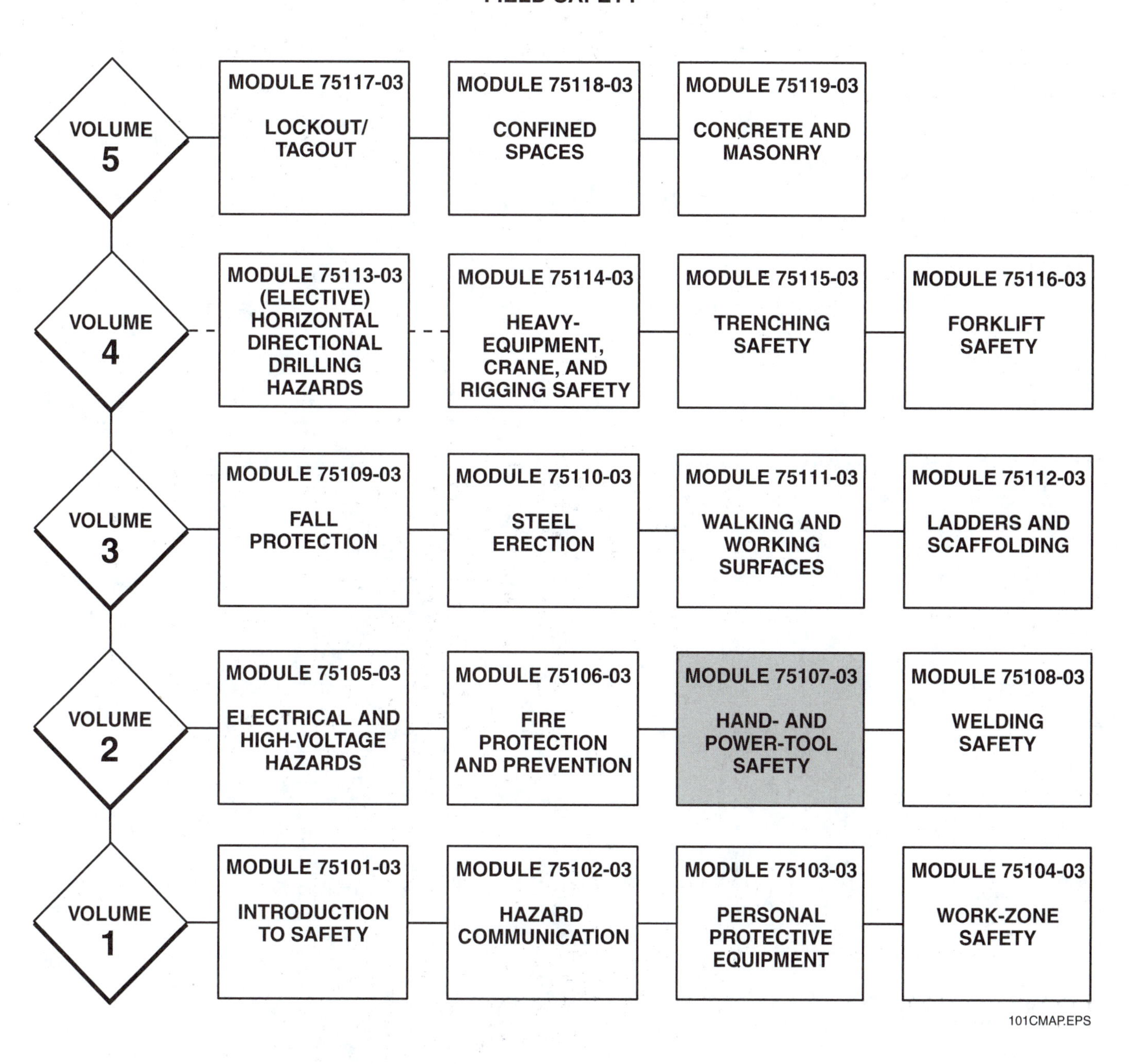

MODULE 75107-03 CONTENTS

Figures

Tables

MODULE 75107-03

Hand- and Power-Tool Safety

Objectives

When you have completed this module, you will be able to do the following:

1. Identify different types of power tools.
2. Describe the uses of hand and power tools.
3. List the five most common power sources for power tools.
4. Describe the risks associated with hand tools.
5. Describe the risks associated with each type of power tool.
6. Explain how to minimize the risks associated with operating hand and power tools.

Prerequisites

Before you begin this module, it is recommended that you successfully complete the following: Field Safety, Modules 75101-03 through 75106-03.

Required Materials

1. Pencil and paper
2. Appropriate personal protective equipment

1.0.0 ◆ INTRODUCTION

Tools are a common part of our lives, so it may be easy to forget that they can pose serious safety risks. All tools are built with safety in mind, but serious tool-related accidents still occur. In one incident, a carpenter apprentice was killed when he was struck in the head by a nail that was fired from a powder-actuated tool.

Hand and **power tools** can cause many different types of injuries. Common hazards include tripping, electrical shock, cutting, and fire. Some of the potential injuries include:

- Burns
- Cuts
- Sprains
- Electrical shock
- Eye injuries
- Hearing loss
- Broken bones

This module describes many of the risks associated with hand and power tools and some of the ways you can protect yourself and those working around you.

2.0.0 ◆ FIRE HAZARDS RELATED TO HAND AND POWER TOOLS

Fire is a significant hazard when working with hand and power tools. Fires occur when sparks come in contact with flammable or combustible materials. Sparks from power tools as well as striking **impact tools** can ignite fumes, oxygen, paper, or solvents being used in the area. To minimize this risk, use the right tools for the job, be aware of your environment, and know what to do if there is a fire.

2.1.0 How Fires Start

For a fire to start, three things are needed in the same place at the same time: fuel, heat, and oxygen. If one of these three is missing, a fire will not start.

Fuel is anything that will combine with oxygen and heat to burn. When pure oxygen is present, such as near a leaking oxygen hose or fitting, material that would not normally be considered fuel (including some metals) will burn.

Heat is anything that will raise a fuel's temperature to the flash point. The flash point is the temperature at which a fuel gives off enough gases (vapors) to burn. The flash points of many fuels are quite low: room temperature or less. When the burning gases raise the temperature of a fuel to the point at which it ignites, the fuel itself will burn and keep burning even if the original source of heat is removed.

Oxygen is always present in the air.

What is needed for a fire to start can be shown as a fire triangle (*Figure 1*). If any one element of the triangle is missing, a fire cannot start. If a fire has started, removing any one element from the triangle will put it out.

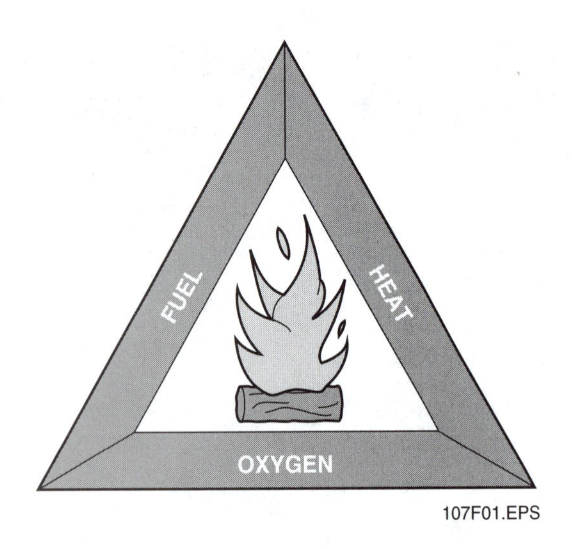

Figure 1 ◆ Fire triangle.

Research has added a fourth side to the fire triangle concept, resulting in the development of a new model called the Fire Tetrahedron. The fourth element involved in the combustion process is referred to as the chemical chain reaction. Specific chemical chain reactions between fuel and oxygen molecules are essential to sustaining a fire once it has begun.

2.2.0 Fire Prevention

Obviously, the best way to provide fire safety is to prevent a fire from starting in the first place. The best way to prevent a fire is to make sure that the three elements needed for fire (fuel, oxygen, and heat) are never present in the same place at the same time.

The following are some basic safety guidelines for fire prevention:

- Always work in a well-ventilated area, especially when you are using flammable materials such as shellac, lacquer, paint stripper, or construction adhesives.
- Never smoke or light matches when you are working with flammable materials.
- Keep oily rags in approved, self-closing metal containers.
- Store combustible materials only in approved containers.
- Know where to find fire extinguishers, what kind of extinguisher to use for different kinds of fires, and how to use the extinguishers.
- Make sure all extinguishers are fully charged. Never remove the tag from an extinguisher; it shows the date the extinguisher was last serviced and inspected.

2.3.0 Fire Fighting

You are not expected to be an expert firefighter. But you may have to deal with a fire to protect your safety and the safety of others. You need to know the location of fire-fighting equipment on your job site. You also need to know which equipment to use on different types of fires. However, only qualified personnel are authorized to fight fires.

Most companies tell new employees where fire extinguishers are kept. If you have not been told, be sure to ask. Also ask how to report fires. The telephone number of the nearest fire department should be clearly posted in your work area. If your company has a company fire brigade, learn how to contact them. Learn your company's fire-safety procedures.

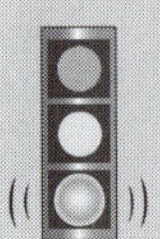

NOTE

For more information on fire protection, refer to the *Fire Protection and Prevention* module in Volume Two.

3.0.0 ◆ HAND-TOOL SAFETY

Hand tools are non-powered tools and may include anything from axes to wrenches (*Figure 2*). Hand tools are dangerous if they are misused or improperly maintained.

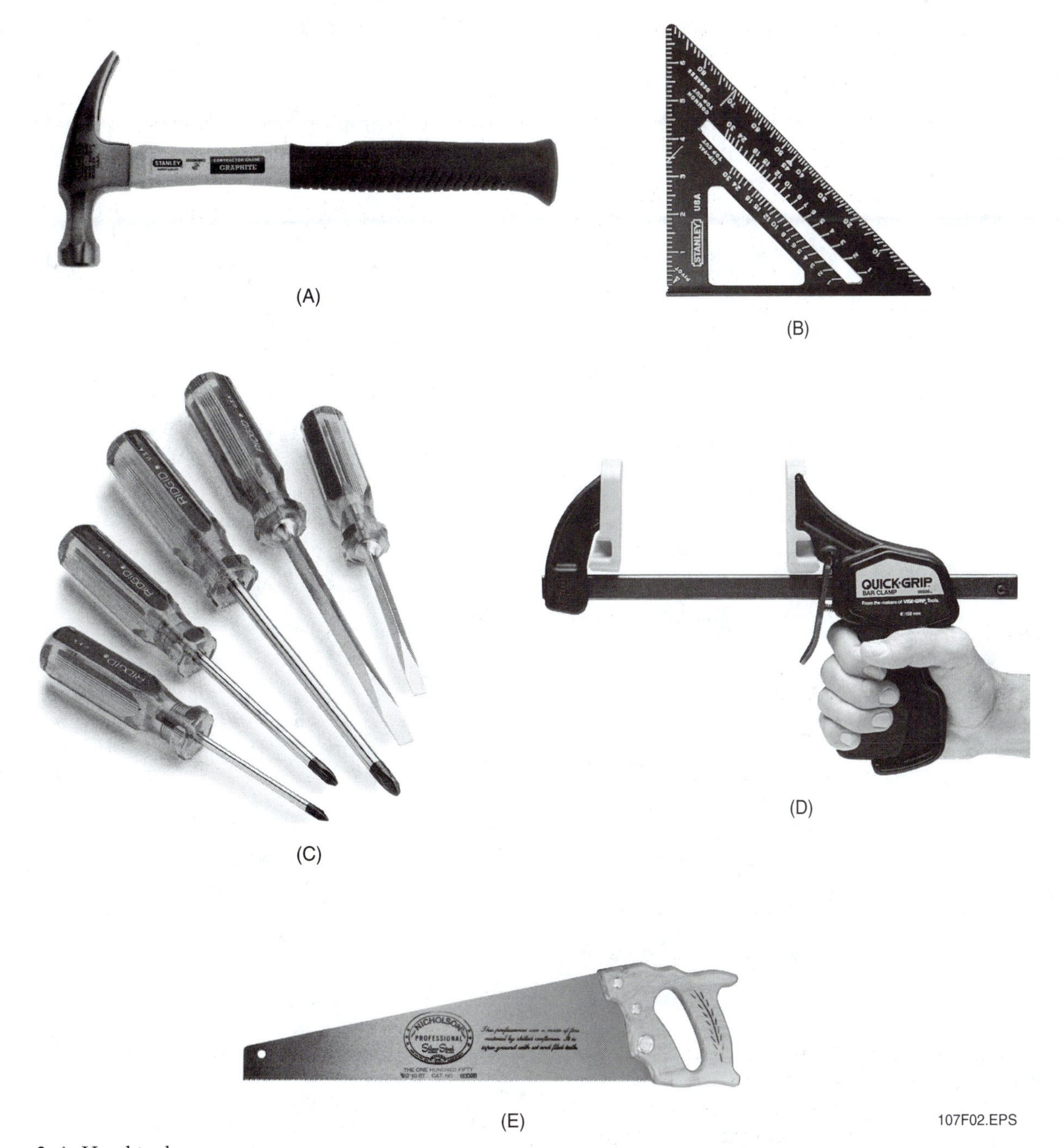

Figure 2 ◆ Hand tools.

The following hazards are associated with hand tools:

- Using the wrong tool can be dangerous. Using a chisel as a screwdriver may cause the tip of the chisel to break and fly off, hitting the user or another person. Only use tools for their designated purpose.
- Always maintain your tools properly. If the wooden handle on a tool such as an ax or hammer is loose, splintered, or cracked, the head of the tool may fly off and strike the user or another person.
- Repair or replace damaged or worn tools. A wrench with its jaws sprung might easily slip, causing hand injuries. If the wrench flies, it may strike the user or another person.
- Impact tools such as chisels, wedges, and drift pins are unsafe if they have mushroomed heads. The heads might shatter when struck, sending sharp fragments flying.

In general, your risks are greatly reduced by inspecting and maintaining tools regularly and always wearing appropriate personal protective equipment such as safety goggles, hard hats, and filtering masks. Also, keep floors dry and clean to prevent accidental slips that can result in injuries caused by the tools you may be using.

3.1.0 Bladed Tools

Bladed tools (*Figure 3*) present the risks of cutting, slicing, snipping, and stabbing. These injuries can be as minor as a small cut or as severe as an amputation or stab wound. To minimize these risks:

- Use bladed tools with the blades and points aimed away from yourself and other people.
- Direct bladed tools away from aisle areas.
- Store bladed tools properly; use the sheath or protective covering if there is one.
- Keep blades sharp and inspect them regularly. Dull blades are difficult to use and control and can be far more dangerous than well-maintained blades.

3.2.0 Impact Tools

Impact tools such as chisels, wedges, and punches are meant to be struck by a hammer (*Figure 4*). The two most common types of injuries associated with impact tools are hammer strikes and eye injuries from flying tool fragments. Getting struck by a hammer is usually the result of carelessness or not paying attention. If you are using a hammer, keep your eyes and your mind on what you're doing.

Injuries caused by flying tool fragments from damaged tools can be prevented by inspecting the tool before using it. Look for mushroomed heads, cracks, chips, or other signs of damage or weakness. Eye injuries can also be prevented by wearing proper eye protection while using impact tools.

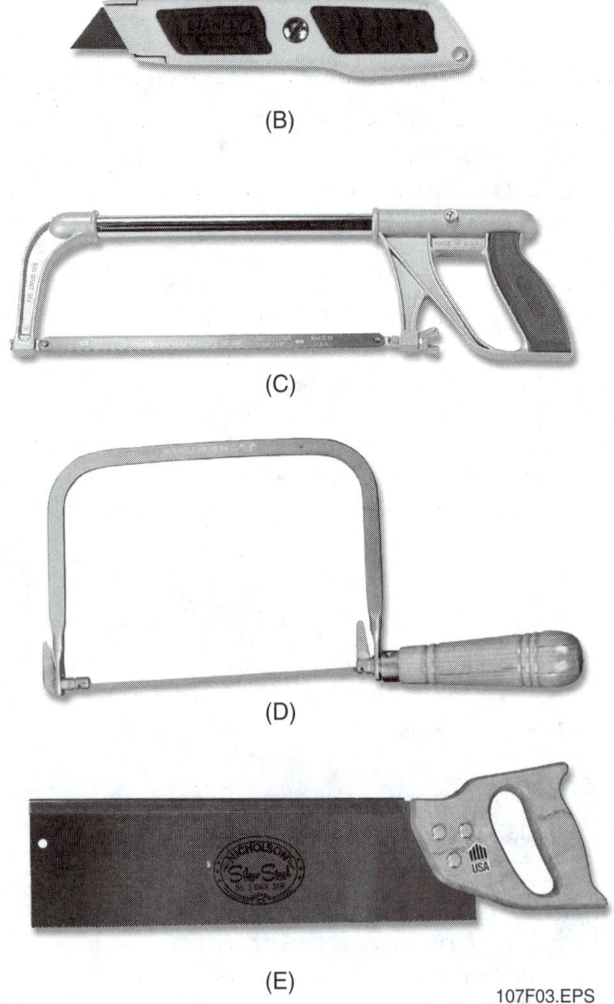

Figure 3 ◆ Bladed tools.

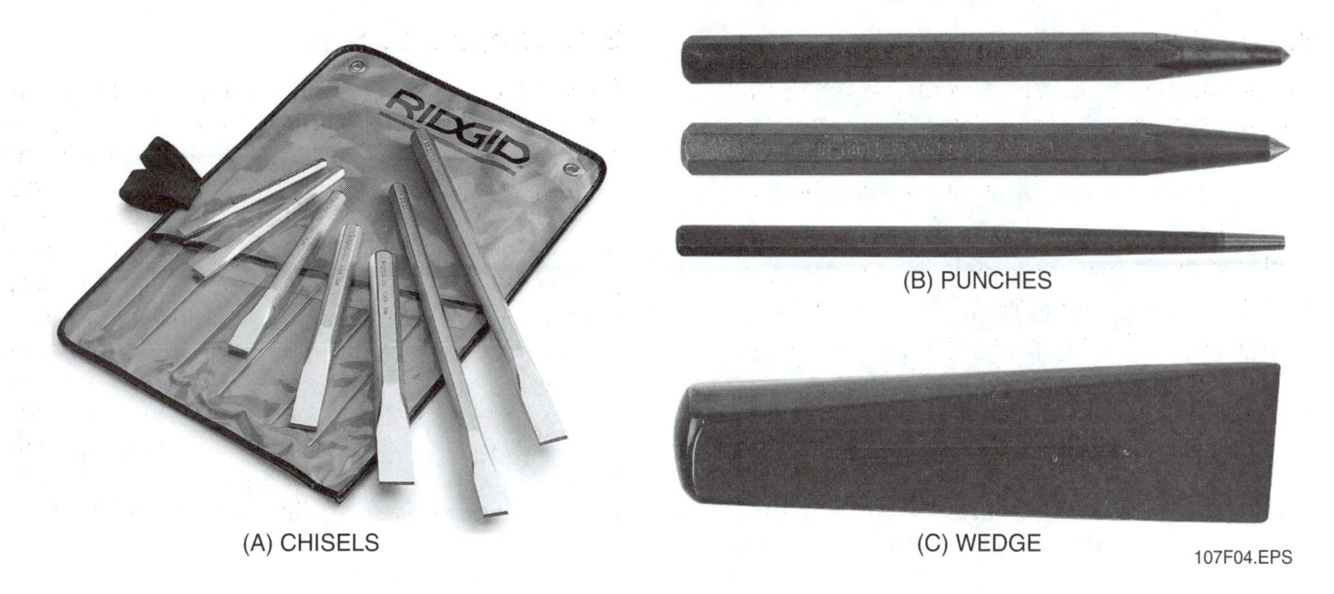

Figure 4 ◆ Impact tools.

3.3.0 Dust and Suspended Particles

Some hand tools create a lot of airborne dust and particles. Sand paper, planes, files, and saws, for example, may create a lot of sawdust or drywall powder (*Figure 5*). Short-term exposure to these particles may cause minor irritation of the eyes, nose, or sinuses. In some individuals, exposure may trigger a more serious reaction such as an asthma attack. Long-term exposure over the course of a career can impact your lungs, possibly causing chronic bronchitis, pneumonia, or other pulmonary disorders. To help prevent these conditions, filtering masks and eye protection should be worn when you are using a tool that creates dust.

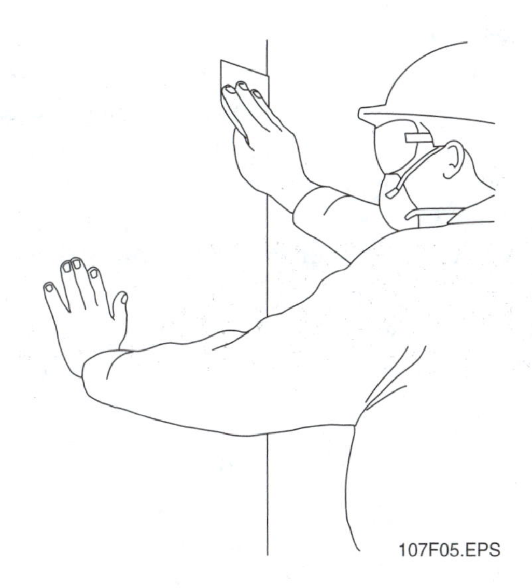

Figure 5 ◆ Sanding drywall.

4.0.0 ◆ POWER TOOLS

Power tools can be hazardous when they are improperly used or not well maintained. Most of the risks associated with hand tools are also risks when using power tools. When you add a power source to a tool, however, the risk factors increase. For example, *Figure 6* shows a radial saw arm. This is far more dangerous than a hand-powered saw.

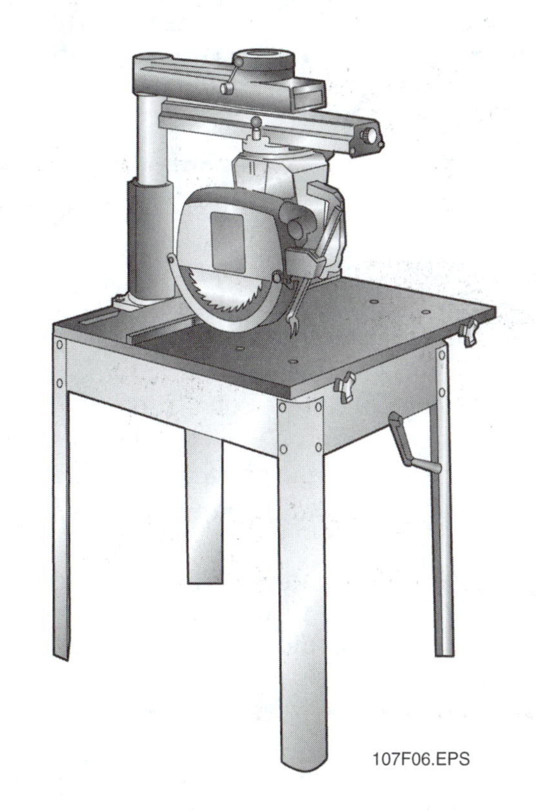

Figure 6 ◆ Radial arm saw.

Power tools are powered by different sources. Some examples of power sources for power tools include:

- Electricity
- Pneumatics (air pressure)
- Liquid fuel
- Hydraulics (fluid pressure)

You should know the safety rules and proper operating procedures for each tool you use. Specific operating procedures and safety rules for using a tool are provided in the operator's/user's manual supplied by the manufacturer. Before operating any power tool for the first time, always read the manual to familiarize yourself with the tool. If the manual is missing, contact the manufacturer for a replacement.

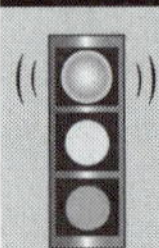

WARNING!

Never use a tool if you are unsure of how to use it correctly.

4.1.0 Safety Guidelines for Power Tools

Before explaining more about power tools, you must understand the general safety rules that apply when using all power tools, regardless of type. Follow these general guidelines in order to prevent accidents and injury:

- Never carry a tool by the cord or hose.
- Keep cords and hoses away from heat, oil, and sharp edges.
- Do not attempt to operate any power tool before being checked out by your instructor on that particular tool.
- Always wear eye protection, a hard hat, and any other required personal protective equipment when operating power tools.
- Wear face and hearing protection when required.
- Wear proper respiratory equipment when necessary.
- Wear the appropriate clothing for the job being done. Wear close-fitting clothing that cannot become caught in moving tools. Roll up or button long sleeves, tuck in shirttails, and tie back long hair. Do not wear any jewelry, including watches or rings.
- Do not distract others or let anyone distract you while operating a power tool.
- Do not engage in horseplay.
- Do not run or throw objects.
- Consider the safety of others, as well as yourself. Observers should be kept at a safe distance away from the work area.
- Never leave a power tool running while it is unattended.
- Assume a safe and comfortable position before using a power tool. Be sure to maintain good footing and balance.
- Secure work with clamps or a vise, freeing both hands to safely operate the tool.
- To avoid accidental starting, never carry a tool with your finger on the switch.
- Be sure that a power tool is properly grounded and connected to a ground fault circuit interrupter (GFCI) before using it (*Figure 7*).

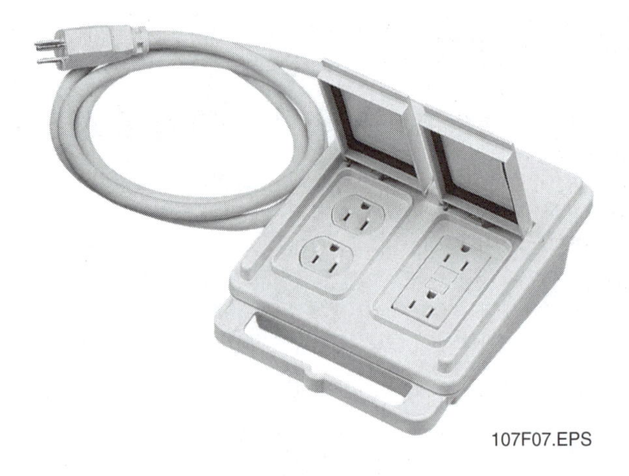

107F07.EPS

Figure 7 ◆ Ground fault circuit interrupter (GFCI).

- Ensure that power tools are disconnected before performing maintenance or changing accessories.
- Do not use dull or broken tools or accessories.
- Use a power tool only for its intended use.
- Keep your feet, fingers, and hair away from the blade and/or other moving parts of a power tool.
- Never use a power tool with **guards** or safety devices removed or disabled.
- Never operate a power tool if your hands or feet are wet.
- Keep the work area clean at all times.
- Become familiar with the correct operation and adjustments of a power tool before attempting to use it.
- Keep tools sharp and clean for the best performance.
- Follow the instructions in the user's manual for lubricating and changing accessories.
- Keep a firm grip on the power tool at all times.

- Use electric extension cords of sufficient size to service the particular power tool you are using.
- Report unsafe conditions to your instructor or supervisor.
- Remove damaged tools from use and tag out with Do Not Use tags.

4.2.0 Electrically Powered Tools

The most serious danger in using **electrically powered tools** is that of electrocution. The *Electrical and High-Voltage Hazards* module in Volume Two covers this topic in great detail; an overview is provided here.

Electricity can cause burns, shocks, explosions, electrocution, and fires. Electrical shocks can be minor and uncomfortable or they can be severe, causing burns or death. Even a small amount of current can cause the heart to stop pumping in rhythm. If not corrected, this condition will result in death. Electrical shock can also cause a loss of balance, muscle control, or consciousness, which could then cause the victim to fall and/or drop a tool. A fall from a ladder or scaffolding can obviously be quite serious.

To prevent electrical shock, tools must provide at least one of the following types of protection:

- *Double insulated* – Double insulation is more convenient than three-wire cords. The user and tools are protected in two ways: by normal insulation on the wires inside, and by a housing that cannot conduct electricity to the operator in the event of a malfunction.
- *Powered by a low-voltage isolation transformer* – If your electrically powered tools do not have either a ground plug or double insulation, check with your supervisor to make sure that you are protected by a low-voltage isolation transformer.
- *Grounded with a three-wire cord* – Three-wire cords have two current-carrying conductors and one grounding conductor. You are probably familiar with the three-prong plug common on electrically powered tools (*Figure 8*). When there is a three-prong cord, that means the tool is powered by a grounded three-wire cord and it should only be plugged into a three-prong, grounded receptacle (*Figure 9*). Any time an adapter is used to accommodate a two-hole receptacle, the adapter wire must be attached to a known ground. Never remove the third prong (grounding conductor) from a plug. If you are using a three-prong extension cord, make sure that it is properly grounded at its source.

Don't Remove the Grounding Prong

An employee was climbing a metal ladder to hand an electric drill to the journeyman installer on a scaffold about 5' above him. When the victim reached the third rung from the bottom of the ladder, he received an electric shock that killed him. The investigation revealed that the extension cord had a missing grounding prong and that a conductor on the green grounding wire was making intermittent contact with the energized black wire, thereby energizing the entire length of the grounding wire and the drill's frame. The drill was not double insulated.

The Bottom Line: Do not disable any safety device on a power tool. A ground fault can be deadly.

Source: The Occupational Safety and Health Administration (OSHA)

Here are some general tips to help you stay safe when using electrically powered tools:

- Electrically powered tools should be used on jobs for which they were intended.
- Wear all appropriate personal protective equipment, such as gloves and safety footwear.
- When not in use, store power tools in a dry place.
- Electrically powered tools should not be used in damp or wet places.
- The work area should be well lit.

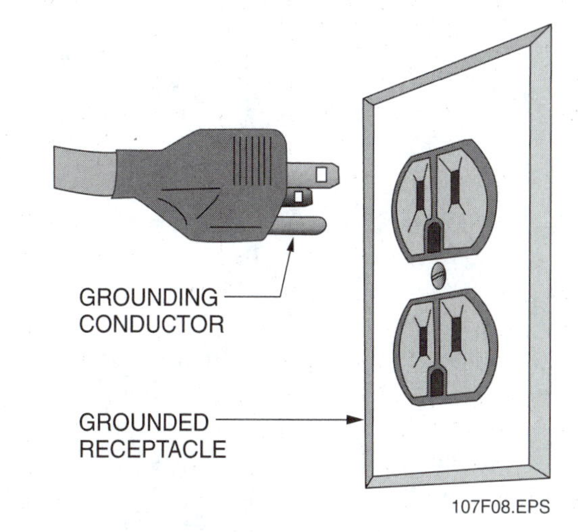

Figure 8 ◆ Grounded three-prong plug.

2-POLE 3-WIRE GROUNDING		15 AMPERE		20 AMPERE		30 AMPERE		50 AMPERE	
		RECEPTACLE	PLUG	RECEPTACLE	PLUG	RECEPTACLE	PLUG	RECEPTACLE	PLUG
5	125V	5-15R	5-15P	5-20R	5-20P	5-30R	5-30P	5-50R	5-50P
6	250V	6-15R	6-15P	6-20R	6-20P	6-30R	6-30P	6-50R	6-50P
7	277V	7-15R	7-15P	7-20R	7-20P	7-30R	7-30P	7-50R	7-50P

107F09.EPS

Figure 9 ◆ Three-prong receptacle configurations.

Because malfunctioning electrically powered tools can cause sparks, these tools can also cause fires and explosions. Make sure you know about any fire hazards in your work area and avoid using electrically powered tools around flammable materials, fumes, and gases. The *Fire Protection and Prevention* module in Volume Two has more information on this topic.

Finally, electrical cords and extension cords pose a tripping hazard. Extension cords should be brightly colored to make them more visible. Cords and cables should be run somewhere other than walkways, or at least along a wall, rather than in the middle of a walkway or across a walkway. Also, avoid running cables and cords across elevated work areas and scaffolding. Occasionally, it may be necessary to run a cord or cable across a walkway. If this cannot be avoided, either tape the cord down and put a carpet over it, or place it in a cord runner designed to minimize the tripping hazard.

4.3.0 Pneumatic Tools

Pneumatic tools (*Figure 10*) such as nail guns, chippers, drills, jackhammers, and sanders are powered by compressed air. The most serious risk in using these tools is the danger of getting hit by one of the tool's attachments or by the fastener it is shooting. Tools that shoot nails, rivets, or staples, and operate at pressures more than 100 pounds per square inch (psi), must be equipped with a safety device that won't allow fasteners to be shot unless the muzzle is pressed against a work surface. Compressed-air guns should never be pointed toward anyone and the muzzle should never be pressed against a person. A safety clip or retainer must be installed on these tools to prevent attachments from being unintentionally shot from the barrel.

The use of pneumatic tools often creates flying debris. If you are using a pneumatic tool, you must wear proper eye protection. Full face protection is highly recommended. Screens must be set up to protect nearby workers from being struck by flying fragments around chippers, riveting guns, staplers, or air drills.

Airless spray guns that atomize paints and fluids at high pressures (1,000 psi or more) must be equipped with automatic or visual safety devices that will prevent the trigger from being pulled until the safety device is manually released.

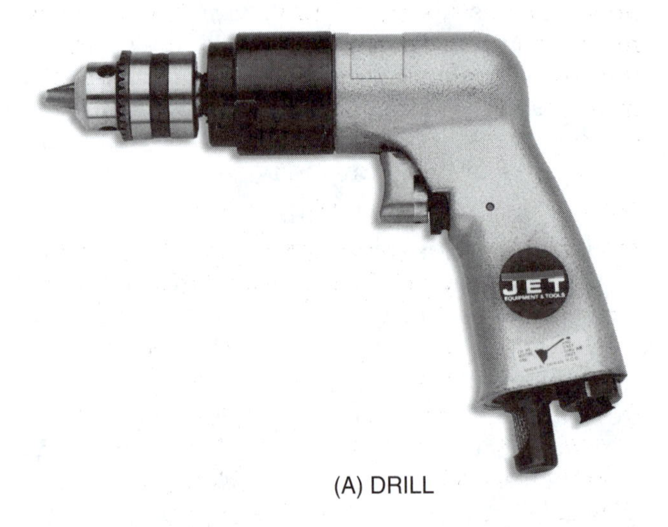

(A) DRILL

(B) NAIL GUN 107F10.EPS

Figure 10 ◆ Pneumatic tools.

4.3.1 Hearing Protection

Some pneumatic tools, like jackhammers, can be very noisy. You must wear ear protection when using these tools. According to OSHA, if you are exposed to sound levels that average at or above 85 **decibels (dB)** over an eight-hour period, (known as an eight-hour time-weighted average or TWA) you should wear hearing protection. 85dB is not very loud. As *Table 1* shows, 85dB is about as loud as the alarm on a typical digital alarm clock. Even at this level, long-term exposure can pose a risk to your hearing. A jackhammer's sound intensity is typically about 100dB; if you are going to be working with or near a jackhammer, you must wear hearing protection.

The competent person on site, who is trained in fitting hearing protectors, must help workers choose the correct size and type of protection for their working environment. The protector selected should be comfortable to wear and reduce noise levels sufficiently to prevent hearing loss.

Hearing protectors must adequately reduce the noise level for each employee's work environment. You must be shown how to use and care for the protectors and must be supervised on the job to make sure you know how and when to wear them.

Table 1 Decibel Levels of Familiar Sounds

Noise Source	Decibel Level
Busy traffic	75dB
Noisy restaurant	80dB
Digital alarm clock	85dB
Average factory	88dB
Screaming child	90dB
Subway train	100dB
Diesel truck	100dB
Helicopter	105dB
Power mower	105dB
Power saw	110dB
Football stadium	117dB
Rock concert	120dB
Car horn	120dB
Jackhammer	130dB
Air raid siren	130dB
PAIN STARTS	**140dB**
Gunshot	140dB
Jet engine	140dB
Rocket launching	180dB
Loudest sound	194dB

A sound's intensity level depends on your distance from the source. The further you are from the sound's source, the quieter it will seem to you. As a rule of thumb, the decibel level may be above the legal limit if you have to raise your voice to be heard one foot away. Other signs of high noise levels are temporary hearing loss or ringing in the ears.

Everyone is different. Some workers will experience hearing loss even if noise is below the legal limit. Since there's no way of telling if you're the one whose hearing will be the first to go, it's best to avoid noise exposure whenever possible. If there's any reason to think the noise level may be too high, the company can have the level measured with instruments. This is called noise monitoring.

For more guidance on when to wear ear protection, what type of ear protection to wear, and how to wear it correctly, see your supervisor or safety technician.

4.3.2 Air Hoses

Pneumatic tools get their air supply through air hoses (*Figure 11*). When using pneumatic tools, always check to see that they are fastened securely to the air hose to prevent them from being disconnected. Most pneumatic tools have a short wire or a positive locking device attaching the air hose to the tool. These will help you make sure that the tool is attached correctly.

If the air hose is more than ½" in diameter, a safety excess flow valve must be installed at the source of the air supply to shut off the air automatically in case the hose breaks. In general, the same safety precautions should be taken with an air hose that are recommended for electric cords. This is because the same types of accidents and injuries can happen with both, including tripping.

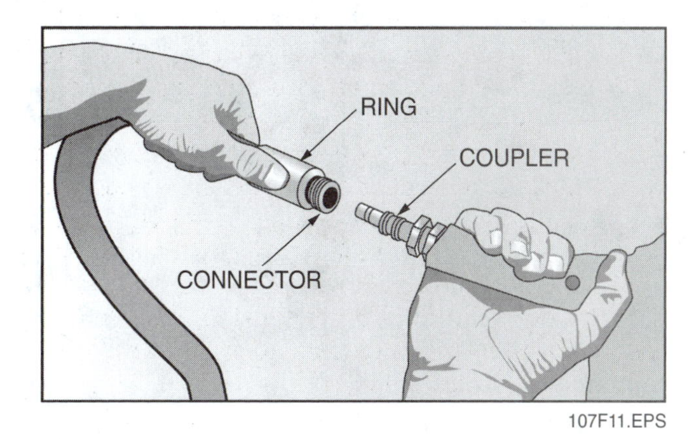

Figure 11 ◆ Pneumatic air hose connection.

Watch Your Step Around Hoses

A worker tripped over a trailing compressed-air hose left in an alleyway and fell hard, breaking his arm and injuring his back. The air tube line had not been put away after use. The alleyway was poorly lit and the incident happened at night. Following an investigation, the company installed floodlighting in poorly lit areas and relocated the air line and hose. Safety awareness and housekeeping training is now given to all staff.

The Bottom Line: Good housekeeping is an important part of safety. Store all tools and hoses after use.

Source: The National Institute for Occupational Safety and Health (NIOSH)

4.3.3 Jackhammers

Besides a damaging sound level of 100dB, jackhammers and other demolition equipment (*Figure 12*) have other dangers that the user should keep in mind. They can create flying debris, so eye protection, preferably full face protection, must be worn. Heavy jackhammers cause muscle fatigue and strains over long periods of use. Heavy-duty rubber grips help to minimize these problems. Wearing gloves and taking reasonable breaks also help to lessen muscle fatigue and strain. Finally, always wear steel-toed safety shoes when using a jackhammer. Safety shoes will minimize injury if the hammer strikes your foot.

4.4.0 Liquid-Fuel Tools

Some power tools, like chainsaws (*Figure 13*), are powered by a liquid fuel, usually gasoline. The most serious hazard with fuel-powered tools comes from fuel vapors that can burn or explode.

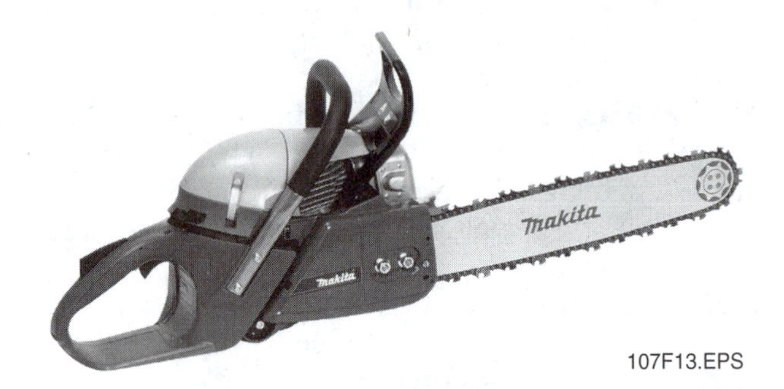

Figure 13 ◆ Chainsaw.

Burning liquid fuel also gives off exhaust fumes, which can be dangerous. Here are a few tips on using **liquid-fuel tools** safely:

- Always wear the appropriate personal protective equipment including eye protection, gloves, and respirators, if necessary.
- Always handle, transport, and store the fuel only in approved flammable-liquid containers.

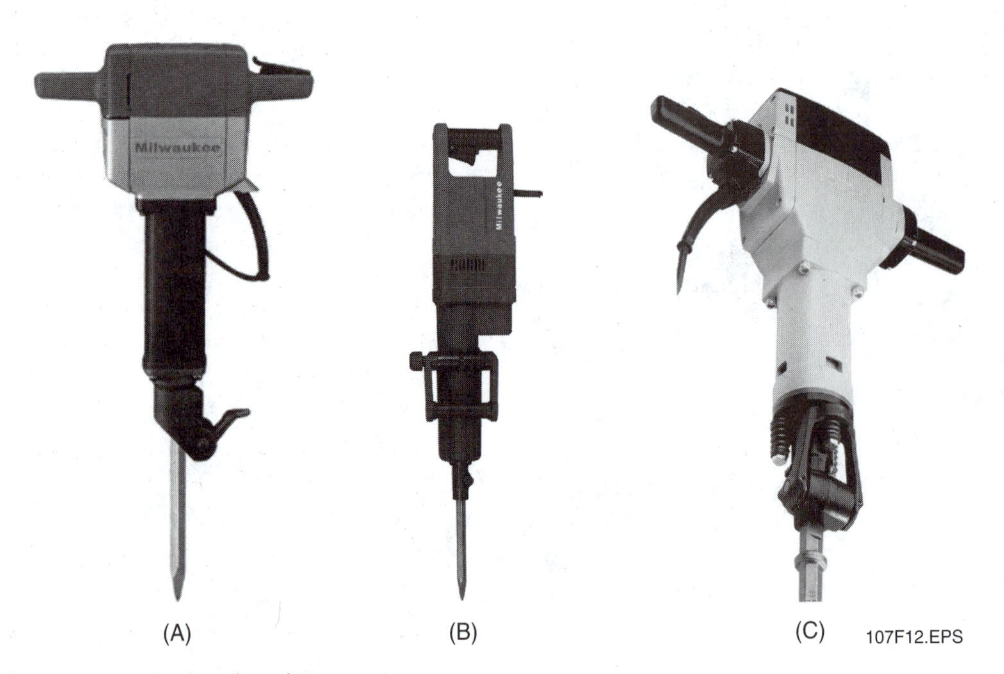

Figure 12 ◆ Jackhammers and other demolition equipment.

Fumes From Liquid-Fuel Tools

A worker at a large, enclosed construction site died of carbon monoxide poisoning after he and six other workers were exposed to high levels of the gas. He died because ventilation on the site was inadequate, and three machines were giving off carbon monoxide: a portable mixer and a trowel, both powered by gasoline, and a forklift powered by propane. This worker would have survived if the work area had been properly ventilated and if he was using the proper personal protective equipment, including a respirator.

The Bottom Line: Keep work areas well ventilated and wear respirators when required. Hazardous air conditions can develop without warning.

Source: The Occupational Safety and Health Administration (OSHA)

- Before the tank for a liquid-fuel powered tool is refilled, shut down the engine and allow it to cool down. This will reduce the risk of a hot tool igniting fuel vapors.
- If you're using a liquid-fueled tool inside a closed area, there must be adequate ventilation and/or respirators in use to avoid breathing dangerous exhaust fumes.
- Fire extinguishers must be available in the area where liquid-fuel powered tools are being used.

4.4.1 Flammable and Combustible Liquids

Liquids can be flammable or combustible. Flammable liquids have a flash point below 100°F. Combustible liquids have a flash point at or above 100°F. Fire can be prevented by:

- *Removing the fuel* – Liquid does not burn. What burns are the gases (vapors) given off as the liquid evaporates. Keeping the liquid in a sealed, approved container prevents evaporation. If there is no evaporation, there is no fuel to burn.
- *Removing the heat* – If the liquid is stored or used away from a heat source, it will not be able to ignite.
- *Removing the oxygen* – The vapor from a liquid will not burn if oxygen is not present. Keeping safety containers tightly sealed prevents oxygen from coming into contact with the fuel.

4.4.2 Fire Extinguishers

All fire extinguishers are labeled, using standard symbols, for the classes of fires on which they can be used. These are the four classes of fires (*Figure 14*).

- *Class A* – Ordinary combustibles such as wood, cloth, and paper
- *Class B* – Flammable liquids such as gasoline, oil, and oil-based paint
- *Class C* – Energized electrical equipment, including wiring, fuse boxes, circuit breakers, machinery, and appliances
- *Class D* – Combustible metals such as magnesium or sodium

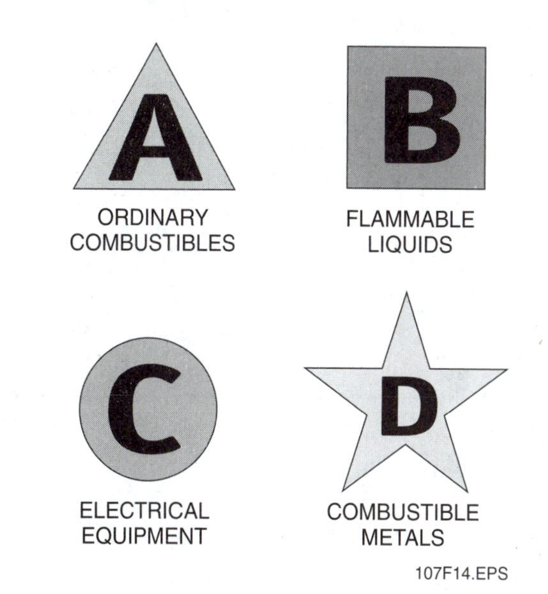

Figure 14 ◆ Fire class symbols.

A red slash through any of the symbols tells you the extinguisher cannot be used on that class of fire. A missing symbol tells you only that the extinguisher has not been tested for a given class of fire, but may be used if an extinguisher labeled for that class of fire is not available.

Remember that the extinguisher must be appropriate for the type of fire. Multi-purpose fire extinguishers, labeled ABC, may be used on all three classes of fire. If you use the wrong type of extinguisher, you can endanger yourself and make the fire worse. It is also very dangerous to use water or an extinguisher labeled only for Class A fires on a gasoline or electrical fire.

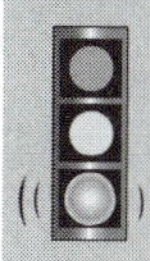

NOTE

For more information about fire protection and safety, refer to the *Fire Protection and Prevention* module in Volume Two.

Watch Out for Combustible Fumes

A worker was using a liquid propane blow torch in the bilge area of a construction barge. The torch flamed out and the worker left the area without turning off the flow of gas. The worker was killed when he returned and lit the torch, igniting the accumulated gas.

The Bottom Line: Always provide adequate ventilation in confined spaces. Never leave a liquid-fuel tool without checking that the flow of gas has been turned off.

Source: The Occupational Safety and Health Administration (OSHA)

4.5.0 Powder-Actuated Tools

Powder-actuated tools operate like loaded guns and should be treated with the same respect and precautions (*Figure 15*). The use of powder-actuated tools requires special training and certification. If you are working around these tools, here are some ways to help you stay safe:

- Never use explosive or flammable materials around powder-actuated tools.
- Never point powder-actuated tools at anybody.
- Never pick up an unattended powder-actuated tool. Instead, tell your supervisor that a powder-actuated tool has been left unattended.
- Never play with powder-actuated tools. Again, these tools are as dangerous as a loaded gun.

Manufacturers use color-coding schemes to identify the strength of a powder load charge. It is extremely important to select the right charge for the job. If you are going to use powder-actuated tools, you must learn the color-coding system that applies to the tool you are using. *Table 2* shows an example of a color-coding system. The power level ranges from 1 to 6, with 1 being the least powerful. This is a general guide. Learn the color code specific to your equipment.

Table 2 Powder Charge Color-Coding System

Power Level	Powder Charge Color
1	Gray
2	Brown
3	Green
4	Yellow
5	Red
6	Purple

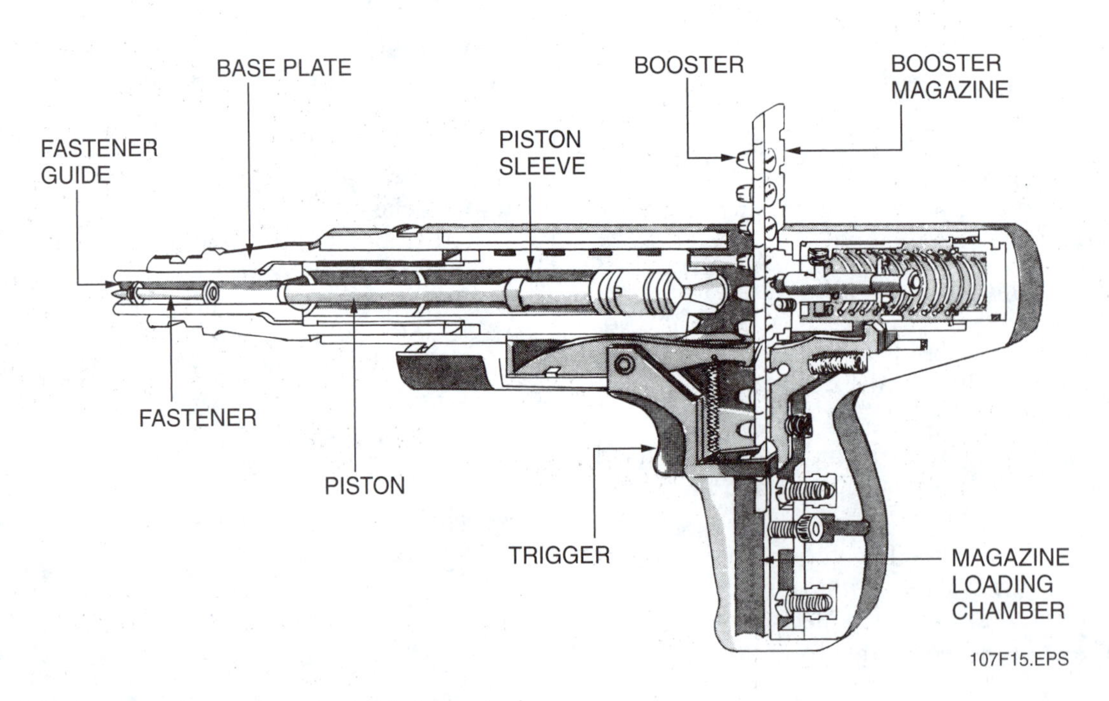

Figure 15 ◆ Major parts of a powder-actuated fastening tool.

4.5.1 Fasteners

When using powder-actuated tools to apply fasteners, dangerous situations can arise. If fasteners are being fired using too much powder or into material that is too thin, it is possible for the fastener to pass right through its target and strike a person or equipment. Fasteners driven into materials like bricks and concrete 3" or less from an edge or corner may create dangerous flying pieces. Fasteners driven less than 1" from the edge or corner of a piece of steel can cause a similar danger of flying shrapnel. Finally, if a fastener is driven into very hard or brittle material, chipping, splattering, or ricocheting could occur. Any of these could result in a life-threatening or seriously debilitating injury. If you're working around powder-actuated tools, always pay attention, watch what you are doing, and take all steps necessary to reduce your risk of injury.

Never use a powder-actuated tool to secure fasteners into easily penetrated material; these tools are designed primarily for installing fasteners into masonry. Do not attempt to use these tools unless you have been properly trained. In addition, all personnel in the area must be aware that the tool is in use and should be wearing appropriate personal protective equipment.

4.6.0 Hydraulic Tools

Hydraulic tools are tools that are operated by the force and movement of liquid forced through an opening such as a tube or hose. For example, portable hydraulic jacks (*Figure 16*) have two basic parts: the pump and the cylinder. The two parts are joined by a high-pressure hydraulic hose. The pump applies pressure to the hydraulic fluid. The cylinder (sometimes called a ram) applies a lifting or pushing force that operates the jack.

107F16.EPS

Figure 16 ◆ Portable hydraulic jack.

The risks and safety rules for hydraulic tools are basically the same as those for pneumatic tools. There is at least one additional item to keep in mind, and that is hydraulic fluid. Hydraulic fluid spilled on the floor is a dangerous slipping hazard. If hydraulic fluid spills, pour an oil-absorbing material onto the spill and clean it up immediately.

4.6.1 Jacks

Jacks are tools used routinely both on and off the job (*Figure 17*). To use a jack safely, remember the following:

- Make sure the jack's base is resting on a firm, level surface.
- To avoid tipping, make sure the jack is centered under the load.

Hydraulic Pressure Can be Dangerous

A laborer for a boring and tunneling company died when he was struck by a hydraulic hose coupling. The victim was part of a crew digging a tunnel. They were using a hydraulic jacking machine to push 54"-diameter casing into the tunnel, one section at a time. The victim was standing next to one of the jacks. When pressure was applied to the hydraulic lines, the male end of the hydraulic hose coupling split in half, and the connection was broken. The sudden release of hydraulic pressure caused the hose with the connector on the end to whip around and strike the victim in the abdomen. An investigation determined that safety straps, also called whip checks, were not installed on the jack and that the worker was standing too close to the jack.

The Bottom Line: Always make sure that all necessary safety devices are in place before using hydraulic equipment. Stand clear of lines and hoses under extreme loads.

Source: The National Institute for Occupational Safety and Health (NIOSH)

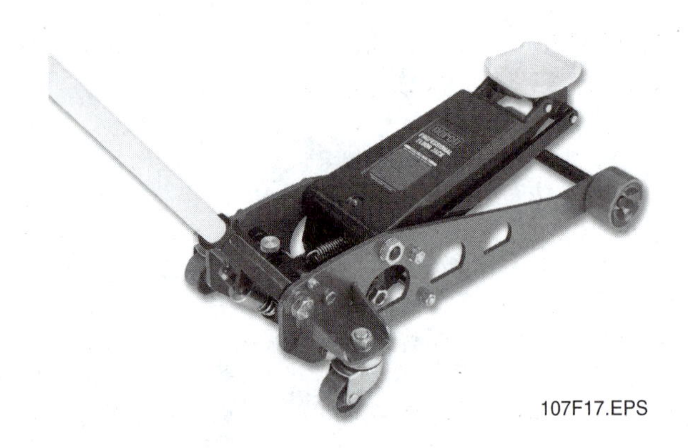

Figure 17 ◆ Jack.

- Make sure the head of the jack is making full and level contact with the load.
- When using a jack, never lift more weight than the jack's weight limit.
- Never jack too high. Jacks are supposed to have a guard that will prevent this from happening. If your jack doesn't have this guard, tell your supervisor, tag it out, and have it repaired.
- A jack should never be used to support a load. Once the load is where you want it, block it up.
- If a hydraulic jack is going to be used in freezing temperatures, it must be filled with antifreeze.

4.7.0 Powered Abrasive-Wheel Tools

Powered abrasive grinding, cutting, polishing, and wire-buffing wheels create special safety problems because they may throw off flying fragments or sparks.

Before an abrasive wheel is mounted, it should be inspected closely and sound- or ring-tested to be sure it is free of cracks or defects. To test a wheel, tap it gently with a light, non-metallic instrument. If it sounds cracked or dead, it could fly apart in operation and should not be used. A good wheel will produce a clear metallic ring when tapped. If you're unsure about how to do this, ask your supervisor to demonstrate the procedure.

To prevent the wheel from cracking, be sure it fits correctly on the spindle. The spindle nut must be tightened enough to hold the wheel in place, without distorting or bending it. Because it is possible for a wheel to explode on startup, never stand directly in front of the wheel until it comes up to full speed.

Portable grinding tools (*Figure 18*) should have safety guards to protect you and your co-workers, not only from the moving wheel surface, but also from flying fragments. If yours doesn't, don't use it. Tag it out and inform your supervisor.

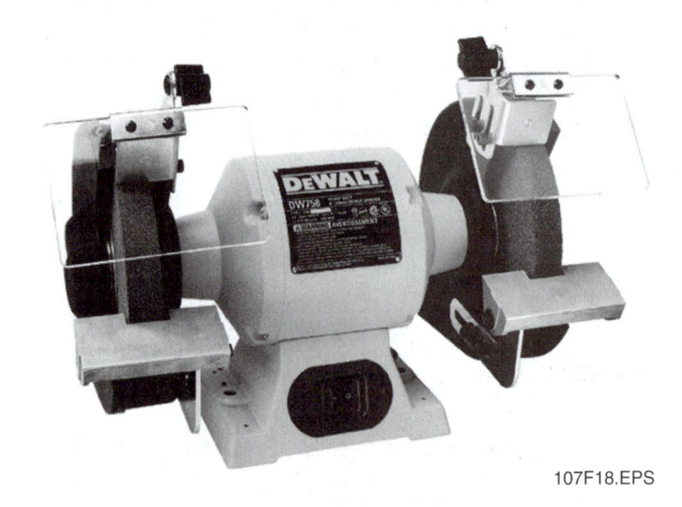

Figure 18 ◆ Bench grinder.

Shore Up Your Load

Two employees were making final adjustments to a large machine in a new paper mill. They were using two hydraulic jacks and two 4" × 4" uprights under one end of a 6,000-pound piece of equipment, which was suspended by four ¾" threaded rods. First, the employees would jack up one end of the piece about an inch. Then, one employee would climb a set of temporary steps to hand-tighten the nuts on the threaded rods. Thus, the 6,000-pound unit was supported solely by the two vertical timbers on the heads of the hydraulic jacks. The timbers were set under a ⅝" side rail without any block or other devices between them. No cribbing, blocking, shoring, or other stabilizing methods were used to secure the load after it was raised. When the end of the piece was jacked up, it fell, crushing one employee and narrowly missing the other.

The Bottom Line: Jacks must not be used to support a load. Always use blocking and bracing to support a raised load.

Source: The National Institute for Occupational Safety and Health (NIOSH)

When using any powered grinder, follow these guidelines:

- Always wear eye protection.
- Turn the power off when the grinder is not in use.
- Don't use a grinder around flammable or explosive materials because the grinder may throw sparks.
- Never clamp a handheld grinder to a vise.

5.0.0 ◆ GUARDS AND SAFETY SWITCHES

Guards are installed on or built into tools to shield dangerous moving parts like belts, gears, shafts, spindles, flywheels, chains, or other reciprocating, rotating, or moving parts of equipment (*Figure 19*). Safety guards should never be removed or disabled in any way. These guards protect you and your co-workers from flying debris, cutting blades, and pinch points. If you find a tool with a missing, broken, or disabled guard, tag it out and tell your supervisor immediately.

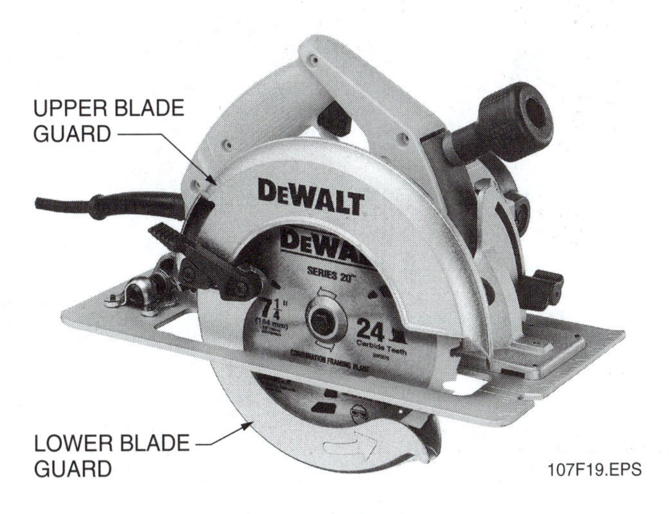

Figure 19 ◆ Circular saw guards.

Safety switches are meant to protect you from a tool accidentally powering up when it shouldn't. Some tools, such as a circular saw, have a constant-pressure switch that will shut off the power when the switch is released. Regardless of what type of switch the tool has, never carry a tool with your finger on the switch.

Summary

If you use hand or power tools, or if you work around people using hand and power tools, you must be aware of the safety hazards. Because tools are so common and are often misused, people don't always know the best or safest way to work with them. Accidents involving tools can cause minor, but painful injuries, or they can have permanent and devastating results like amputation or death. To stay safe, make sure that you are familiar with a tool before you use it and that the tool you are about to use is in good working condition. Also, wear all of the necessary personal protective equipment and clothing for the tools you are using and practice all of the safeguards associated with your work.

Review Questions

1. A bladed hand tool is safest if it is _____.
   a. slightly dull to prevent serious cuts
   b. inspected and maintained regularly
   c. well oiled
   d. kept warm to prevent shattering

2. A wrench with sprung jaws is dangerous because _____.
   a. the space between the jaws and the nut or bolt is a pinch point
   b. the loose jaws could cause the wrench to slip off and hit someone
   c. the jaws could easily break off after they are sprung
   d. sprung jaws can bend the bolt

3. Bladed tools can directly cause all of the following injuries *except* _____.
   a. stab wounds
   b. punctures
   c. burns
   d. amputations

4. When inspecting an impact tool, _____ indicates that the tool may be dangerous.
   a. a mushroomed head
   b. a missing guard
   c. a two-prong plug
   d. no insulation

5. The two most common types of injuries associated with impact tools are hammer strikes and eye injuries.

   a. True
   b. False

6. You may safely plug a grounded plug into a two-prong receptacle if you carefully remove the grounding prong.

   a. True
   b. False

7. Since they are only powered by air, pneumatic tools do not pose any safety risks.

   a. True
   b. False

8. Liquid-fuel tools could start a fire when _____.

   a. flammable fuel vapor comes into contact with the hot tool
   b. the power cord shorts out and causes sparks
   c. the operator is using the tool while standing in water
   d. the air hose is not grounded

9. All of the following are risks typically associated with hydraulic tools *except* _____.

   a. slipping
   b. flying objects
   c. tipping
   d. tripping

10. Powered abrasive wheels can explode on startup.

    a. True
    b. False

GLOSSARY

Trade Terms Introduced in This Module

Bladed tools: Tools that use sharp edges to accomplish their tasks. Bladed tools include saws, knives, scissors, tin snips, and wire cutters.

Decibels (dB): A measure of sound intensity or loudness. The higher the decibel level, the louder and more potentially damaging the sound is.

Electrically powered tools: Tools that use electrical current to operate.

Guards: Devices that protect tool operators from dangerous moving parts such as blades, gears, and pulleys.

Hydraulic tools: Tools that use a pressurized fluid, typically hydraulic oil, to operate.

Impact tools: Tools that must strike or be struck to accomplish their task. Impact tools include hammers, chisels, and taps.

Liquid-fuel tools: Tools that use a liquid fuel, such as gasoline or liquid propane, to operate.

Pneumatic tools: Tools that use pressurized air to operate.

Power tools: Tools that require a power source, such as electricity, hydraulics, or pneumatics, to operate.

ACKNOWLEDGMENTS

Figure Credits

Cooper Tools	107F02 (E) 107F03 (D) and (E), 107F04 (C)
Jensen Tools, Inc.	107F02 (D)
The Stanley Works	107F02 (A) and (B), 107F03 (B), 107F04 (B), 107F10 (B)
Ridge Tool Company	102F02 (C), 107F03 (C), 107F04 (A)
Coleman Cable, Inc.	107F07
Jet Tools and Equipment	107F10 (A)
Milwaukee Electric Tool Company	107F12 (A) and (B)
Bosch Power Tools	107F12 (C)
Makita USA, Inc.	107F13
Lincoln Automotive	107F16
Norco Industries, Inc.	107F17
DeWalt Industrial Tool Company	107F18, 107F19

NCCER CURRICULA — USER UPDATE

NCCER makes every effort to keep its textbooks up-to-date and free of technical errors. We appreciate your help in this process. If you find an error, a typographical mistake, or an inaccuracy in NCCER's curricula, please fill out this form (or a photocopy), or complete the online form at **www.nccer.org/olf**. Be sure to include the exact module ID number, page number, a detailed description, and your recommended correction. Your input will be brought to the attention of the Authoring Team. Thank you for your assistance.

Instructors – If you have an idea for improving this textbook, or have found that additional materials were necessary to teach this module effectively, please let us know so that we may present your suggestions to the Authoring Team.

NCCER Product Development and Revision

13614 Progress Blvd., Alachua, FL 32615

Email: curriculum@nccer.org
Online: www.nccer.org/olf

❑ Trainee Guide ❑ AIG ❑ Exam ❑ PowerPoints Other ______________________________

Craft / Level: Copyright Date:

Module ID Number / Title:

Section Number(s):

Description:

Recommended Correction:

Your Name:

Address:

Email: Phone:

Module 75108-03

Welding Safety

COURSE MAP

This course map shows all of the modules in Field Safety. The suggested training order begins at the bottom and proceeds up. The local Training Program Sponsor may adjust the training order.

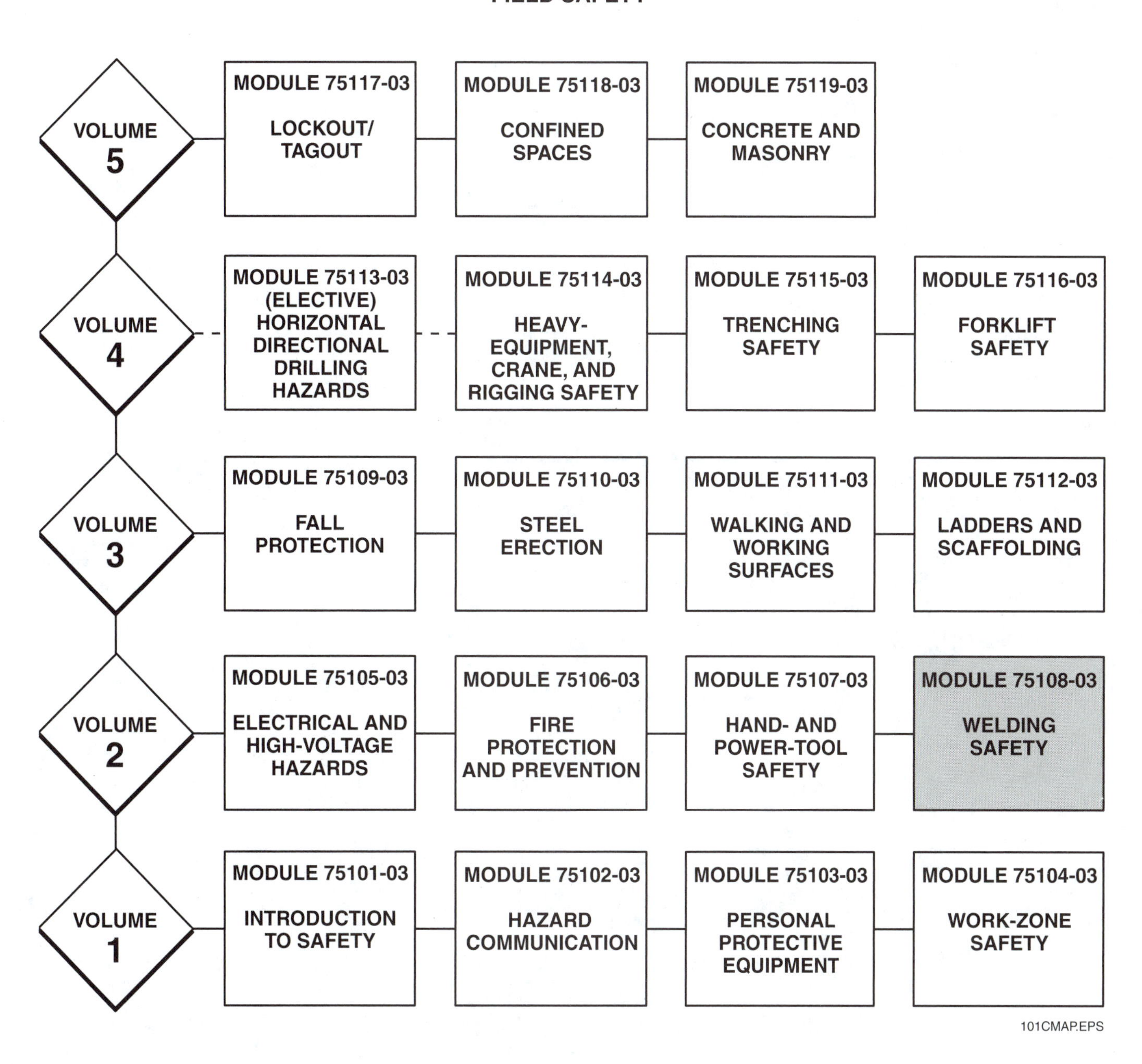

MODULE 75108-03 CONTENTS

Figures

Tables

MODULE 75108-03

Welding Safety

Objectives

When you have completed this module, you will be able to do the following:

1. Identify and describe the safety hazards associated with welding and metal cutting operations.
2. Identify the purpose and characteristics of a hot work permit.
3. Describe the fire hazards associated with welding operations.
4. Identify the hazards associated with handling and storing compressed gases.
5. Identify the hazards associated with toxic fumes generated by welding and cutting processes and the methods used to avoid these hazards.
6. Describe the methods used to prevent injury to workers, including use of appropriate personal protective equipment.

Prerequisites

Before you begin this module, it is recommended that you successfully complete the following: Field Safety, Modules 75101-03 through 75107-03.

Required Materials

1. Pencil and paper
2. Appropriate personal protective equipment

1.0.0 ◆ INTRODUCTION

Welding, **brazing**, and metal cutting are common tasks performed at a construction site. Oxyfuel cutting, also called flame cutting or burning, uses a cutting torch powered by an oxygen/acetylene (oxyacetylene) combination (*Figure 1*). Construction workers use torches to cut concrete reinforcing bars (rebar) and metal framing components. Plumbers and HVAC technicians use brazing torches powered by liquid propane or other **flammable** gases to join copper pipe. Welders perform a variety of tasks, including joining piping (*Figure 2*), attaching **curtain wall** sections to the building framework, attaching metal framing components, and repairing construction equipment. Welding processes produce an intense flame or electric arc that can cause severe burns or start a fire. Some welding processes also use compressed **inert gases** such as argon to shield the welding arc. All welding processes work on the principle of melting steel and other metals.

NOTE

This module provides information and instructions for construction workers who are working in areas where welding or metal cutting may take place. Workers actually performing welding and cutting operations must have more extensive safety training. Such training is provided in the following Contren™ modules:

- Module 29101-03, *Welding Safety*
- Module 29102-03, *Oxyfuel Cutting*

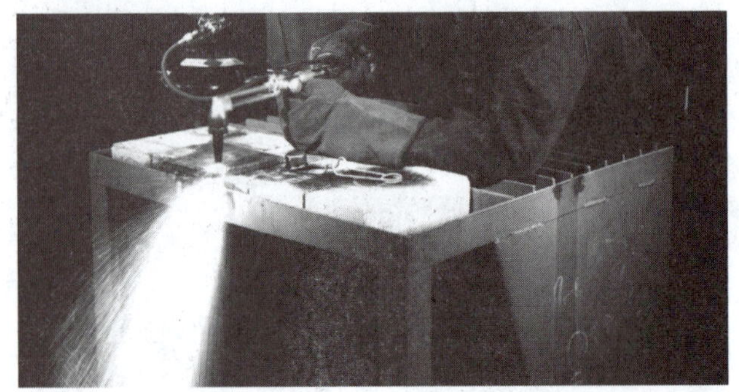

108F01.EPS

Figure 1 ◆ Metal cutting.

108F02.EPS

Figure 2 ◆ Welder joining pipe sections.

Regardless of the gases involved, welding activities always carry a high risk of severe burns and fire hazards. Being aware of these risks is of great importance and will help keep you safe.

Persons working with or around welding, brazing, or metal cutting equipment can be severely burned from the intense heat and flames generated in these processes. In addition, the gases used in these processes pose an additional danger because they are stored under pressure in metal cylinders (*Figure 3*).

Many hazards are involved in the handling, storage, and use of compressed gases. To understand these hazards, we must realize that compressed gases store potential energy. It takes energy to compress and confine the gas. That energy is stored until purposely released to perform useful work or until accidentally released by container failure or other causes.

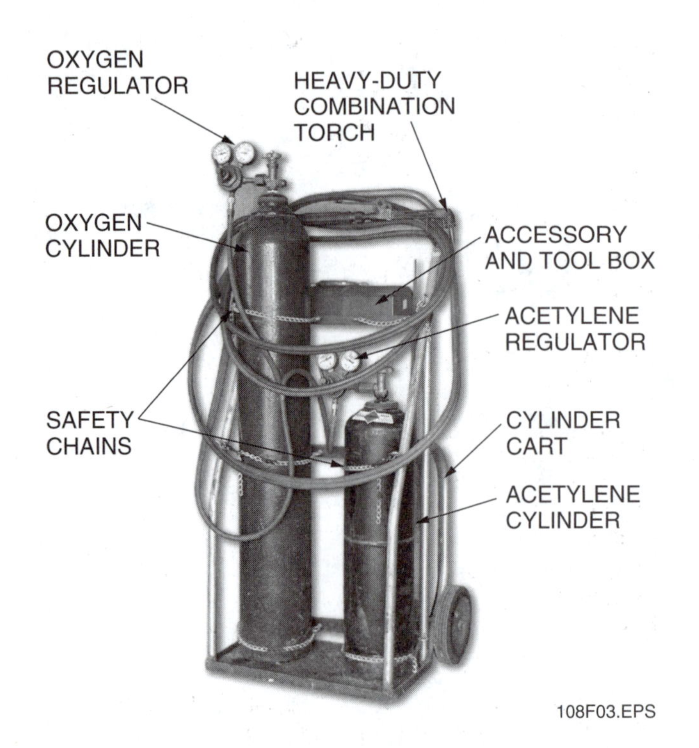

108F03.EPS

Figure 3 ◆ Compressed-gas cylinders used for oxyfuel cutting.

Some compressed gases, acetylene for example, are highly flammable. Therefore, flammable compressed gases have additional stored energy besides simple compression-released energy. Other compressed gases, such as nitrogen, can cause **asphyxiation** because they displace oxygen. Another compressed gas, oxygen, will explode if it comes into contact with grease or oil. In case of a fire, escaping oxygen will make the fire more intense.

When welding or working around welding, workers should keep in mind the many risks associated with welding. They include:

- Fire and explosion
- Fumes and smoke
- Noise
- Heat
- Intense light

Because of these hazards, working around welding operations requires a high level of awareness and use of the proper safety equipment and procedures. This module provides details of the health and safety hazards to which workers may be exposed when working around welding operations, and covers the equipment and methods you can use to reduce the risks presented by those hazards.

2.0.0 ◆ FIRE AND EXPLOSION HAZARDS

Welding involves extremely high temperatures. Any material that comes in contact with a welding flame or arc may catch fire or explode. The welding process also creates hot sparks that can start a fire. Molten metal known as slag may drop from the material being welded, or be thrown out from the weld area.

Oxyfuel cutting also creates a potential fire hazard due to the high-intensity flame produced by the cutting torch and the sparks and slag that are by-products of the process. The force of the oxygen stream can cause sparks to be thrown for a considerable distance, so it is important to move or protect any nearby flammable material. In processes that produce dust and fumes, it is important to protect against igniting the dust or fumes. High concentrations of grain dust, sawdust, and some chemical fumes can ignite or explode when exposed to a flame or intense heat.

2.1.0 Fire Prevention

The key to fire prevention on a work site where welding is being done is to prevent any material from coming in contact with the high temperatures, sparks, and slag generated by the welding process.

The work area for welding should be shielded from the rest of the work site. When possible, welding should be done in a work area built for welding, with firebrick walls and a fire-resistant floor and ceiling. When the welding is done at a job location, people and material around the welding area must be shielded from the welding process.

All flammable material within a 35' radius must be either removed or protected from heat, sparks, or slag generated by the welding process. Walls and floors must be fireproof, and should have no cracks or openings. Fireproof shielding must be placed to protect or block any such openings. Where the walls or shielding are made of metal, and there is combustible material directly on the other side, that material should be moved away to prevent accidental ignition due to heat radiation. These areas should be made safe for welding and cutting operations with concrete floors, arc filter screens, protective drapes, curtains or blankets, and fire extinguishers. No combustibles should be stored nearby.

When it is necessary to perform welding at the work site, welding shields must be erected around the work area (*Figure 4*). These shields must be positioned to prevent any sparks or slag from contacting flammable material. They must also be placed to prevent any workers in the area from seeing the arc of the welding equipment.

108F04.EPS

Figure 4 ◆ Welding shield.

Never use any type of container that has held a combustible or flammable material as a platform for the material to be welded unless it has been thoroughly cleaned or filled with an inert gas. Traces of flammable or combustible materials can ignite or explode, or the heat can cause them to give off toxic fumes, which can travel over a large area. If the contents of a container are unknown, always assume the container is unsafe for use in a welding operation.

Never perform welding or cutting on drums, barrels, tanks, vessels, or other containers until they have been emptied and cleaned thoroughly. These containers may contain a residue of flammable materials and substances such as detergents, solvents, greases, tars, or acids. These substances might produce flammable, toxic, or explosive vapors when heated. Clean containers only in well-ventilated areas. Vapors can accumulate during cleaning, causing explosions or injury. Containers must be cleaned by steam cleaning, flushing with water, or washing with detergent until all traces of the material have been removed.

After cleaning the container, fill it with water or an inert gas such as nitrogen, argon, or carbon dioxide (CO_2) for additional safety (*Figure 5*). Air, which contains oxygen, is displaced from inside the container by the water or inert gas. Without oxygen, combustion cannot take place. When using water, position the container to minimize the air space. When using an inert gas, provide a vent hole so the inert gas can purge the air containing oxygen.

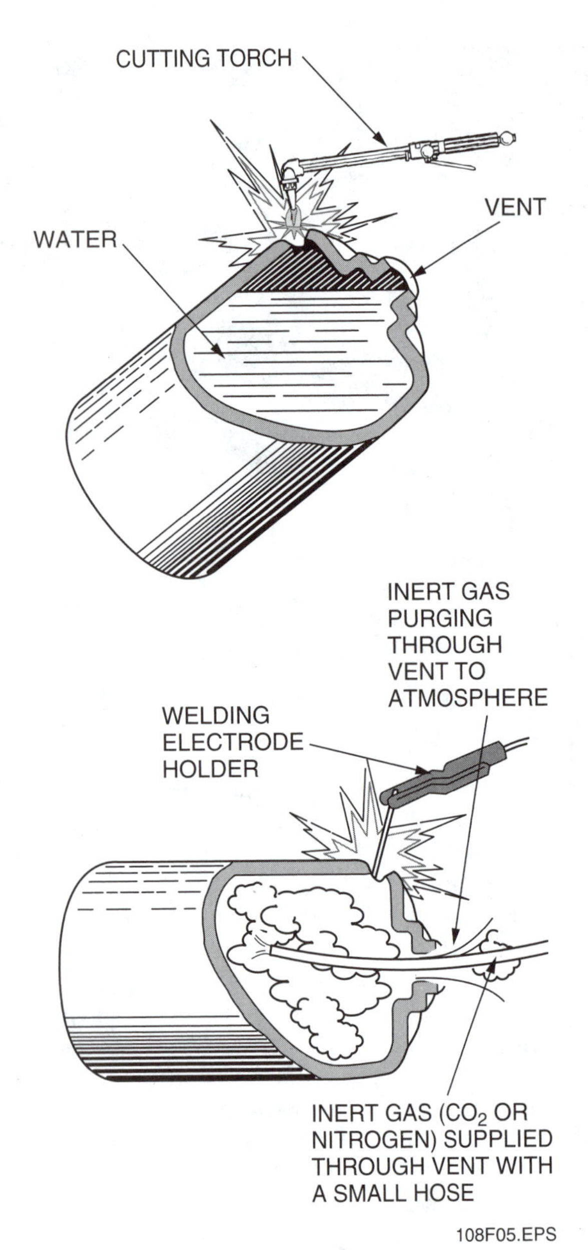

Figure 5 ◆ Purging a container.

A fire watch must be posted during the welding operation, and for at least 30 minutes after the welding work is finished. The fire watch is there to ensure that sparks and slag from the welding process do not ignite any materials in the area. The fire watch also watches for any other sort of emergency situation that the welder cannot see because of the welding helmet. For more information on fire safety, refer to the *Fire Prevention and Protection* module in Volume Two.

2.2.0 Hot Work Permits

A hot work permit (*Figure 6*) is an official authorization from the site manager to perform work that may pose a fire hazard. The permit includes information such as the time, location, and type of work being done. The hot work permit system promotes the development of standard fire safety guidelines. Permits also help managers keep records of who is working where and at what time. This information is essential in the event of an emergency or other times when personnel need to be evacuated.

A fire watch is used during jobs requiring hot work permits. During a fire watch, a person other than the welder or cutting operator must constantly scan the work area for fires. Fire watch personnel must have ready access to fire extinguishers and alarms and know how to use them. Cutting operations must never be performed without a fire watch. Whenever oxyfuel cutting equipment is used, there is a great danger of fire. Hot work permits and fire watches are used to minimize this danger.

Most sites require the use of hot work permits and fire watches. When these requirements are violated, severe penalties may be imposed.

Welding Where Flammable Materials Have Been Stored

Two employees were welding brackets onto a 55,000-gallon oil storage tank. The half-full tank contained explosive vapors from a waste chemical and waste oil products from automobile and truck service stations. One worker was killed and another injured when the tank exploded and the top was blown off.

The Bottom Line: Had these employees followed proper procedures for welding on or near containers where flammable liquids have been stored, the tank would have been emptied and cleaned thoroughly, or filled with water before they started welding.

Source: The Occupational Safety and Health Administration (OSHA)

HOT WORK PERMIT

For Cutting, Welding, or Soldering with Portable Gas or ARC Equipment

(References: *1997 Uniform Fire Code Article 49* & *National Fire Protection Association Standard NFPA 51B*)

Job Date____________ Start Time____________ Expiration____________ WO #____________

Name of Applicant______________________________ Company____________ Phone____________

Supervisor______________________________ Phone____________

Location / Description of Work ____

IS A FIRE WATCH REQUIRED?

1. _______ (yes or no) Are combustible materials in building construction closer than 35' to the point of operation?
2. _______ (yes or no) Are combustibles more than 35' away but would be easily ignited by sparks?
3. _______ (yes or no) Are wall or floor openings within a 35' radius exposing combustible material in adjacent areas, including concealed spaces in floors or walls?
4. _______ (yes or no) Are combustible materials adjacent to the other side of metal partitions, walls, ceilings, or roofs which could be ignited by conduction or radiation?
5. _______ (yes or no) Does the work necessitate disabling a fire detection, suppression, or alarm system component?

YES to any of the above indicates that a qualified fire watch is required.

Fire Watcher Name(s) ______________________________ Phone____________

NOTIFICATIONS

Notify the following groups at least 72 hours prior to work and 30 minutes after work is completed.
Write in names of persons contacted.

Notify in person OR by phone ONLY if question #5 above is answered "yes":

- Facilities Management Fire Alarm Supervisor

Notify by phone or in person: (If by phone, write down name of person and send them a completed copy of this permit.)

- Facilities Management Fire Protection Group
- Environmental Health & Safety Industrial Hygiene Group

SIGNATURES REQUIRED

Project Manager______________________________ Date ____________ Phone____________

I understand and will abide by the conditions described in this permit. I will implement the necessary precautions which are outlined on both sides of this permit form. Thirty minutes after each hot work session, I will reinspect work areas and adjacent areas to which sparks and heat might have spread to verify that they are fire safe, and contact Facilities Management Alarm Technicians to have any disabled fire-protection systems reactivated.

____________________ ____________________ Date ____________ Phone____________
Permit Applicant Company or Department

1/17/03

108F06.EPS

Figure 6 ◆ Hot work permit.

Before a hot work permit is issued, the following precautions must be taken:

- Operations and contractor personnel must make a joint inspection of the work area to ensure that no flammable or combustible materials are present. This inspection includes the areas above and below where the work will be performed as well as the surrounding area.
- Any combustible materials that cannot be moved must be covered with a fire-resistant blanket or other material. In some cases, they may also be wet down with water.
- Open sewers in the area must be covered and sealed.
- A gas test must be performed to be sure no flammable gases or vapors are present.
- An inspection of the work area must be performed.
- Fire watches must be in place. It is recommended that at least one fire watch be posted with an extinguisher to watch for possible fires.
- Fire extinguishers must be available.
- Communication procedures must be established.
- Heat, flames, and sparks must be controlled.

If a fire starts, the first step is to activate the alarm and get everyone out of the work area. If you are trained in fire fighting, you may then attempt to put out a small fire using the appropriate fire extinguisher.

2.3.0 Compressed Gases

Most welding-related fires occur during oxyfuel gas welding or cutting. Anyone performing this work must be well trained in the function and operation of each part of oxyfuel gas welding or cutting equipment. They must also be trained and tested in the correct methods of starting, testing for leaks, and shutting down oxyfuel gas welding or cutting equipment. They must know the hazards involved in the use of fuel gas, oxygen, and shielding gas cylinders and how these cylinders are stored safely.

2.3.1 Handling and Storage of Compressed-Gas Cylinders

Oxygen and fuel gas cylinders or other flammable materials must be stored separately (*Figure 7*). The storage areas must be separated by 20' or by a wall 5' high with at least a one-half hour burn rating. The purpose of the distance or wall is to keep the heat of a small fire from causing the oxygen cylinder safety valve to release. If the safety valve releases the oxygen, a small fire will quickly become a raging inferno.

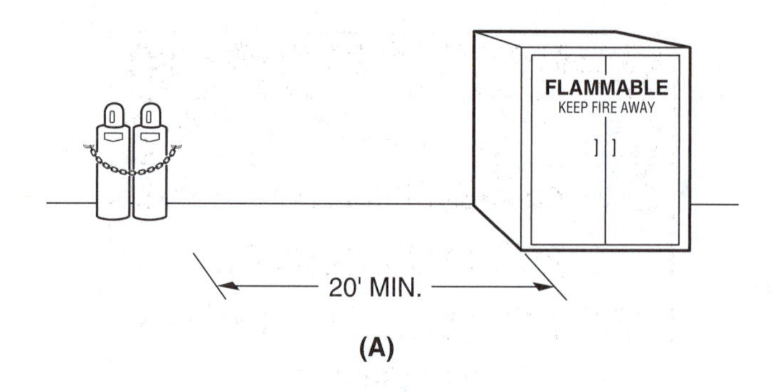

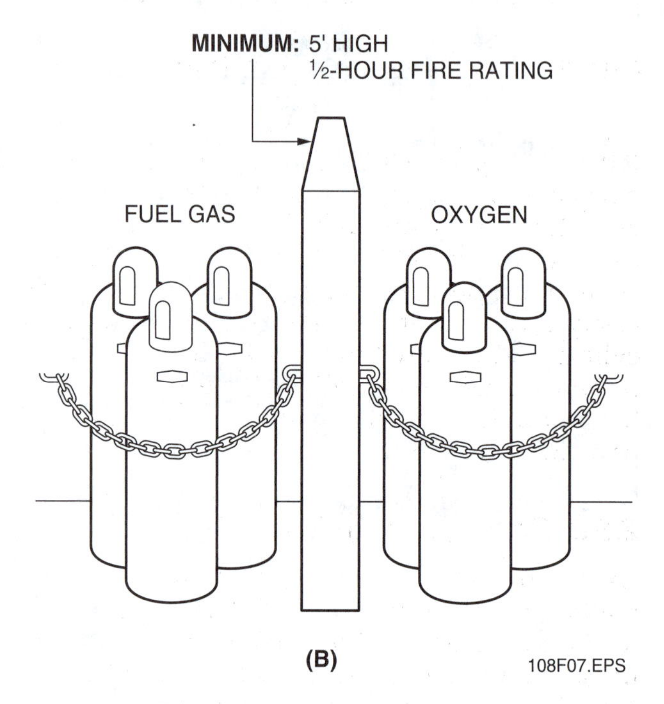

Figure 7 ◆ Cylinder storage requirements.

Inert gas cylinders may be stored separately or with either fuel cylinders or oxygen cylinders. Empty cylinders must be stored separately from full cylinders, although they may be stored in the same room or area. All cylinders must be stored upright and have the protective caps screwed on firmly. Cylinders must be secured with a chain or other device so that they cannot be knocked over accidentally.

Cylinder storage areas must be located away from halls, stairwells, and exits so that in case of an emergency they will not block an escape route. Storage areas should also be located away from heat, radiators, furnaces, and welding sparks. The location of storage areas should be such that unauthorized people cannot tamper with the cylinders. A warning sign such as the one shown in *Figure 8* must be posted in the storage area.

Cylinders equipped with a valve protection cap must have the cap in place unless the cylinder is in use. The protection cap prevents the valve from being broken off if the cylinder is knocked over.

108F08.EPS

Figure 8 ◆ Sample warning sign for cylinder storage areas.

If the valve of a full high-pressure cylinder (argon, oxygen, CO_2, or mixed gases) is broken off, the cylinder can fly through the air like a missile, causing damage and personal injury. Never lift a cylinder by the safety cap or valve. The valve can easily break off or be damaged. When moving cylinders, the valve protection cap must be in place, especially if the cylinders are mounted on a truck or trailer. Cylinders must never be dropped or handled roughly.

2.3.2 General Precautions

Use warm water (not boiling) to loosen cylinders that are frozen to the ground. Any cylinder that leaks, has a bad valve, or has gas-damaged threads must be identified and reported to the supplier. Use a piece of soapstone to write the problem on the cylinder. If closing the cylinder valve cannot stop the leak, the cylinder should be moved outdoors to a safe location, away from any source of ignition, and the supplier notified. Post a warning sign, and then slowly release the cylinder pressure.

In its gaseous form, acetylene is extremely unstable and explodes easily. For this reason it must remain at pressures below 15 psi. If an acetylene cylinder is tipped over, stand it upright and wait at least 30 minutes before using it. If liquid acetone is withdrawn from a cylinder, it will gum up the safety check valves and regulators and decrease the stability of the acetylene stored in the cylinder. For this reason, acetylene must never be withdrawn at a per-hour rate that exceeds one-tenth of the volume of the cylinder(s) in use. Acetylene cylinders in use should be opened no more than one-and-one-half turns and, preferably, no more than three-fourths of a turn.

Other precautions include:

- Use only compressed-gas cylinders containing the correct gas for the process used and properly operating regulators designed for the gas and pressure used.
- Keep cylinders in the upright position and securely chained to a fixed support.
- All hoses, fittings, and other parts must be suitable and maintained in good condition.
- Keep combustible cylinders in one area of the building for safety. Cylinders must be at a safe distance from arc welding or cutting operations and any other source of heat, sparks, or flame.
- Never allow the welding electrode, electrode holder, or any other electrically hot parts to come in contact with the cylinder.
- When opening a cylinder valve, stand to one side of the valve outlet to avoid any debris that might be expelled by the valve.
- Always use valve protection caps on cylinders that are not in use or are being transported.
- Never lift a cylinder by its protective cap. Use a lifting cage (*Figure 9*) designed for the purpose.

Use Caution with High-Pressure Cylinders

A pipe fitter was setting up an argon bottle in preparation for tungsten inert gas (TIG) welding. The pipe fitter knew a pressure regulator was needed on the argon bottle before installing the manifold. He failed to pay attention to detail during the setup process, and attached a shop-fabricated distribution manifold directly to the argon bottle without installing a pressure regulator. The manifold contained two flow meters, one for the TIG torch and one for a purge line for welding stainless steel pipe. When he opened the valve on the bottle, one of the meters broke because of the gas. Fortunately, no one was injured, the only damage was to the equipment.

The Bottom Line: All hoses and regulators must be connected properly and double-checked for safety.

Source: Department of Energy Case History Information Services

Figure 9 ◆ Cylinder lifting cage.

3.0.0 ◆ INJURY AND ILLNESS HAZARDS

Because of the flames, extreme heat, and intense light generated by welding and metal cutting, these processes are inherently dangerous. The following paragraphs describe the hazards associated with welding and metal cutting, and how to avoid them.

3.1.0 Heat Hazards

The welding process generates intense heat and sparks, both of which can cause severe burns. Eye injuries and burns will result from contact with hot welding rods and slag, as well as sparks and metal chips.

In addition to acute injuries, welders and those in the area can suffer from heat exhaustion or heat stroke due to exposure to excessive heat. Conditions associated with overheating include heat exhaustion, heat cramps, and heat stroke.

You should always observe the following guidelines when working in a hot work space, in order to prevent heat exhaustion, cramps, or heat stroke:

- Drink plenty of fluids.
- Do not overexert yourself.
- Take frequent, short, work breaks.

3.1.1 Heat Exhaustion

Heat exhaustion can cause pale, clammy skin; heavy sweating with dizziness; nausea and possible vomiting; fast, weak pulse; slightly elevated body temperature; and possible fainting.

If a person appears to be suffering from heat exhaustion, the following response is indicated:

- Remove the victim from the heated area.
- Apply cool, wet cloths.
- Fan the victim, but stop if shivering or goose bumps occur.
- Have the victim lie down.
- Give the victim about one-half glass of water every 15 minutes if conscious.
- Seek medical attention.

3.1.2 Heat Cramps

Heat cramps can come after an attack of heat exhaustion. Loss of salt from heavy sweating can cause these cramps. Heat cramps cause abdominal pain, nausea, dizziness, muscular twitching, and severe muscle cramps. The skin becomes pale and heavy sweating occurs.

If you are suffering from heat cramps, the following response is indicated:

- Drink about one-half glass of water every 15 minutes.
- Sit or lie down in a cool area.
- Gently stretch and massage cramped muscles.
- Seek medical attention.

3.1.3 Heat Stroke

Heat stroke is an immediate, life-threatening emergency that requires urgent medical attention. It is characterized by headache, nausea, and vision problems. Body temperature can reach as high as 106°F. Hot, flushed, dry skin; slow, deep breathing; possible convulsions; and loss of consciousness will accompany these symptoms.

If a person appears to be suffering from heat stroke, the following response is indicated:

- Get medical help immediately.
- Have the victim lie down to prevent shock.
- Remove the victim from the heated area.
- Give nothing by mouth.
- Cool the victim by immersion in cool water or use cool compresses.

3.2.0 Light and Radiation Hazards

Arc welding produces an extremely intense light (*Figure 10*) that can cause permanent damage to the retina of the eye. The damage is caused by both the visible light and invisible ultraviolet (UV) radiation generated by the arc. A condition known as arc eye or welder's flash can result from even a brief exposure, typically less than one minute. The symptoms of arc eye usually become apparent several hours after exposure. These symptoms include a feeling of sand or grit in the eye, blurred vision, intense pain, tearing, burning, and headaches.

It is not just the welder who is at risk of arc flash. The arc can reflect off materials in the work area and affect co-workers who are in the surrounding area.

108F10.EPS

Figure 10 ◆ Arc welding light.

Use the Correct Eye Filter

A company had used oxyacetylene welding in their operations for years. They purchased an arc welder and put it into operation in their shop, using their existing safety equipment. The worker who used the arc welder suffered welder's flash for several days following his first use of the arc welder. Investigation found that the eye filters, which were sufficient for use with the oxyacetylene torch, did not provide proper protection from the arc flash created by the electrical arc.

The Bottom Line: Always use the appropriate personal protective equipment.

About half of the recorded cases of welder's flash involved co-workers who were not welding. Proper eye protection is necessary for all people working in the area where welding is being performed. Those who work around welding equipment without proper eye protection can suffer permanent eye damage.

The ultraviolet radiation generated by the welding arc can also cause burns to exposed skin, similar to sunburn. This increases the risk of skin cancer for those who are exposed. Infrared radiation generated by the arc can damage the cornea of the eye and lead to the formation of cataracts many years later.

3.3.0 Noise Hazards

Noise is an often overlooked hazard in many workplaces; the welding site is no exception. The roar of a cutting flame, the snap of a welding arc coming to life, and the heavy clang of metal plate dropping from a cut can all have an impact on the worker's hearing. Exposure to continual extreme noise can also cause increased blood pressure and stress levels, leading to possible heart disease. Workers exposed to constant or frequent noise can become nervous or irritable, and tend to tire more easily.

Noise that averages in excess of 90 decibels over an eight-hour period is considered a workplace hazard under the OSHA standard for construction work. Keep in mind that maximum noise levels are stated in relation to time, as shown in *Table 1*. Under these conditions, employers are required to provide hearing protection (*Figure 11*) and annual hearing tests for workers. For more information on noise hazards and hearing protection, refer to the *Hand- and Power-Tool Safety* module in Volume Two.

Table 1 Maximum Noise Levels

Sound Level (decibels)	Maximum Hours of of Continuous Exposure per Day	Examples
90	8	Power lawn mower
92	6	Belt sander
95	4	Tractor
97	3	Hand drill
100	2	Chain saw
102	1.5	Impact wrench
105	1	Spray painter
110	0.5	Power shovel
115	0.25 or less	Hammer drill

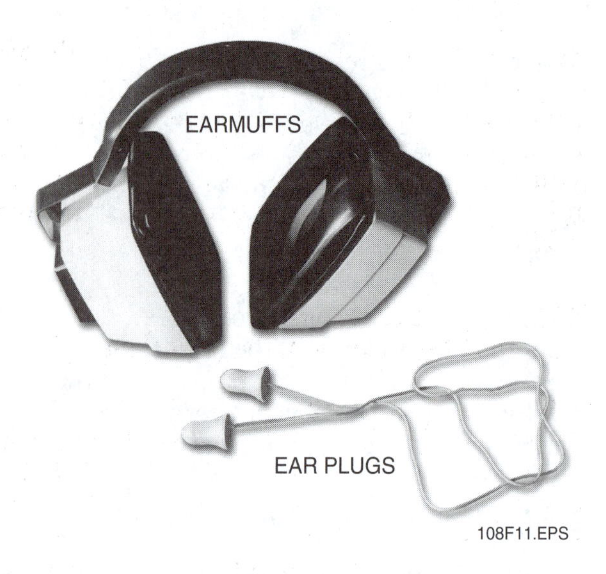

Figure 11 ◆ Hearing protection.

3.4.0 Electrical Hazards

An arc welding setup consists of a welding machine that produces the welding current, an electrode holder, an electrode, and a workpiece clamp. When the equipment and workpiece are connected, as shown in *Figure 12*, the voltage is strong enough to cause an electric arc between the welding electrode and the workpiece, and an electrical current will flow. The presence of an electrical current creates the danger of electric shock. This danger increases when the welding is done in areas that are wet or cramped. Even a minor electrical shock can be dangerous if it causes you to lose your balance and fall.

The gloves worn for welding (*Figure 13*) should be dry so that electricity cannot be conducted through them. Boots should have rubber soles for the same reason. Metal and other conductive work surfaces should have an insulating layer, such as a rubber mat, for the worker's protection.

Figure 13 ◆ Protective gloves.

All electrical equipment should be properly grounded. The equipment should be inspected daily for breaks or cracks in the insulation on cables and electrode holders, or even more frequently if conditions warrant. The cables and electrode holders must be kept dry at all times, and electrodes should never be changed with bare hands, wet gloves, or while standing on a wet floor or grounded surface.

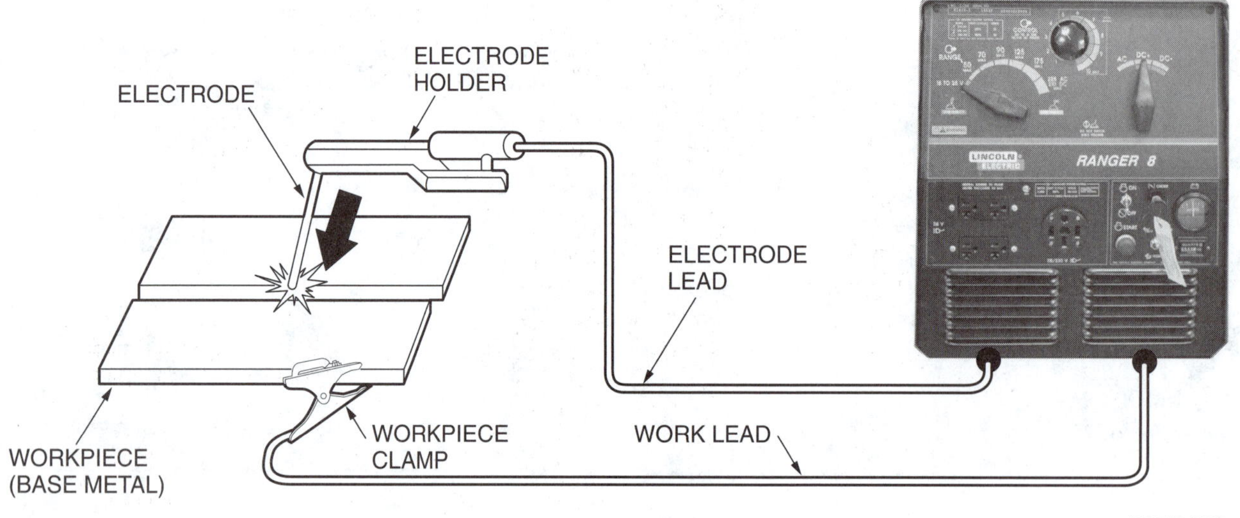

Figure 12 ◆ Arc welding setup.

For more information on electrical and high-voltage hazards, refer to the *Electrical and High-Voltage Hazards* module in Volume Two.

3.5.0 Physical Injury Hazards

Welding is a very physical process. Whether it is the moving of materials to be welded, moving into and out of confined spaces for welding, moving the equipment needed for welding, or just holding a torch in place for long periods of time, welding takes a toll on the body. Those working around a welding operation may be asked to help out with some of these activities, and must take care to prevent injuries to themselves and others. Common workplace injuries in these circumstances include back injuries, shoulder pain, tendonitis, knee-joint diseases, carpal tunnel syndrome, and reduced muscle strength. Most of these injuries can be prevented by following procedures for proper lifting, positioning work at convenient locations and heights, allowing yourself to change position frequently, and minimizing vibration of equipment and work surfaces.

4.0.0 ◆ ACCIDENT AND INJURY PREVENTION

The hazards of working around welding can be reduced, but they can never be totally eliminated. The best way to stay safe is to take steps to prevent accidents and injuries. This is done through good work practices, dressing properly (*Figure 14*), and using appropriate personal protective equipment, as previously described.

Figure 14 ◆ Proper attire for welding.

INSIDE TRACK Flame-Resistant Clothing

Flame-resistant (FR) coveralls used by welders during cutting and welding normally resist ignition and burning. However, some welder coveralls in facility supplies may no longer be flame resistant and their use may put workers at risk from serious burn injuries. Most FR material is cotton that has been treated with chemicals to provide the fire-resistant properties required for these garments. Some of the compounds used to provide fire resistance lose their effectiveness over time. For example, one type of protective compound is certified to retain its original flame resistance through 25 washings in temperatures up to 165°F. After 25 washings, clothing treated with this compound is no longer usable as fire protection.

It is important to understand that treated cotton FR garments are not fireproof—they will ignite if exposed to sufficient thermal energy, such as a flash fire or an electric arc. When the source of ignition is removed, the FR material will not continue to burn.

Source: Department of Energy Case History Information Services

Those working around the welding process should wear clothing that reduces the risk of injury. For example:

- Wear a welding helmet or snug-fitting cutting goggles that include filter lenses providing shade 5 or 6 protection, as well as a clear lens (*Figure 15*). The filter level should be appropriate for the type of welding being performed (*Table 2*).
- Wear a long-sleeved work shirt with the collar buttoned, and no pockets. The shirt should be worn outside the pants to prevent sparks from getting caught in the waistband.
- A cap worn with the bill turned backwards will prevent sparks from falling down the back of the shirt collar.
- Earplugs provide noise protection and also protect the ears from sparks.
- Wear trousers without cuffs that cover the ankles and tops of boots.
- Wear high-top leather boots.
- Wear all-leather, gauntlet-style welder's gloves.
- Use a respirator if the materials used require it.

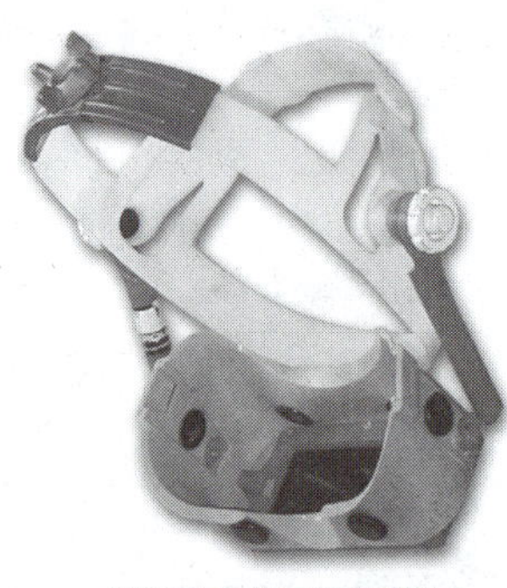

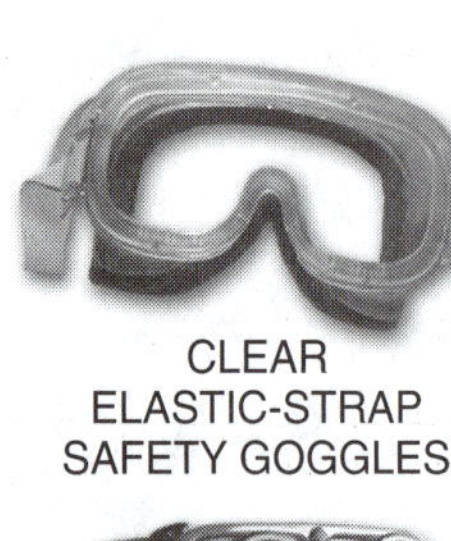

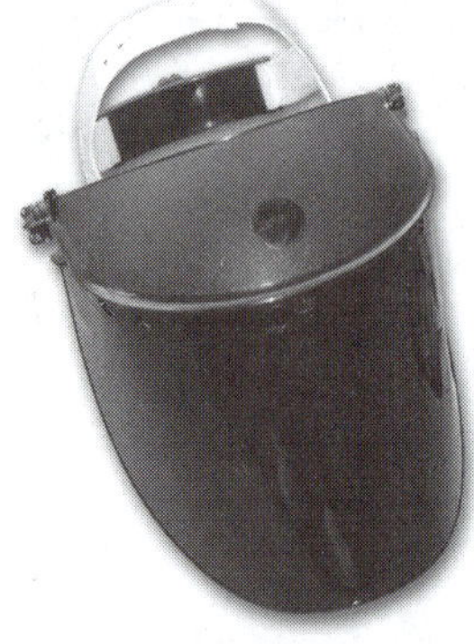

Figure 15 ◆ Eye protection.

Those who are working in an area where welding is being performed should wear flash goggles to prevent exposure to the sudden, intense light associated with the welding process.

5.0.0 ◆ VENTILATION

The gases, dust, and fumes caused by welding and cutting operations are often toxic. Studies have linked these fumes to a variety of serious illnesses and medical conditions. For that reason, mechanical ventilation must be provided to remove fumes produced by welding or cutting processes. The following general rules can be used to determine if there is adequate ventilation:

- The welding area must contain at least 10,000 cubic feet of air for each welder.
- There must be proper air circulation.
- Partitions, structural barriers, or equipment must not block air circulation.

Table 2 Appropriate Filter Levels for Various Types of Welding

Filter Lenses for Protection Against Radiant Energy

Operations	Electric Size 1/32	Arc Current	Minimum Protective Shade
Shielded metal arc welding	Less than 3	Less than 60	7
	3-5	60-160	8
	5-8	160-250	10
	More than 8	250-550	11
Gas metal arc welding and flux arc welding		less than 60	7
		60-160	10
		160-250	10
		250-500	10
Gas Tungsten arc welding		less than 50	8
		50-150	8
		150-500	10
Air carbon Arc cutting	(Light)	less than 500	10
	(Heavy)	500-1000	11
Plasma arc welding		less than 20	6
		20-100	8
		100-400	10
		400-800	11
Plasma arc cutting	(light)**	less than 300	8
	(medium)**	300-400	9
	(heavy)**	400-800	10
Torch brazing			3
Torch soldering			2
Carbon arc welding			14

Filter Lenses for Protection Against Radiant Energy

Operations	Plate thickness-inches	Plate thickness-mm	Minimum* Protective Shade
Gas Welding:			
Light	Under 1/8	Under 3.2	4
Medium	1/8 to 1/2	3.2 to 12.7	5
Heavy	Over 1/2	Over 12.7	6
Oxygen cutting:			
Light	Under 1	Under 25	3
Medium	1 to 6	25 to 150	4
Heavy	Over 6	Over 150	5

* As a rule of thumb, start with a shade that is too dark to see the weld zone. Then go to a lighter shade which gives sufficient view of the weld zone without going below the minimum. In oxyfuel gas welding or cutting where the torch produces a high yellow light, it is desirable to use a filter lens that absorbs the yellow or sodium line in the visible light of the (spectrum) operation.

** These values apply where the actual arc is clearly seen. Experience has shown that lighter filters may be used when the arc is hidden by the workplace.

108T01.TIF

Even when there is adequate ventilation, workers must try to avoid inhaling fumes and smoke from welding or cutting activities. The heated fumes and smoke generally rise straight up. Observe the column of smoke and position yourself to avoid it. A small fan may also be used to divert the smoke. Care must be taken to keep the fan from blowing directly on the work area because the fumes and gases must be present at an electric arc in order to shield the molten metal from the air.

5.1.0 Fume Hazards

Welding and metal cutting work create fumes. Fumes are solid particles consisting of the base metal, electrodes or welding wire, and any coatings applied to them. Most fumes are not considered dangerous as long as there is adequate ventilation. Adequate ventilation can be a problem in tight or cramped working quarters. To ensure adequate room ventilation, local exhaust ventilation should be used to capture fumes (*Figure 16*).

The exhaust hood should be kept 4" to 6" away from the source of the fumes. Welders should recognize that fumes of any type, regardless of their source, should not be inhaled. The best way to avoid problems is to provide adequate ventilation. If this is not possible, breathing protection must be used. Protective devices for use in poorly ventilated or confined spaces are shown in *Figures 17* and *18*.

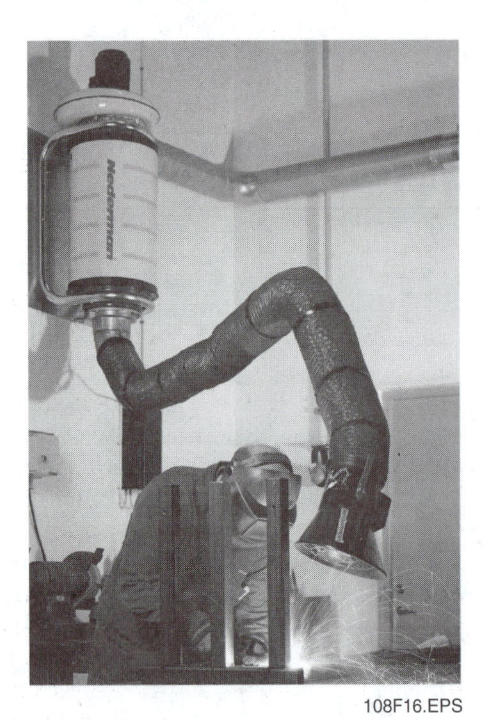

108F16.EPS

Figure 16 ◆ Flexible exhaust pickup.

If respirators are used, your employer should offer medical screenings and worker training on respirator fitting to make sure you know how to use a respirator properly.

> **WARNING!**
> Fumes and gases can be dangerous. Overexposure can cause nausea, headaches, dizziness, metal fume fever, and severe toxic effects which can be fatal. Reports have indicated irritation to the eyes, skin, and respiratory system and more severe complications. In confined spaces the gases may cause asphyxiation.

108F17.EPS

Figure 17 ◆ Typical respirator.

108F18.EPS

Figure 18 ◆ Belt-mounted respirator.

5.2.0 Respirators

Metals such as zinc, cadmium, beryllium, and lead give off toxic fumes when heated (see the *Appendix* at the back of this module for more information). Such metals require the use of respirators for protection from harmful fumes. Respirators are grouped into three main types based on how they work to protect the wearer from contaminants. The types are:

- Air-purifying respirators
- Supplied-air respirators (SAR)
- Self-contained breathing apparatus (SCBA)

5.2.1 Air-Purifying Respirators

Air-purifying respirators provide the lowest level of protection. They are made for use only in atmospheres that have enough oxygen to sustain life (at least 19.5%). Air-purifying respirators use special filters and cartridges to remove specific gases, vapors, and particles from the air. The respirator cartridges contain charcoal, which absorbs certain toxic vapors and gases. When the wearer detects any taste or smell, the charcoal's absorption capacity has been reached. This means the cartridge can no longer remove the contaminants. The respirator filters remove particles such as dust, mists, and metal fumes by trapping them within the filter material. Filters should be changed when breathing becomes difficult.

Depending on the contaminants, cartridges can be used alone or in combination with a filter/prefilter and filter cover. Air-purifying respirators should be used for protection against only the types of contaminants listed on the filters and cartridges. Also refer to the National Institute for Occupational Safety and Health (NIOSH) approval label affixed to each respirator and filter carton. Respirator manufacturers typically classify air-purifying respirators into four groups:

- No maintenance
- Low maintenance
- Reusable
- Powered air-purifying respirators (PAPRs)

No-maintenance and low-maintenance respirators are typically used for residential or light commercial work that does not call for constant and heavy respirator use. No-maintenance respirators are typically half-mask respirators with permanently attached cartridges or filters. The entire respirator is discarded when the cartridges or filters are spent. Low-maintenance respirators are generally half-mask respirators that use replaceable cartridges and filters. However, they are not designed for constant use.

Reusable respirators (*Figure 19*) are made in half-mask and full-face piece styles. These respirators require the replacement of cartridges, filters, and respirator parts. Their use also requires a complete respirator maintenance program.

108F19.EPS

Figure 19 ◆ Reusable half-mask air-purifying respirator.

Powered air-purifying respirators (PAPRs) are made in half-mask, full-face piece, and hood styles. They use battery-operated blowers to pull outside air through the cartridges and filters attached to the respirator. The blower motors can be either mask- or belt-mounted. Depending on the cartridges used, they can filter particles, dust, fumes, and mists along with certain gases and vapors. PAPRs like the one shown in *Figure 20* have a belt-mounted, powered air-purifier unit connected to the mask by a breathing tube. Many models also have an audible and visual alarm that is activated when the airflow falls below the required minimum level. This feature gives an immediate indication of a loaded filter or low battery charge

Know What You're Cutting or Welding

Cutting or welding operations involving materials, coatings, or electrodes containing cadmium, mercury, lead, zinc, chromium, and beryllium result in toxic fumes. For cutting or welding of such materials, always use proper area ventilation and wear an approved full-face, supplied-air respirator (SAR) that uses breathing air from outside of the work area. For occasional, very short-term exposure to fumes from zinc- or copper-coated materials, a high-efficiency particulate arresting (HEPA)-rated or metal-fume filter may be used on a standard respirator.

condition. Units with the blower mounted in the mask do not use a belt-mounted powered air purifier connected to a breathing tube.

5.2.2 Supplied-Air Respirators

Supplied-air respirators (*Figure 21*) provide a supply of air for extended periods of time through a high-pressure hose connected to an external source of air, such as a compressor, compressed-air cylinder, or pump. They provide a higher level of protection in atmospheres where air-purifying respirators are not adequate. Supplied-air respirators are typically used in toxic atmospheres. Some can be used in atmospheres that are immediately dangerous to life and health (IDLH) as long as they are equipped with an air cylinder for emergency escape. An atmosphere is considered IDLH if it poses an immediate hazard to life or produces immediate, irreversible, and debilitating effects on health. There are two types of supplied-air respirators: continuous-flow and pressure-demand.

A continuous-flow supplied-air respirator provides air to the user in a constant stream. One or two hoses are used to deliver the air from the air source to the face piece. Unless the compressor or pump is specially designed to filter the air, or a portable air-filtering system is used, the unit must be located where there is breathable air. Continuous-flow respirators are made with tight-fitting half-masks or full-face pieces. They are also made with hoods. The flow of air to the user may be adjusted either at the air source (fixed flow) or on the unit's regulator (adjustable flow). A pressure-demand supplied-air respirator is similar to the continuous-flow type except that it supplies air to the user's face piece via a pressure-demand valve as the user inhales and fresh air is required. It typically has a two-position exhalation valve that allows the worker to switch between pressure-demand and negative-pressure modes to facilitate entry into, movement within, and exit from a work area.

108F20.EPS

Figure 20 ◆ Powered air-purifying respirator.

108F21.EPS

Figure 21 ◆ Supplied-air respirator.

5.2.3 Self-Contained Breathing Apparatus (SCBA)

SCBAs (*Figure 22*) provide the highest level of respiratory protection. They can be used in oxygen-deficient atmospheres (below 19.5% oxygen), in poorly ventilated or confined spaces, and in IDLH atmospheres. These respirators provide a supply of air for about 30 to 60 minutes from a compressed-air cylinder worn on the user's back. An emergency escape breathing apparatus (EEBA) is a smaller version of a SCBA cylinder. EEBAs are used for escape from hazardous environments and generally provide a five- to ten-minute supply of air.

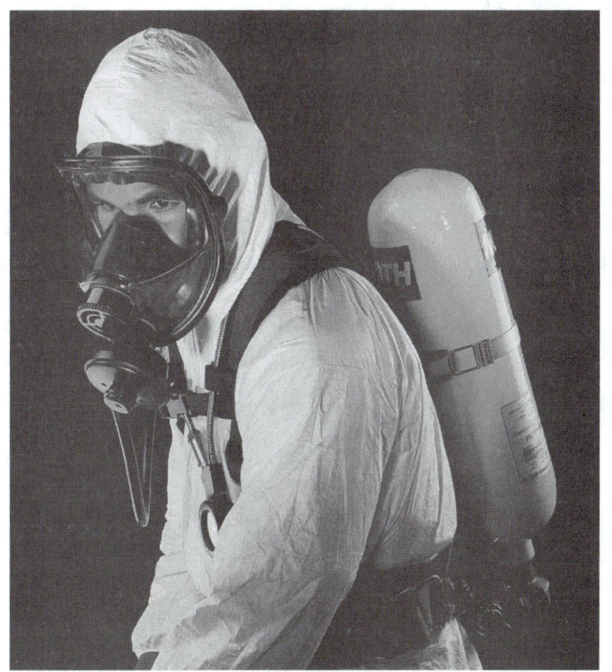

108F22.EPS

Figure 22 ◆ Self-contained breathing apparatus.

5.3.0 Respiratory Program

A respirator must be properly selected for the contaminant present and its concentration level (*Figure 23*). It must be properly fitted and used in accordance with the manufacturer's instructions. It must be worn during all times of exposure. Employers must have a respiratory protection program consisting of the following:

- Standard operating procedures for selection and use
- Employee training
- Regular cleaning and disinfecting
- Sanitary storage
- Regular inspection
- Annual fit testing
- Pulmonary function testing

BEFORE USING A RESPIRATOR YOU MUST DETERMINE THE FOLLOWING:

1. THE TYPE OF CONTAMINANT(S) FOR WHICH THE RESPIRATOR IS BEING SELECTED
2. THE CONCENTRATION LEVEL OF THE CONTAMINANT(S)
3. WHETHER THE RESPIRATOR CAN BE PROPERLY FITTED ON THE WEARER'S FACE

ALL RESPIRATOR INSTRUCTIONS, WARNINGS, AND USE LIMITATIONS CONTAINED ON EACH PACKAGE MUST ALSO BE READ AND UNDERSTOOD BY THE WEARER BEFORE USE.

108F23.EPS

Figure 23 ◆ Use the correct respirator.

As an employee, you are responsible for wearing respiratory protection when needed. When it comes to vapors or fumes, both can be eliminated (in certain concentrations) by the use of air-purifying devices as long as the oxygen levels are acceptable. Examples of fumes are smoke billowing from a fire or the fumes generated when welding. Always check the cartridge on your respirator to make sure it is the correct type to use for the air conditions and contaminants found on your job site.

When selecting a respirator to wear while working with specific materials, you must first determine the hazardous ingredients contained in the material and their exposure levels, and then choose the proper respirator to protect yourself at those levels. Always read the product's MSDS. It identifies the hazardous ingredients and should list the type of respirator and cartridge recommended for use with the product.

Limitations that apply to all half-mask (air-purifying) respirators are as follows:

- These respirators do not completely eliminate exposure to contaminants, but they will reduce the exposure to an acceptable level.
- These respirators do not supply oxygen and must not be used in areas where the oxygen level is below 19.5%.
- These respirators must not be used in areas where chemicals have poor warning signs, such as no taste or odor.

If breathing becomes difficult, if you become dizzy or nauseated, if you smell or taste the chemical, or if you experience other noticeable effects of exposure, leave the area immediately, return to a fresh air area, and seek any necessary assistance.

5.3.1 Positive and Negative Fit Checks

Respirators are useless unless properly fit-tested to each individual. To obtain the best protection from your respirator, you must perform positive and negative fit checks each time you wear it. These fit checks must be repeated until you have obtained a good face seal.

To perform a positive fit check, do the following:

Step 1 Adjust the face piece for the best fit, then adjust the head and neck straps to ensure good fit and comfort.

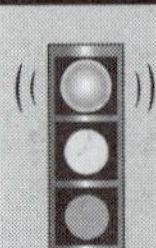

WARNING!
Do not overtighten the head and neck straps. Tighten them only enough to stop leakage. Overtightening can cause face-piece distortion and dangerous leaks.

Step 2 Block the exhalation valve with your hand or other material.

Step 3 Breathe out into the mask.

Step 4 Check for air leakage around the edges of the face piece.

Step 5 If the face piece puffs out slightly for a few seconds, a good face seal has been obtained.

To perform a negative fit check, do the following:

Step 1 Block the inhalation valve with your hand or other material.

Step 2 Attempt to inhale.

Step 3 Check for air leakage around the edges of the face piece.

Step 4 If the face piece caves in slightly for a few seconds, a good face seal has been obtained.

A respirator must be clean, in good condition, and all of its parts must be in place for it to give you the proper protection. Respirators must be cleaned every day. Failure to do so will limit their effectiveness and offer little or no protection. For example, suppose you wore the respirator for two weeks and did not clean it. The bacteria from breathing into the respirator, plus the airborne contaminants that managed to enter the face piece, have made the inside of your respirator very unsanitary. Continued use may cause you more harm than good. Remember, only a clean and complete respirator will provide you with the necessary protection. Follow these guidelines.

- Inspect the condition of your respirator before and after each use.
- Do not wear a respirator if the face piece is distorted or if it is worn and cracked. You will not be able to get a proper face seal.
- Do not wear respirators if any part is missing. Replace worn straps or missing parts before use.
- Do not expose respirators to excessive heat or cold, chemicals, or sunlight.

Dangerous Welding By-Products

A project involved welders using oxyacetylene torches to cut scrap metal into manageable pieces. Before the project started, samples of the scrap metal were analyzed to determine what hazards existed. The results of these analyses did not identify arsenic as a hazard.

Routine sampling was conducted as the project progressed. The samples were analyzed and found to exceed the OSHA exposure level for arsenic. The personnel protective measures, which were based on the original analysis, were inadequate to protect from this arsenic hazard. The source of the arsenic could not be determined and therefore could not be eliminated. An arsenic compliance plan was developed and implemented, detailing how operations could continue while still protecting the workforce.

The Bottom Line: Had testing of the scrap material not been conducted after the work started, the workers would have suffered life-threatening exposure to arsenic.

Source: Department of Energy Case History Information Services

- Clean and wash your respirator each day. Remove the cartridge and filter, hand wash the respirator using mild soap and a soft brush, and let it air dry overnight.
- Sanitize your respirator each week. Remove the cartridge and filter, then soak the respirator in a sanitizing solution for at least two minutes. Thoroughly rinse with warm water and let it air dry overnight. Store the clean and sanitized respirator in its resealable plastic bag. Do not store the respirator face down. This will cause distortion of the face piece.

6.0.0 ◆ WELDING AND CUTTING OPERATIONS IN CONFINED SPACES

When welding or cutting is being performed in any confined space, the gas cylinders and welding machines are left on the outside. Before any work is started, heavy equipment mounted on wheels must be securely blocked to prevent accidental movement. Where a welder must enter a confined space through a manhole or other opening, means must be provided for quickly removing the worker in case of emergency (*Figure 24*). When safety belts and lifelines are used for this purpose, they are attached to the welder's body before entry. An attendant with a pre-planned rescue procedure is stationed outside to observe the welder at all times and must be capable of putting the rescue operations into effect.

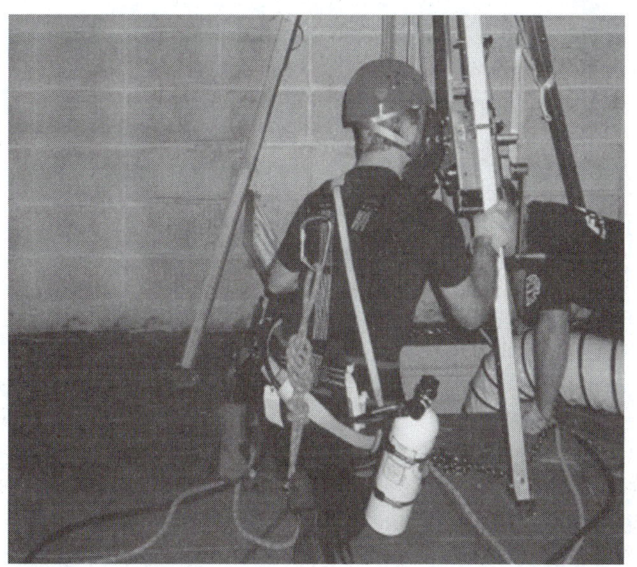

108F24.EPS

Figure 24 ◆ Confined-space entry.

When welding operations are suspended for any substantial period of time, such as during lunch or overnight, all electrodes must be removed from the holders. The holders must be carefully placed so that accidental contact cannot occur. The welding machines must also be disconnected from the power source.

Leaks or improperly closed valves can allow gas to escape during periods of inactivity. Special procedures are also required when oxyacetylene equipment will not be used for a substantial period of time: the gas and oxygen supply valves must be closed, the regulators must be released, the gas and oxygen lines must be bled, and the valves on the torch must be shut off.

Summary

Welding and metal cutting are integral parts of the construction process. If you are working near someone who is welding or using a cutting torch, you must be mindful of the related health and safety hazards. Be sure to wear all necessary personal protective equipment such as flash goggles, flame-resistant clothing, and protective gloves. Also, make sure that you are aware of what is going on if you are working around welding. Extreme heat, bright flashes of light, toxic fumes, and high voltage are just some of the hazards associated with welding. Being careless around welding can lead to tragic results.

Compressed-gas cylinders must be handled carefully. Protective caps must be used when the cylinders are stored or being moved. Safety chains must be used to prevent cylinders from falling over during use, storage, and transport. If the valve breaks off, the pressure release from the cylinder may cause it to act like an unguided missile. The gases used in oxyfuel cutting and welding are highly flammable. Under the right conditions, they will explode. For example, if pure oxygen comes into contact with grease or oil, an explosion is likely.

Anyone working in an environment where welding or metal cutting is being performed must be properly trained to recognize and avoid these hazards.

Review Questions

1. Before it can be used as a welding platform, a container that held a combustible chemical must be thoroughly cleaned, then filled with _____.
   a. pure oxygen
   b. solvent
   c. water or an inert gas
   d. concrete

2. A hot work permit is not required if the welder has access to a fire extinguisher.
   a. True
   b. False

3. A hot work permit is usually issued by _____.
   a. the county fire marshal
   b. a welder
   c. the site manager
   d. the welding foreman

4. If a gas cylinder is frozen to the ground, it is best to loosen it with boiling water.
   a. True
   b. False

5. Dizziness, nausea, and abdominal pain are signs of excessive _____ exposure.
   a. noise
   b. vibration
   c. heat
   d. light

6. Welder's flash is a condition caused by _____.
   a. visible light alone
   b. ultraviolet radiation alone
   c. a combination of ultraviolet radiation and visible light
   d. infrared radiation

7. Cornea damage and cataracts are caused by _____.
   a. visible light alone
   b. ultraviolet radiation alone
   c. a combination of ultraviolet radiation and visible light
   d. infrared radiation

8. The maximum continuous noise level allowed on a construction work site over an 8-hour period is _____.
   a. 85 decibels
   b. 90 decibels
   c. 95 decibels
   d. 100 decibels

9. Electrical welding equipment should be inspected for breaks or cracks in the insulation, cables, or electrode holders at least every _____.
   a. hour
   b. day
   c. two days
   d. week

10. When welding, your shirt should be worn _____.
    a. tucked into the pants, with the collar buttoned
    b. tucked into the pants, with the collar unbuttoned
    c. hanging out over the pants, with the collar buttoned
    d. hanging out over the pants, with the collar unbuttoned

GLOSSARY

Trade Terms Introduced in This Module

Asphyxiation: A deadly condition caused by interference with breathing, resulting in a lack of oxygen and an excess of carbon dioxide in the blood.

Brazing: A method of joining metals with a nonferrous filler metal at a temperature above 800°F, but lower than the melting point of the metals being joined.

Curtain wall: A nonbearing exterior wall that is set into, and attached to, the steel or concrete structure of a building.

Flammable: Easily set on fire.

Inert gas: A colorless, odorless, tasteless gas that has very little chemical reactivity.

APPENDIX

Welding Health Hazards

Hazardous Chemical Agents

Beryllium – Beryllium is sometimes used as an alloying element with copper and other base metals. Acute exposure to high concentrations of beryllium can result in chemical pneumonia. Long-term exposure can result in shortness of breath, chronic cough, and significant weight loss accompanied by fatigue and general weakness.

Cadmium – Cadmium is frequently used as a rust-preventive coating on steel and also as an alloying element. Acute exposures to high concentrations of cadmium fumes can produce severe lung irritation, pulmonary edema, and in some cases, death. Long-term exposure to low levels of cadmium in air can result in emphysema (a disease affecting the ability of the lungs to absorb oxygen) and can damage the kidneys.

Carbon monoxide – Carbon monoxide is a gas usually formed by the incomplete combustion of various fuels. Welding and cutting may produce significant amounts of carbon monoxide. In addition, welding operations that use carbon dioxide as the inert gas shield may produce hazardous concentrations of carbon monoxide in poorly ventilated areas. This is caused by a breakdown of shielding gas. Carbon monoxide is odorless, colorless, and cannot be detected. Common symptoms of overexposure include pounding of the heart, a dull headache, flashes before the eyes, dizziness, ringing in the ears, and nausea.

Chlorinated hydrocarbon solvents – Various chlorinated hydrocarbons are used in degreasing and other cleaning operations. The vapors of these solvents are a concern in welding and cutting because the heat and ultraviolet radiation from the arc will decompose the vapors and form highly toxic and irritating phosgene gas. (See *Phosgene*.)

Fluorides – Fluoride compounds are found in the coatings of several types of fluxes used in welding. Exposure to these fluxes may irritate the eyes, nose, and throat. Repeated exposure to high concentrations of fluorides in air over a long period may cause pulmonary edema (fluid in the lungs) and bone damage. Exposure to fluoride dusts and fumes may also produce skin rashes.

Iron oxide – Iron is the principal alloying element in steel manufacture. During the welding process, iron oxide fumes arise from both the base metal and the electrode. The primary acute effect of this exposure is irritation of nasal passages, throat, and lungs. Although long-term exposure to iron oxide fumes may result in iron pigmentation of the lungs, most authorities agree that these iron deposits in the lungs are not dangerous.

Lead – The welding and cutting of lead-bearing alloys or metals whose surfaces have been painted with lead-based paint can generate lead oxide fumes. Inhalation and ingestion of lead oxide fumes and other lead compounds will cause lead poisoning. Symptoms include a metallic taste in the mouth, loss of appetite, nausea, abdominal cramps, and insomnia. Over time, anemia and general weakness, chiefly in the muscles of the wrists, develop. Lead adversely affects the brain, central nervous system, circulatory system, reproductive system, kidneys, and muscles.

Mercury – Mercury compounds are used to coat metals to prevent rust or inhibit foliage growth (marine paints). Under the intense heat of the arc or gas flame, mercury vapors are produced. Exposure to the vapors may produce stomach pain, diarrhea, kidney damage, or respiratory failure. Long-term exposure may produce tremors, emotional instability, and hearing damage.

Nitrogen oxides – The ultraviolet light of the arc can produce nitrogen oxides from the nitrogen and oxygen in the air. Nitrogen oxides are produced by gas metal arc welding (GMAW or short-arc), gas tungsten arc welding (GTAW or heli-arc), and plasma arc cutting. Even greater quantities are formed if the shielding gas contains nitrogen. Nitrogen dioxide, one of the oxides formed, has the greatest health effect. This gas is irritating to the eyes, nose, and throat, but dangerous concentrations can be inhaled without any immediate discomfort. High concentrations can cause shortness of breath, chest pain, and fluid in the lungs (pulmonary edema).

Ozone – Ozone is produced by ultraviolet light from the welding arc. Ozone is produced in greater quantities by gas metal arc welding (GMAW or short-arc), gas tungsten arc welding (GTAW or heli-arc), and plasma arc cutting. Ozone is a highly active form of oxygen and can cause great irritation to all mucous membranes. Symptoms of ozone exposure include headache, chest pain, and dryness of the upper respiratory tract. Excessive exposure can cause fluid in the lungs (pulmonary edema). Both nitrogen dioxide and ozone are thought to have long-term effects on the lungs.

Phosgene – Phosgene is formed by the decomposition of chlorinated hydrocarbon solvents by ultraviolet radiation. It reacts with moisture in the lungs to produce hydrogen chloride, which in turn destroys lung tissue. For this reason, any use of chlorinated solvents should be well away from welding operations or any operation in which ultraviolet radiation or intense heat is generated.

Zinc – Zinc is used in large quantities in the manufacture of brass, galvanized metals, and various other alloys. Inhalation of zinc oxide fumes can occur when welding or cutting on zinc-coated metals. Exposure to these fumes is known to cause metal fume fever. Symptoms of metal fume fever are very similar to those of common influenza. They include fever (rarely exceeding 102°F), chills, nausea, dryness of the throat, cough, fatigue, and general weakness and aching of the head and body. The victim may sweat profusely for a few hours, after which the body temperature begins to return to normal. The symptoms of metal fume fever have rarely, if ever, lasted beyond 24 hours. The subject can then appear to be more susceptible to the onset of this condition after a period of nonexposure, such as on a Monday or on a weekday following a holiday, than they are on other days.

Hazardous Physical Agents

Infrared radiation – Exposure to infrared radiation, produced by the electric arc and other flame cutting equipment, may heat the skin surface and the tissues immediately below the surface. Except for this effect, which can progress to thermal bums in some situations, infrared radiation is not dangerous to welders. Most welders protect themselves from infrared and ultraviolet radiation with a welder's helmet or glasses and protective clothing.

Intense visible light – Exposure of the human eye to intense visible light can produce adaptation, pupillary reflex, and shading of the eyes. Such actions are protective mechanisms to prevent excessive light from being focused on the retina. In the arc welding process, eye exposure to intense visible light is prevented for the most part by the welder's helmet. However, some individuals have sustained retinal damage due to careless viewing of the arc. At no time should the arc be observed without eye protection.

Ultraviolet radiation – Ultraviolet radiation is generated by the electric arc in the welding process. Skin exposure to ultraviolet radiation can result in severe bums, in many cases without prior warning. Ultraviolet radiation can also damage the lens of the eye. Many arc welders are aware of the condition known as arc eye, a sensation of sand in the eyes. This condition is caused by excessive eye exposure to ultraviolet radiation. Ultraviolet rays also increase the skin effects of some industrial chemicals, such as coal tar and creosol compounds.

Figure Credits

Controls Corporation of America	108F01
Mathey Dearman, Inc.	108F02
Gerald Shannon	108F03, 108F12, 108F13
Sellstrom Manufacturing Co., Inc.	108F04
SAF-T-CART	108F09
Charles Rogers	108F10, 108F11, 108F14, 108F15
Nederman, Inc.	108F16
Hornell, Inc.	108F17, 108F18, 108F21
Scott Health and Safety	108F19, 108F20
North Safety Products	108F22
Brad Krauel	108F24

NCCER CURRICULA — USER UPDATE

NCCER makes every effort to keep its textbooks up-to-date and free of technical errors. We appreciate your help in this process. If you find an error, a typographical mistake, or an inaccuracy in NCCER's curricula, please fill out this form (or a photocopy), or complete the online form at **www.nccer.org/olf**. Be sure to include the exact module ID number, page number, a detailed description, and your recommended correction. Your input will be brought to the attention of the Authoring Team. Thank you for your assistance.

Instructors – If you have an idea for improving this textbook, or have found that additional materials were necessary to teach this module effectively, please let us know so that we may present your suggestions to the Authoring Team.

NCCER Product Development and Revision

13614 Progress Blvd., Alachua, FL 32615

Email: curriculum@nccer.org

Online: www.nccer.org/olf

❑ Trainee Guide ❑ AIG ❑ Exam ❑ PowerPoints Other ____________________

Craft / Level: Copyright Date:

Module ID Number / Title:

Section Number(s):

Description:

Recommended Correction:

Your Name:

Address:

Email: Phone:

Module 75109-03

Fall Protection

COURSE MAP

This course map shows all of the modules in Field Safety. The suggested training order begins at the bottom and proceeds up. The local Training Program Sponsor may adjust the training order.

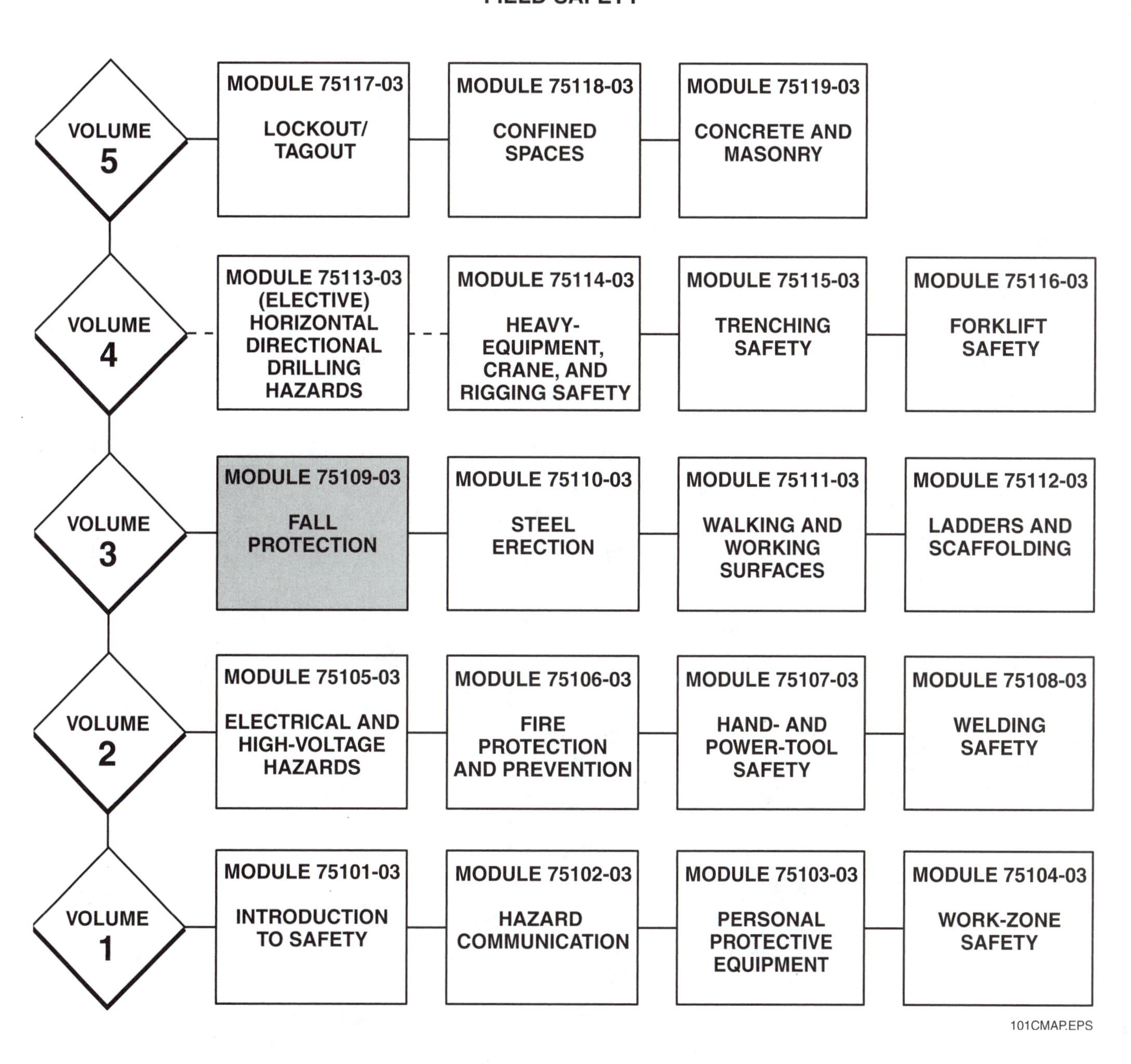

MODULE 75109-03 CONTENTS

Figures

MODULE 75109-03

Fall Protection

Objectives

When you have completed this module, you will be able to do the following:

1. Explain and identify safety hazards associated with working at elevated heights.
2. Demonstrate and explain how to properly use fall-protection equipment.

Prerequisites

Before you begin this module, it is recommended that you successfully complete the following: Field Safety, Modules 75101-03 through 75108-03.

Required Materials

1. Pencil and paper
2. Appropriate personal protective equipment

1.0.0 ◆ INTRODUCTION

Falls are the leading cause of death in the construction industry. In fact, more than one third of all deaths in the industry are the result of a fall. Fall protection is required when workers are exposed to falls from work areas with elevations that are 6' or higher. The types of work areas that put the worker at risk include:

- Scaffolding
- Ladders
- **Leading edges**
- Ramps or runways
- Wall or floor openings
- Roofs
- Excavations, pits, and wells
- Concrete forms
- Unprotected sides and edges

Falls happen because of the inappropriate use or lack of fall-protection systems (*Figure 1*). They also happen because of worker carelessness. It is your responsibility to learn how to set up, use, and maintain fall-protection equipment. Not only will this keep you alive and uninjured, it could save the lives of your co-workers.

2.0.0 ◆ FALLING HAZARDS AND SAFEGUARDS

Falls are classified into two groups: falls from an elevation and falls on the same level. Falls from an elevation can happen when someone is doing work from scaffolding, work **platforms**, decking, concrete forms, ladders, or excavations. Falls from elevations are almost always fatal. This is not to say that falls on the same level aren't also extremely dangerous. When a worker falls on the same level, usually from tripping or slipping, head injuries often occur. Sharp edges and pointed objects such as exposed rebar could cut or stab the worker.

The following safe practices can help prevent slips and falls:

- Wear safe, strong work boots that are in good repair.
- Watch where you step. Be sure your footing is secure.
- Do not allow yourself to get in an awkward position. Stay in control of your movements at all times.
- Maintain clean, smooth walking and working surfaces. Fill holes, ruts, and cracks. Clean up slippery material and litter.
- Install cables, extension cords, and hoses so that they will not become tripping hazards.
- Do not run on scaffolding, work platforms, decking, roofs, or other elevated work areas.

Obey Safety Warnings

A crew was installing the final structural steel beam (bar joist) in the roof of a new cold storage warehouse under construction. After a crane lifted the beam into place, it was not quite straight. The foreman on the job wanted to use a hammer to straighten it. He was standing on a portion of roof decking that had already been completed. To get to the beam, he reached his left foot out over an open, undecked area of the roof. He rested his left foot on the nearest joist girder. As he was preparing to strike a blow with the hammer, his foot slipped off the girder. His hands caught the bar joist, but he couldn't hold on and fell. The area where the foreman needed to work had been barricaded with wire rope safety lines on all four sides, but he removed these lines to gain access. He was not using fall-protection equipment. He was killed when he fell 38' to the floor below.

The Bottom Line: Never cross a safety line. Always wear fall-protection equipment when working on elevated surfaces.

Source: Electronic Library of Construction Occupational Safety and Health

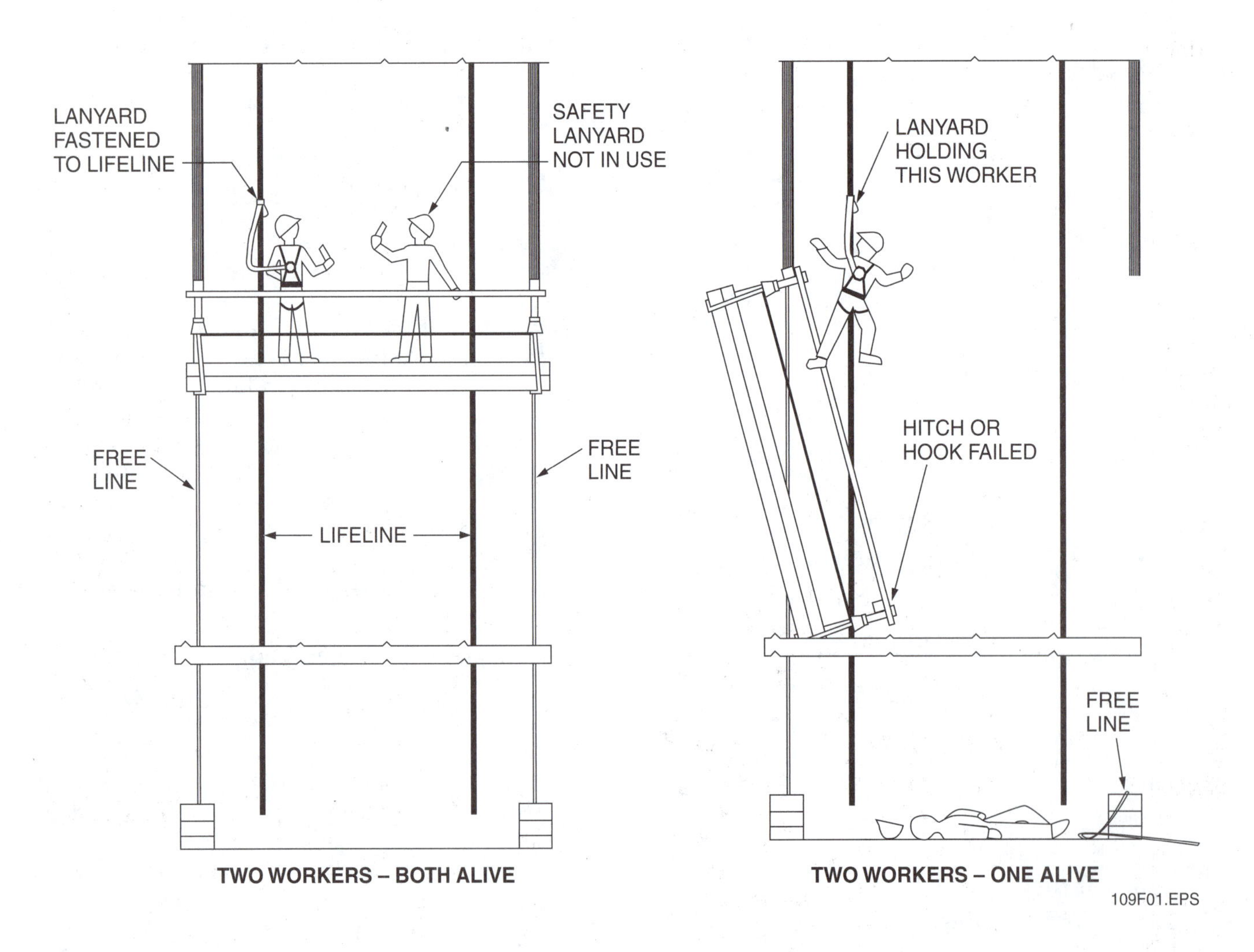

Figure 1 ◆ Proper and improper safety harness use.

The best way to survive a fall from an elevation is to use fall-protection equipment. The three most common types of fall-protection equipment are guardrails, **personal fall-arrest systems**, and safety nets.

2.1.0 Guardrails

Guardrails are the most common type of fall protection (*Figure 2*). They protect workers by providing a barrier between the work area and the ground or lower work areas. They may be made of wood, pipe, steel, or wire rope and must be able to support 200 pounds of force applied to the top rail.

(A)

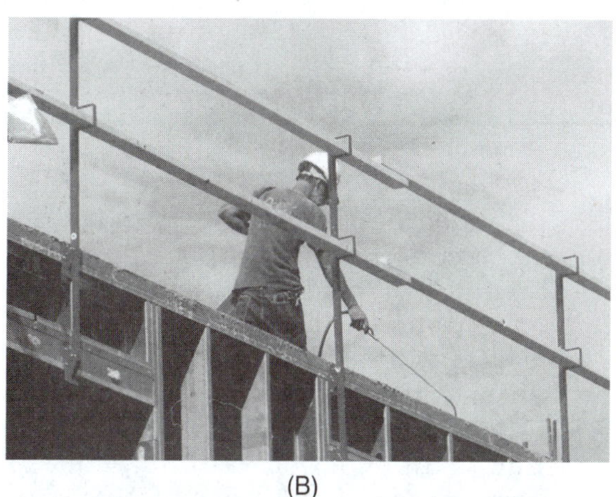

(B)

(C)

109F02.EPS

Figure 2 ◆ Guardrails.

2.2.0 Personal Fall-Arrest Systems

Personal fall-arrest systems catch workers after they have fallen. They are designed and rigged to prevent a worker from **free falling** a distance of more than 6' and hitting the ground or a lower work area. When describing personal fall-arrest systems, these terms must be understood:

- *Free-fall distance* – The vertical distance a worker moves after a fall before a deceleration device is activated.
- *Deceleration device* – A device such as a shock-absorbing **lanyard** or **self-retracting lifeline** that brings a falling person to a stop without injury.
- *Deceleration distance* – The distance it takes before a person comes to a stop when falling. The required deceleration distance for a fall-arrest system is a maximum of 3½'.
- *Arresting force* – The force needed to stop a person from falling. The greater the free-fall distance, the more force is needed to stop or arrest the fall.

Personal fall-arrest systems use specialized equipment. This equipment includes:

- **Body harnesses** and belts
- Lanyards
- Deceleration devices
- **Lifelines**
- Anchoring devices and equipment **connectors**

2.2.1 Body Harnesses

Full-body harnesses with sliding back D-rings (*Figure 3*) are used in personal fall-arrest systems. They are made of straps that are worn securely around the user's body. This allows the arresting force to be distributed throughout the body, including the shoulders, legs, torso, and buttocks.

Icy Scaffolding Can Be Deadly

A laborer was working on the third level of a tubular welded-frame scaffold. It was covered with ice and snow. Planking on the scaffold was weak, and a guardrail had not been set up. The worker slipped and fell head first approximately 20' to the pavement below.

The Bottom Line: Don't work on a wet or icy scaffold. Make sure all scaffolding is sturdy and includes the proper guardrails.

Source: The Occupational Safety and Health Administration (OSHA)

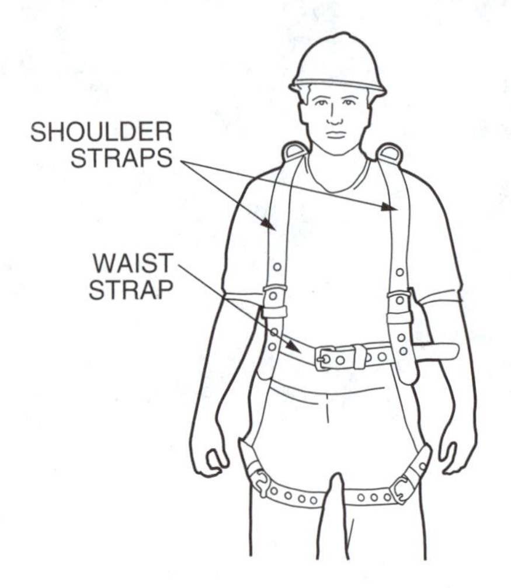

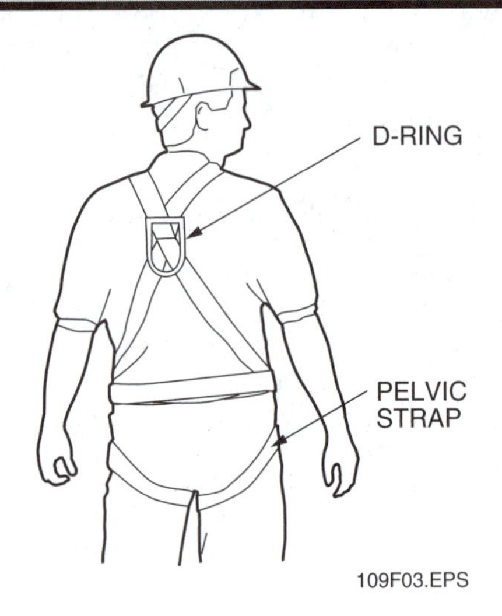

Figure 3 ◆ Full-body harnesses with sliding back D-rings.

This distribution decreases the chance of injury. When a fall occurs, the sliding D-ring moves to the nape of the neck. This keeps the worker in an upright position and helps to distribute the arresting force. The worker then stays in a relatively comfortable position while waiting for rescue.

Selecting the right full-body harness depends on a combination of job requirements and personal preference. Harness manufacturers normally provide selection guidelines in their product literature. Some types of full-body harnesses can be equipped with front chest D-rings, side D-rings, or shoulder D-rings. Harnesses with front chest D-rings are typically used in ladder climbing and personal-positioning systems (*Figure 4*). Those with side D-rings are also used in personal-positioning systems. Personal-positioning systems allow workers to hold themselves in place, keeping their hands free to accomplish a task.

A personal-positioning system should not allow a worker to free fall more than 2'. The **anchorage** that it's attached to should be able to support at least twice the impact load of a worker's fall or 3,000 pounds, whichever is greater.

It Really Works

A construction worker was moving material on a roof. The worker was standing on a metal roof panel when his right foot slipped. In an attempt to correct himself, the worker caught his left foot in a corrugation rib. The worker then fell forward to his knees on an underlying bottom insulation sheet and broke through the layer of transite. After a short free fall, the clutch mechanism of his retractable lanyard engaged, gently slowing the worker's descent to a complete stop.

The Bottom Line: Personal fall-protection equipment can save your life.

Source: U.S. Department of Energy

109F04.EPS

Figure 4 ◆ Harness with front chest D-ring.

Sliding back D-rings are the most common type of D-ring.The procedure for putting on a sliding back D-ring full-body harness is as follows:

Step 1 Hold the harness by the back D-ring.

Step 2 Shake the harness, allowing all the straps to fall into place.

Step 3 Unbuckle and release the waist and/or leg straps.

Step 4 Slip the straps over your shoulders so that the D-ring is located in the middle of your back (*Figure 5*).

Step 5 Fasten the waist strap. It should be tight but not binding.

Step 6 At each leg, pull the straps between the legs and buckle.

Step 7 After all the straps have been buckled, tighten all friction buckles so that the harness fits snugly but allows full range of movement.

Step 8 Pull the chest strap around the shoulder straps and fasten it in the mid-chest area. Tighten it enough to pull the shoulder straps taut.

109F05.EPS

Figure 5 ◆ Properly positioning the D-ring.

2.2.2 Lanyards

Lanyards are short, flexible lines with connectors on each end (*Figure 6*). They are used to connect a body harness or **body belt** to a lifeline, deceleration device, or anchorage point. There are many kinds of lanyards made for different uses and climbing situations. All must have a minimum breaking strength of 5,000 pounds. They come in both fixed and adjustable lengths and are made out of steel, rope, or nylon webbing. Some have a shock absorber (*Figure 7*) which absorbs up to 80% of the arresting force when a fall is being stopped. When choosing a lanyard for a particular job, always follow the manufacturer's recommendations.

NOTE

In the past, body belts were often used instead of a full-body harness as part of a fall-arrest system. As of January 1, 1998, however, they were banned from such use. This is because body belts concentrate all of the arresting force in the abdominal area, causing the worker to hang in an uncomfortable and potentially dangerous position while awaiting rescue.

CAUTION

When activated during the fall-arresting process, a shock-absorbing lanyard stretches in order to reduce the arresting force. This potential increase in length must always be taken into consideration when determining the total free-fall distance from an anchor point.

109F06.EPS

Figure 6 ◆ Typical lanyard.

109F07.EPS

Figure 7 ◆ Lanyard with a shock absorber.

2.2.3 Deceleration Devices

Deceleration devices limit the arresting force to which a worker is subjected when the fall is stopped suddenly. Rope grabs and self-retracting lifelines are two common deceleration devices (*Figure 8*). A rope grab connects to a lanyard and attaches to a lifeline. In the event of a fall, the rope grab is pulled down by the attached lanyard, causing it to grip the lifeline and lock in place. Some rope grabs have a mechanism that allows the worker to unlock the device and slowly descend down the lifeline to the ground or surface below.

Self-retracting lifelines provide unrestricted movement and fall protection while climbing and descending ladders and similar equipment or when working on multiple levels. Typically, they have a 25' to 100' galvanized-steel cable that automatically takes up the slack in the attached lanyard, keeping the lanyard out of the worker's way. In the event of a fall, a centrifugal braking mechanism engages to limit the worker's free-fall distance. Self-retracting lifelines and lanyards that limit the free-fall distance to 2' or less must be able to support a minimum tensile load of 3,000 pounds. Those that do not limit the free fall to 2' or less must be able to hold a tensile load of at least 5,000 pounds.

2.2.4 Lifelines

Lifelines are ropes or flexible steel cables that are attached to an anchorage. They provide a means for tying off personal fall-protection equipment.

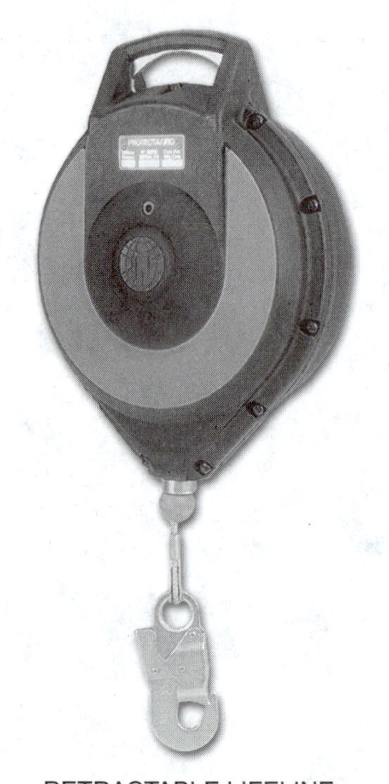

RETRACTABLE LIFELINE

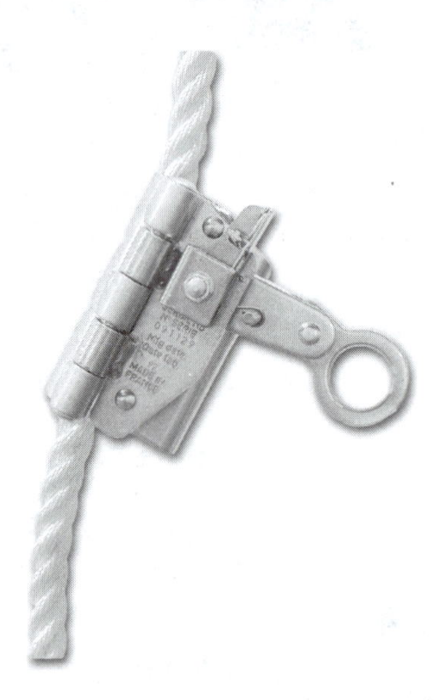

ROPE GRAB

109F08.EPS

Figure 8 ◆ Rope grab and retractable lifeline.

Vertical lifelines (*Figure 9*) are suspended vertically from a fixed anchorage. A fall-arrest device such as a rope grab is attached to the lifeline. Vertical lifelines must have a minimum breaking strength of 5,000 pounds. Each worker must use his or her own line. This is because if one worker falls, the movement of the lifeline during the fall arrest may also cause the other workers to fall. A vertical lifeline must be connected in a way that will keep the worker from moving past its end, or it must extend to the ground or the next lower working level.

Myths and Facts About Falls in Construction

Myth 1: In the construction industry, falls are not a leading cause of death.
Fact: One third of all deaths in the construction industry are caused by falls.

Myth 2: You have to fall a long distance to kill yourself.
Fact: Half of the construction workers who die in falls fall from a height of 21' or less. If you hit your head hard enough, you can die at any height. Even if you survive a fall, you may be laid up for some time with an injury.

Myth 3: Experienced workers don't fall.
Fact: The average age of construction workers who have fallen to their death is 47. That's not exactly young and inexperienced.

"It just happens so fast. It's when you think you're safe that you need to be more careful."
– *Gene, Builder*

Myth 4: Working safely is costly.
Fact: Some fall-protection equipment is inexpensive, such as ladder stabilizers, guardrail holders, and fall-protection kits. Other items such as harnesses, lifelines, and safe scaffolding are more costly. Injury and death, however, are much more expensive in the end.

"I fell three stories and was out of work for 8 weeks. I was subcontracting and didn't have comp. This was a long time ago, but I probably lost around $5,000. A harness would have cost me $50 back then." – *Dan, General Contractor*

Myth 5: Fall-protection equipment is more of a hindrance than a help.
Fact: Nothing is more of a hindrance than a lifelong disability you may experience due to a fall.

Source: Electronic Library of Construction Occupational Safety and Health

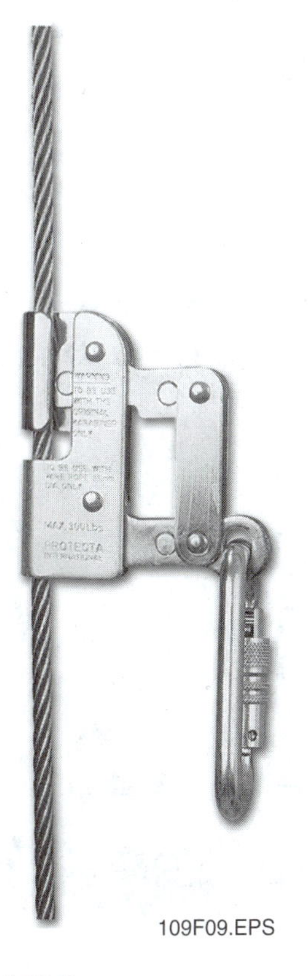

109F09.EPS

Figure 9 ◆ Vertical lifeline.

Horizontal lifelines (*Figure 10*) are connected horizontally between two fixed anchorages. Horizontal lifelines must be designed, installed, and used under the supervision of a qualified and competent person. The more workers tied off to a single horizontal line, the stronger the line and anchors must be.

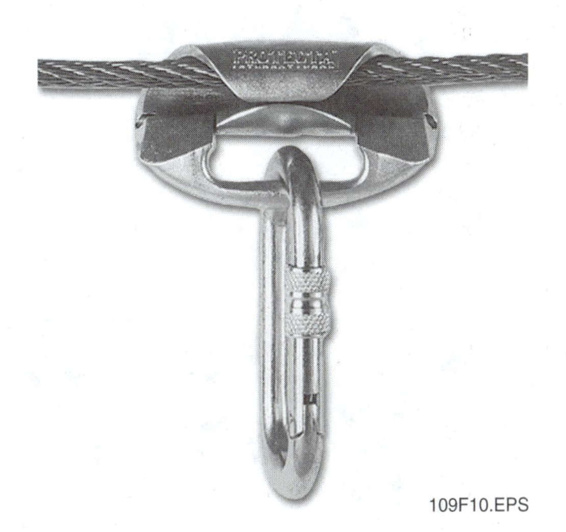

109F10.EPS

Figure 10 ◆ Horizontal lifeline.

Death Due to Unguarded Protruding Steel Bar

A laborer fell approximately 8' through a roof opening to a foundation that had about 20 half-inch rebars protruding straight up. The laborer was impaled on one of the bars and died.

The Bottom Line: Even a short-distance fall can be fatal. Use a personal fall-protection system and check the area for potential hazards.

Source: The Occupational Safety and Health Administration (OSHA)

2.2.5 Anchoring Devices and Equipment Connectors

Anchoring devices, commonly called tie-off points, support the entire weight of the fall-arrest system. The anchorage must be capable of supporting 5,000 pounds for each worker attached. Eye bolts (*Figure 11*) and overhead beams are considered anchorage points.

The D-rings, buckles, carabiners, and snaphooks (*Figure 12*) that fasten and/or connect the parts of a personal fall-arrest system are called connectors. There are regulations that specify how they are to be made, and that require D-rings and snaphooks to have a minimum tensile strength of 5,000 pounds.

NOTE

As of January 1, 1998, only locking-type snaphooks are permitted for use in personal fall-arrest systems.

109F11.EPS

Figure 11 ◆ Eye bolt.

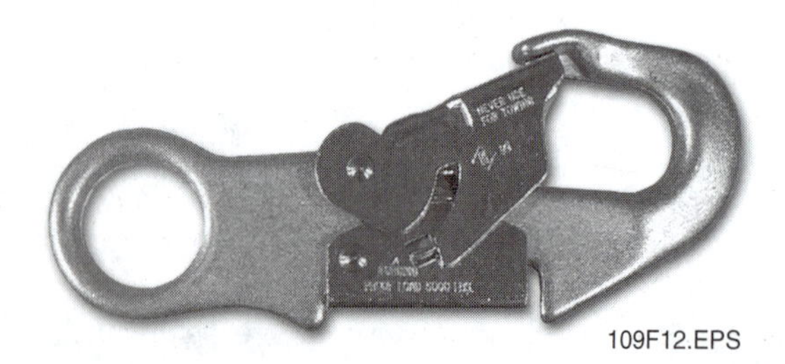

109F12.EPS

Figure 12 ◆ Double locking snaphook.

2.2.6 Selecting an Anchor Point and Tying Off

Connecting the body harness either directly or indirectly to a secure anchor point is called tying off. Tying off is always done before you get into a position from which you can fall. Follow the manufacturer's instructions on the best tie-off methods for your equipment. Use the following general guidelines in addition to the manufacturer's instructions when selecting an anchorage and tying off.

When selecting an anchorage point, it should be:

- Directly above the worker
- Easily accessible
- Damage-free and capable of supporting 5,000 pounds per worker
- High enough so that no lower level is struck should a fall occur
- Separate from work basket tie-offs

Be sure to check the manufacturer's equipment labels and allow for any equipment stretch and deceleration distance.

What's Wrong With This Picture?

109SA01.EPS

Guardrails Can Save Your Life

A construction subcontractor project engineer and a crane operator were performing final checks and making crane adjustments from the top of a container about 20' high. The engineer stepped out onto an 8' × 8' temporary work platform. The crane operator turned toward the platform area just in time to see the project engineer trip and fall from the platform.

The engineer was not wearing fall-protection equipment when he stepped out onto the platform. The platform didn't have guardrails. The engineer died from head and internal injuries.

The Bottom Line: Make sure any work platform has proper guardrails installed. Always wear the required fall-protection equipment.

Source: U.S. Department of Energy

2.2.7 Using Personal Fall-Arrest Equipment

Before using fall-protection equipment on the job, you must know the basics of fall protection and the proper use of the equipment. All equipment supplied by your employer must meet established standards for strength. Before each use, always read the instructions and warnings on any fall-protection equipment. Inspect the equipment using the following guidelines:

- Examine harnesses and lanyards for mildew, wear, damage, and deterioration.
- Make sure no straps are cut, broken, torn, or scraped.
- Check for damage from fire, chemicals, or corrosives.
- Make sure all hardware is free of cracks, sharp edges, and burrs.
- Check that snaphooks close and lock tightly and that buckles work properly.
- Check ropes for wear, broken fibers, pulled stitches, and discoloration.
- Make sure lifeline anchors and mountings are not loose or damaged.

Do not mix or match equipment from different manufacturers. All substitutions must be approved by your supervisor. All damaged or defective parts must be taken out of service immediately and tagged as unusable or destroyed. If the equipment was used in a previous fall, remove it from service until it can be inspected by a qualified person.

2.3.0 Safety Net Systems

Safety nets are used for fall protection on bridges and similar projects. They must be installed as close as possible, not more than 30', beneath the work area. There must be enough clearance under a safety net to prevent a worker who falls into it from hitting the surface below. There must also be no obstruction between the work area and the net.

Depending on the actual vertical distance between the net and the work area, the net must extend 8' to 13' beyond the edge of the work area. Mesh openings in the net must be limited to 36 square inches and 6" on the side. The border rope must have a 5,000-pound minimum breaking strength, and connections between net panels must be as strong as the nets themselves. Safety nets must be inspected at least once a week and after any event that might have damaged or weakened them. Worn or damaged nets must be removed from service.

3.0.0 ◆ RESCUE AFTER A FALL

Every elevated job site should have an established rescue and retrieval plan. Planning is especially important in remote areas where help is not readily available. Before beginning work, make sure that you know what your employer's rescue plan calls for you to do in the event of a fall. Find out what rescue equipment is available and where it is located. Learn how to use equipment for self-rescue and the rescue of others.

Drop Testing Safety Nets

Safety nets should be drop-tested at the job site after the initial installation, whenever relocated, after a repair, and at least every six months if left in one place. The drop test consists of a 400-pound bag of sand of 29" to 31" in diameter that is dropped into the net from at least 42" above the highest walking/working surface at which workers are exposed to fall hazards. If the net is still intact after the bag of sand is dropped, it passed the test.

Communicate With Your Co-Workers

A crew was demolishing the roof of the warehouse portion of a commercial building. The work was being done at night because the coal tar on the roof would release hazardous gases if disturbed in the heat of the day. The site had adequate halogen lighting, but none of the workers on the job were using fall protection.

After the roofing material was removed, 4' × 8' sheets of plywood were exposed. Any damaged sheets needed to be replaced. A helper was to follow the workers who were replacing the plywood, and pick up the damaged sheets and put them in a chute.

One evening, a worker removed a sheet of damaged plywood, but had run out of nails to attach the replacement plywood. He walked away to get more nails. The opening where the damaged plywood had been was left unguarded. No one knew this.

The helper came along shortly after the worker left, picked up the sheet of damaged plywood, and headed for the chute. He stepped into the opening and was killed when he fell approximately 27'.

The Bottom Line: Always use the appropriate fall-protection equipment.

Source: Electronic Library of Construction Occupational Safety and Health

Location is Everything

An worker was placing metal bridge decking onto the stringers of a bridge deck to be welded. After the first decking was placed down on stringers, the employee stepped onto it in order to put down the next decking. The decking was not secured in place and shifted.

Although safety nets were being used under another section of the bridge, they had not been moved forward as the crew moved to another area. The worker fell approximately 80' into the river and was killed.

The Bottom Line: Make sure that all safety equipment is in place before beginning any job.

Source: The Occupational Safety and Health Administration (OSHA)

If a fall occurs, any employee hanging from the fall-arrest system must be rescued safely and quickly. Your employer should have previously determined the method of rescue for fall victims, which may include equipment that lets the victim rescue himself or herself, a system of rescue by co-workers, or a way to alert a trained rescue squad. If a rescue depends on calling for outside help such as the fire department or rescue squad, all the needed phone numbers must be posted in plain view at the work site. In the event a co-worker falls, follow your employer's rescue plan. Call any special rescue service needed. Communicate with the victim and monitor him or her constantly during the rescue.

Summary

Falls are a very common and serious type of accident on construction sites. You have a responsibility to yourself and others to work safely and use fall-protection equipment. Personal injury and damage to equipment is less likely to happen when fall-protection equipment is used. Fall-protection equipment must be used correctly every time you work at an elevated site. It is important that everyone on a job site knows and follows their company's fall-protection procedures. If these procedures are not followed, or if protection equipment is not used, it is likely that falls will result in serious injuries or death.

Review Questions

1. Fall protection is required when working at elevations over _____.
   a. 6'
   b. 8'
   c. 10'
   d. 12'

2. Falls represent _____ of all deaths in the construction industry.
   a. ¼
   b. ⅓
   c. ½
   d. ⅔

3. Guardrails must be able to support _____ pounds of force applied to the top rail.
   a. 75
   b. 100
   c. 150
   d. 200

For Questions 4 through 7, match the correct term to the definition.

4. _____ Arresting force

5. _____ Deceleration device

6. _____ Deceleration distance

7. _____ Free-fall distance

   a. The distance it takes before a person comes to a stop when falling
   b. A device such as a shock-absorbing lanyard or self-retracting lifeline that brings a falling person to a stop without injury
   c. The vertical distance a worker moves after a fall before a deceleration device is activated
   d. The force needed to stop a person from falling
   e. A device that supports the entire weight of the fall-arrest system

8. Short, flexible lines with connectors on each end are called _____.
   a. anchorages
   b. deceleration devices
   c. lanyards
   d. lifelines

9. D-rings, buckles, and snaphooks are examples of deceleration devices.
   a. True
   b. False

10. Safety nets should be placed no more than _____ beneath the work area.
    a. 10'
    b. 20'
    c. 30'
    d. 35'

PROFILE IN SUCCESS

P.D. Frey, CSP

Corporate Safety Health and Environmental Director
Austin Industries

What has been your greatest career-related achievement?
My most significant accomplishment has been recognition as a Certified Safety Professional (CSP). This required the highest degree of commitment, covered the broadest spectrum of information, and is the most recognized and valued credential in the safety field.

How did you choose a career in the Safety Industry?
My interest in safety began when I was involved in a lost-time injury while attending college. I lost my footing while attempting to throw water from a five-gallon bucket onto a concrete slab. The bucket of water dropped on my thumb. I lost the tip of my thumb and thumb nail, and as a result, had to have a skin graft. This accident got me interested in safety.

What types of training have you been through?
I have been through all the required OSHA safety training: traffic control, fall protection, excavation, scaffolding, etc. The most important information is always the same: Why do we do it that way and how can I apply it to my work situation? The best training activities usually include hands-on activities and take-home materials for further review and explanation.

What kinds of work have you done in your career?
I have been focused primarily in the construction arena. I have been involved in heavy highway, commercial and petro-chemical related construction activities, and even had one millwork/lumber yard experience. I enjoy the dynamic, challenging world of construction where there is an ever-changing array of improvement opportunities.

Tell us about your present job.
A Corporate Safety Director position provides more freedom, more responsibility, and more travel than other positions. In some ways, it moves me further away from direct safety activities. The focus shifts toward policy and procedure development, legal and insurance implications, global vs. internal issues. I am focused less on regulatory and more on liability issues. It is rewarding to see issues resolved, programs implemented, and safety improvement achieved.

What factors have contributed most to your success?
I have to stress the importance of education. It is the foundation to any career. The importance of good writing skills and the ability to speak to groups of people become more important as you advance in the safety field.

What advice would you give to those new to Safety industry?
Don't depend only upon knowledge gained from books. Talk to people, ask for their input and suggestions; you will be surprised that in many cases, they will come up with a better solution than you would have recommended! Safety is best practiced when it is a shared learning experience.

GLOSSARY

Trade Terms Introduced in This Module

Anchorage: A secure point of attachment for lifelines, lanyards, or deceleration devices.

Body belt: A strap with means both for securing it about the waist and for attaching it to a lanyard, lifeline, or deceleration device.

Body harness: Straps that may be secured about the worker in a manner that will distribute the fall-arrest forces over at least the thighs, pelvis, waist, chest, and shoulders, with means for attaching it to other components of a personal fall-arrest system.

Connector: A device that is used to couple (connect) parts of the personal fall-arrest system and positioning device systems together. It may be an independent component of the system, such as a carabiner. It may also be an integral component of part of the system, such as a buckle or D-ring sewn into a body belt or body harness, or a snaphook spliced or sewn to a lanyard.

Free fall: The act of falling before a personal fall-arrest system begins to apply force to arrest the fall.

Lanyard: A flexible line of rope, wire rope, or strap that generally has a connector at each end for connecting the body belt or body harness to a deceleration device, lifeline, or anchorage.

Leading edge: The edge of a floor, roof, or formwork for a floor or other walking/working surface (such as the deck) which changes location as additional floor, roof, decking, or formwork sections are placed, formed, or constructed. A leading edge is considered to be an unprotected side and edge during periods when it is not actively and continuously under construction.

Lifeline: A component consisting of a flexible line connected vertically to an anchorage at one end (vertical lifeline), or connected horizontally to an anchorage at both ends (horizontal lifeline), and which serves as a means for connecting other components of a personal fall-arrest system to the anchorage.

Personal fall-arrest system: A system used to stop an employee in a fall from a working level. It consists of an anchorage, connectors, and a body harness, and may include a lanyard, deceleration device, lifeline, or suitable combinations of these. As of January 1, 1998, using a body belt for fall arrest is prohibited.

Platform: A work surface elevated above lower levels. Platforms can be constructed using individual wood planks, fabricated planks, fabricated decks, and fabricated platforms.

Self-retracting lifeline: A deceleration device containing a drum-wound line that can be slowly extracted from, or retracted onto, the drum under slight tension during normal employee movement, and which, after the onset of a fall, automatically locks the drum and arrests the fall.

Figure Credits

Construction Safety Association of Ontario	109F02 (A)
Fall Protection Systems, Inc.	109F02 (B) and (C)
Protecta International	109F04, 109F06, 109F07, 109F08, 109F09, 109F10, 109F11, 109F12
Gary Wilson	109SA01

NCCER CURRICULA — USER UPDATE

NCCER makes every effort to keep its textbooks up-to-date and free of technical errors. We appreciate your help in this process. If you find an error, a typographical mistake, or an inaccuracy in NCCER's curricula, please fill out this form (or a photocopy), or complete the online form at **www.nccer.org/olf**. Be sure to include the exact module ID number, page number, a detailed description, and your recommended correction. Your input will be brought to the attention of the Authoring Team. Thank you for your assistance.

Instructors – If you have an idea for improving this textbook, or have found that additional materials were necessary to teach this module effectively, please let us know so that we may present your suggestions to the Authoring Team.

NCCER Product Development and Revision

13614 Progress Blvd., Alachua, FL 32615

Email: curriculum@nccer.org
Online: www.nccer.org/olf

❑ Trainee Guide ❑ AIG ❑ Exam ❑ PowerPoints Other ______________________

Craft / Level: Copyright Date:

Module ID Number / Title:

Section Number(s):

Description:

Recommended Correction:

Your Name:

Address:

Email: Phone:

Module 75110-03

Steel Erection

COURSE MAP

This course map shows all of the modules in Field Safety. The suggested training order begins at the bottom and proceeds up. The local Training Program Sponsor may adjust the training order.

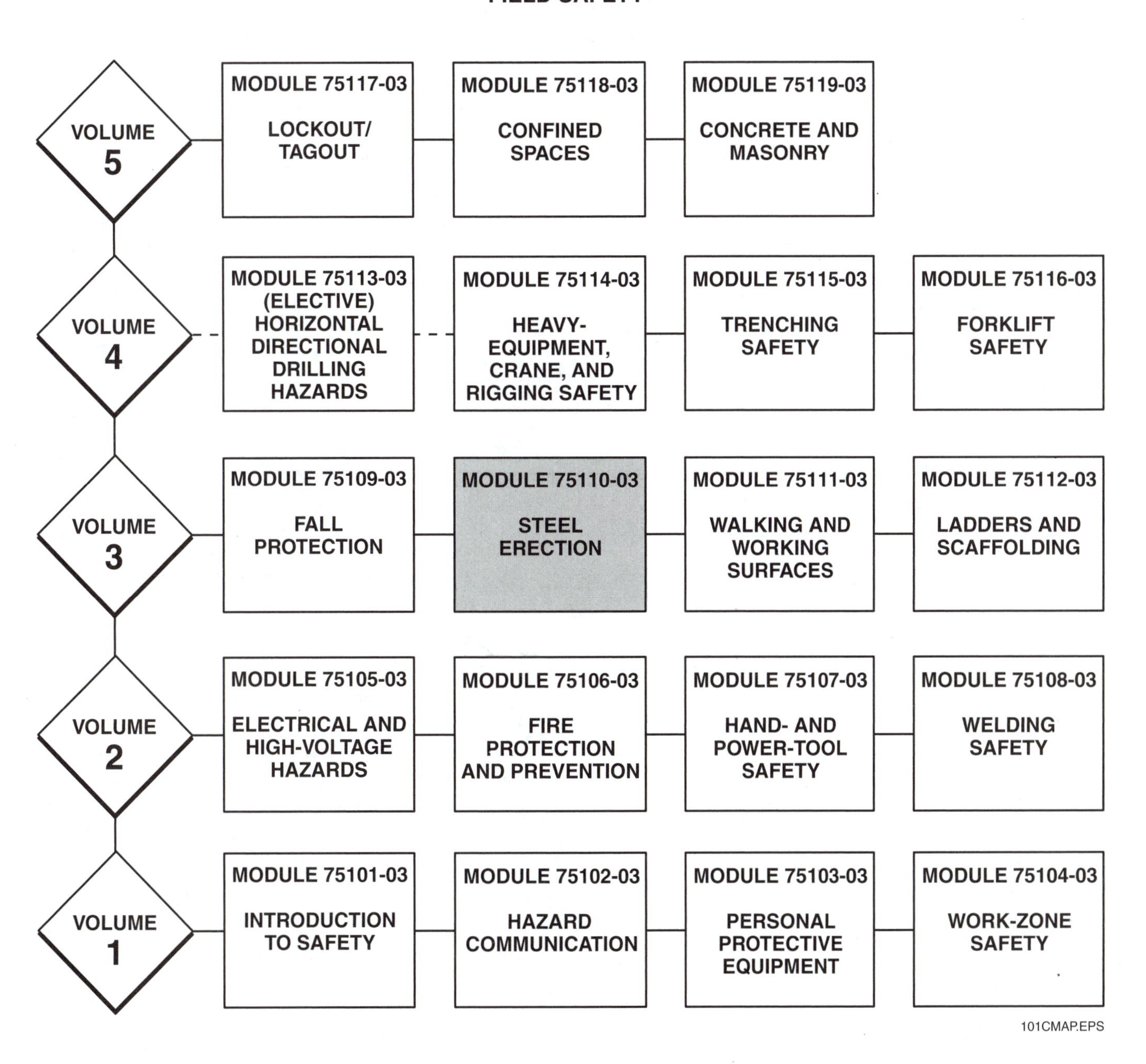

MODULE 75110-03 CONTENTS

Figures

MODULE 75110-03

Steel Erection

Objectives

When you have completed this module, you will be able to do the following:

1. Describe the steel-erection process.
2. Identify common safety hazards associated with steel-erection jobs.
3. Explain the safeguards that are required during a job to prevent personal injury and damage to equipment and property.

Prerequisites

Before you begin this module, it is recommended that you successfully complete the following: Field Safety, Modules 75101-03 through 75109-03.

Required Materials

1. Pencil and paper
2. Appropriate personal protective equipment

1.0.0 ◆ INTRODUCTION

Steel erection is used during the construction of many types of structures including high-rise buildings, metal decking, bridges, office buildings, schools, medical facilities, and retail stores. Steel-erection jobs can be extremely dangerous. The most common type of construction-site accident is falling from elevations. In fact, more workers die from falling than any other accident on steel-erection jobs. Workers can also be struck or crushed by falling loads, electrocuted by power lines, or run over or trapped by cranes (*Figure 1*).

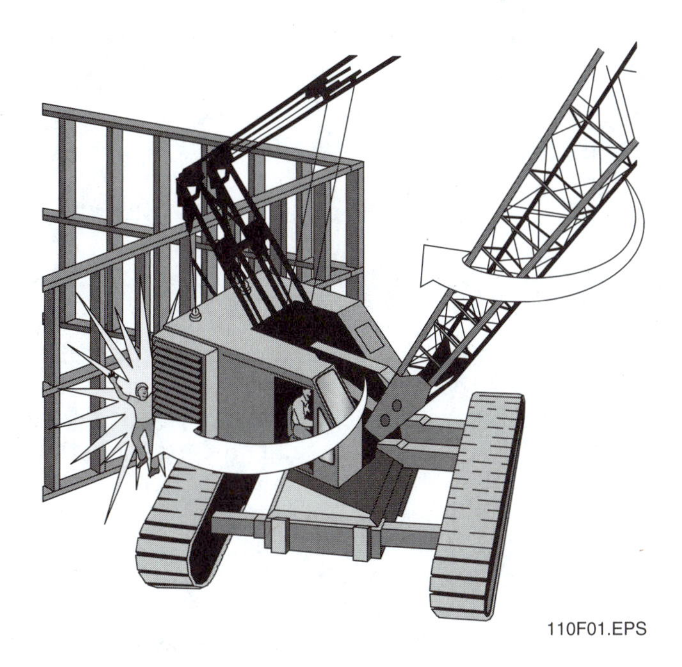
110F01.EPS

Figure 1 ◆ Worker struck by moving crane.

It's important to know the hazards and safety procedures for every job you work on regardless of the work you are doing. You must also understand that your first responsibility on a job is safety. This includes your own safety, the safety of others on the site, and the safe use of equipment on the site.

2.0.0 ◆ RIGGING

Rigging and hoisting of steel members and materials are essential parts of the steel-erection process. Rigging operations can be extremely complicated and dangerous, especially when moving heavy material like steel. Some rigging

Guard Floor Openings

A worker had just completed bolting up a vertical beam at a manufacturing facility. He was attempting to disconnect the hoisting line when he fell 60' through an unguarded floor opening in the metal decking on which he was standing.

The Bottom Line: This worker would not have fallen if guardrails had been in place. Always ensure that the area is safe before you begin working.

Source: The Occupational Safety and Health Administration (OSHA)

operations require cranes, while others use a loader to move materials around the job site. Regardless of whether rigging operations involve a simple vertical lift, a powered **hoist**, or a crane, only experienced workers who have been properly trained can do it without supervision.

The basic equipment used in rigging operations includes the following items:

- **Slings**
- **Hitches**
- **Rigging hardware**
- Hoists

2.1.0 Slings

During a rigging operation, the load that is being moved must be connected to the device that is doing the moving. In steel erection, that device will most likely be a crane. The connector that links the load to the lifting device is often a length of synthetic, chain, or wire rope material. This material is called the sling (*Figure 2*).

All slings are required to have identification tags. An identification tag must be securely attached to each sling and clearly marked with the information required for each type of sling. For all three types of slings, that information will include the manufacturer's name or trademark, and the rated capacity of the type of hitch used with that sling. The rated capacity is the maximum load weight that the sling is designed to carry.

Always check the identification tag before using a sling. If the load to be lifted weighs more than the rated load capacity of the sling, do not use it and immediately tell your supervisor.

2.2.0 Hitches

The way a sling is arranged to hold the load is called the rigging configuration or hitch (*Figure 3*). Hitches can be made using just the sling or by also using connecting hardware. One of the most important parts of rigging is making sure that the load is held securely and the movement of the load is controlled.

Different types of hitches are used for different types of loads. The type of hitch the rigger uses depends on the type of load to be lifted. For example, different hitches are used to secure loads of steel beams, pipes, concrete slabs, or heavy machinery.

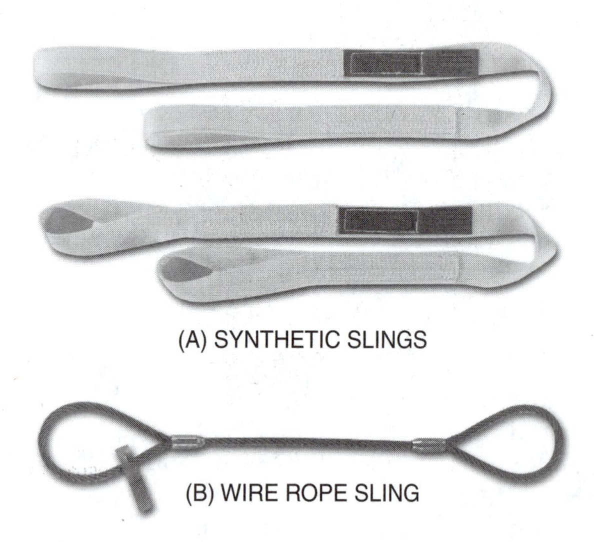

Figure 2 ◆ Examples of slings.

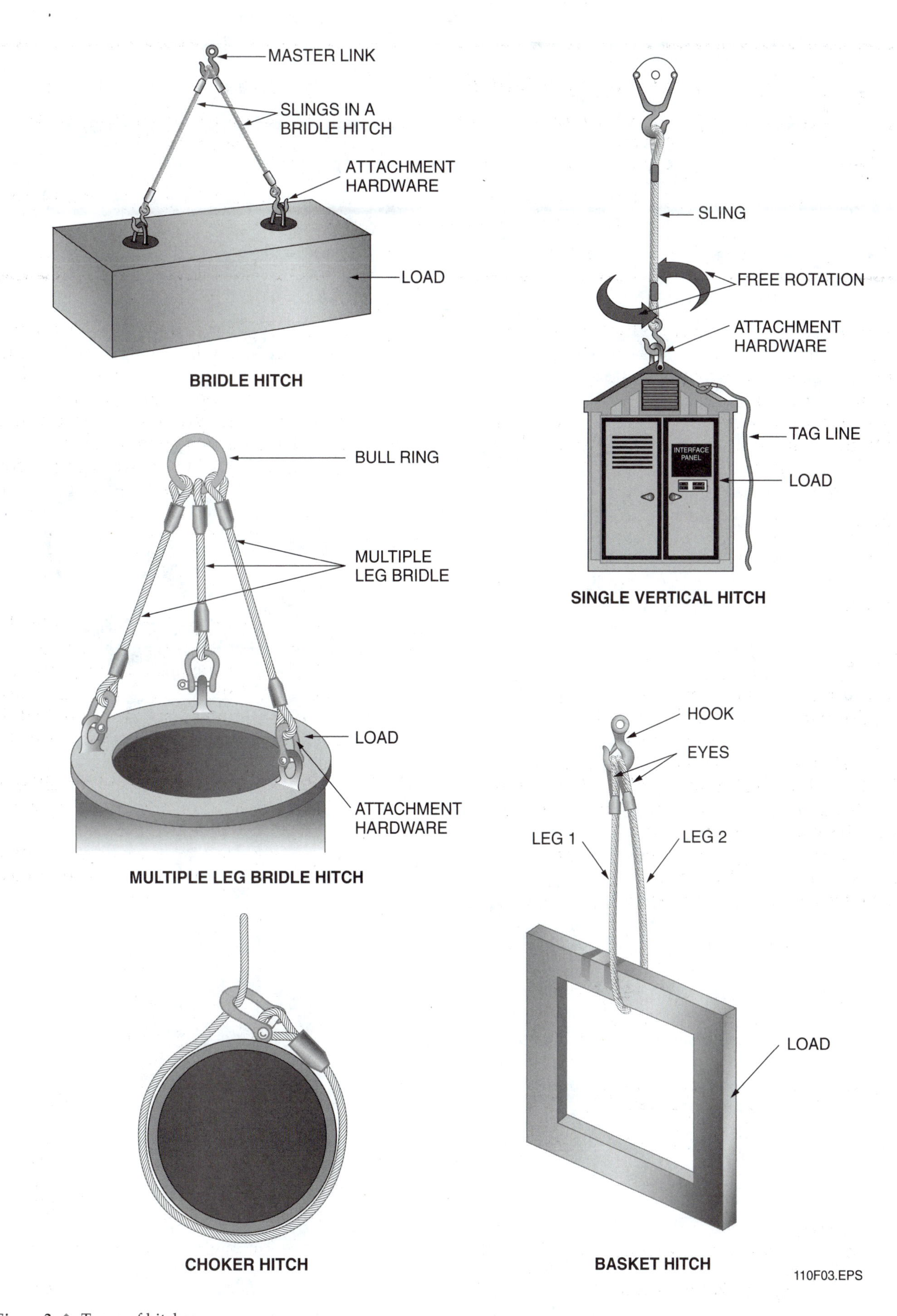

Figure 3 ◆ Types of hitches.

Lack of Training is Deadly

Two carpenters were attempting to set a 30' I-beam on a 15'-high concrete block wall. The carpenters were not adequately trained to do steel-erection work. They released the spanner connection from the hoist line before the I-beam was secured to the bearing plates. The beam tipped and fell to the concrete floor, along with the two carpenters. Both workers died from their injuries.

The Bottom Line: Always ensure that workers are properly trained for the job.

Source: The Occupational Safety and Health Administration (OSHA)

Controlling the movement of the load once the lift is in progress is another extremely important part of rigging. Riggers must consider the intended movement of the load when choosing a hitch. For example, some loads are lifted straight up and then straight down. Other loads are lifted up, turned 180 degrees in midair, and then set down in a completely different location. It's important to have a plan for rigging operations and follow it. Otherwise, accidents, injuries, and death can occur.

2.3.0 Rigging Hardware

Rigging hardware is as crucial as the crane, the slings, or any specially designed lifting frame. Depending on the job, rigging hardware (*Figure 4*) can consist of shackles, eyebolts, lifting clamps, and rigging hooks. If the hardware that connects the slings to either the load or the master link (*Figure 5*) were to fail, the load would drop just as it would if the crane, hoist, or slings were to fail. Hardware failure related to improper attachment, selection, or inspection contributes to a great number of the deaths, serious injuries, and property damage in rigging accidents. It is important that the proper hardware is selected, maintained, and inspected on a regular basis.

2.4.0 Hoists

Hoists (*Figure 6*) are devices that give you a mechanical advantage for lifting a load when cranes are not being used and when you cannot lift objects manually. All hoists use a pulley system to transmit power and lift a load. Some hoists are mounted on trolleys and use electricity or compressed air for power; others are operated by hand.

(A) WEB SLING SHACKLE

(B) RIGGING HOOK WITH SAFETY LATCH

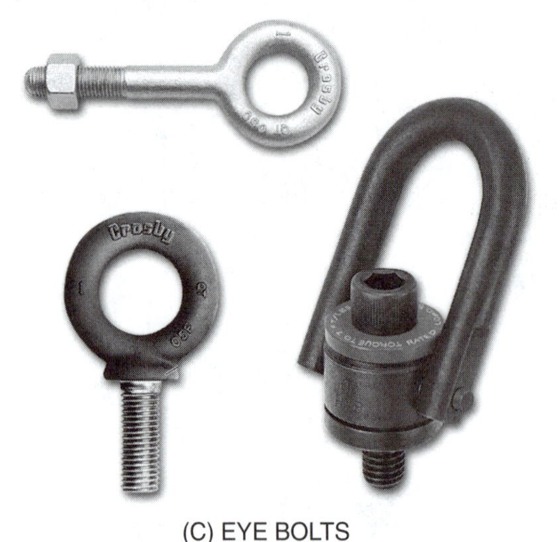

(C) EYE BOLTS

(D) STANDARD LIFTING CLAMP

110F04.EPS

Figure 4 ◆ Rigging hardware.

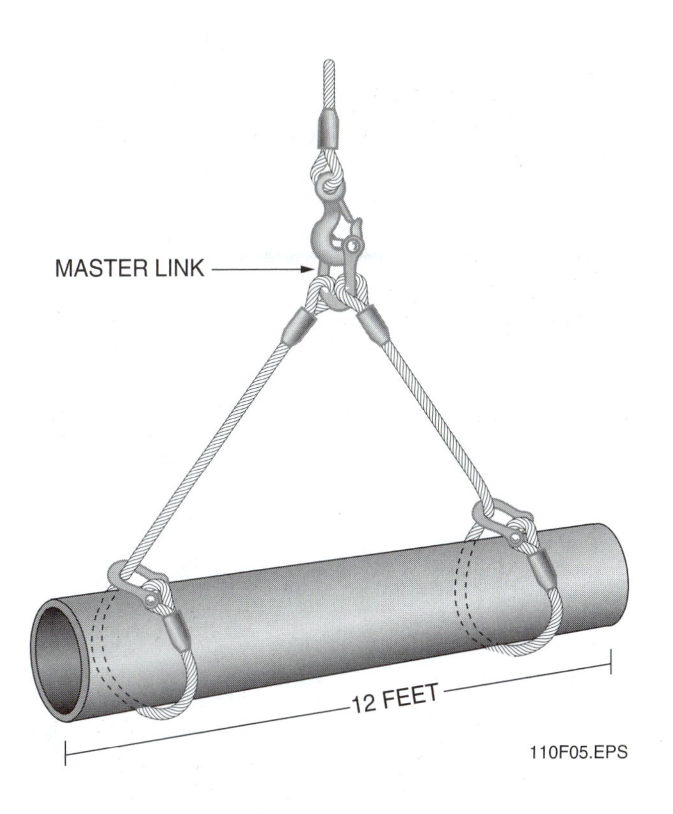

Figure 5 ◆ Master link.

3.0.0 ◆ HAZARDS AND SAFEGUARDS

The improper use of equipment or carelessness of workers can result in accidents, injury, or death. Anyone working on a steel-erection job can minimize risks. This is done by being aware of the hazards involved and following the proper safety procedures and guidelines. The main hazards involved with steel erection are falls, falling loads and objects, and material handling.

3.1.0 Falls

More people are injured in falls than in any other type of accident. On average, between 150 and 200 construction workers are killed each year in fall-related accidents. In one instance, a worker was walking on a 5½"-wide steel beam that was over 37' off the ground. The beam wobbled and the worker lost his balance and fell to his death. He was not wearing any fall-protection equipment.

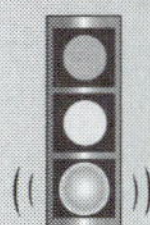

NOTE
For more information on fall protection, refer to the *Fall Protection* module in Volume Three.

Figure 6 ◆ Various chain hoists.

Don't Blame the Weather

Two steel workers were erecting lightweight steel I-beams on the third floor of a 12-story building, 54' above the ground. They were not using fall-arrest equipment. One employee removed a choker sling from a beam and then attempted to place the sling onto a lower hook on a series of stringers. While the crawler tower crane was booming away from the steel, the wind moved the stringer into the beam the employee was standing on. The beam moved while the employee was trying to disengage the hook, causing him to lose his balance and fall to his death.

The Bottom Line: Always wear the appropriate personal protective equipment for the job.

Source: The Occupational Safety and Health Administration (OSHA)

Falls are classified into two groups: falls from an elevation and falls on the same level. Falls from an elevation can occur when someone is doing work from scaffolding, work platforms, decking, concrete forms, ladders, or around excavations. Falls from elevations are almost always fatal. This is not to say that falls on the same level aren't also extremely dangerous. When a worker falls on the same level, usually from tripping or slipping, sharp edges and pointed objects can cut or stab the worker.

The following safe practices can help prevent slips and falls:

- Wear safe work boots that are in good repair.
- Watch where you step. Be sure your footing is secure.
- Do not allow yourself to get in an awkward position. Stay in control of your movements at all times.
- Maintain clean, smooth walking and working surfaces. Fill holes, ruts, and cracks. Clean up slippery material and litter.
- Install cables, extension cords, and hoses so that they will not become tripping hazards.
- Do not run on scaffolding, work platforms, decking, roofs, or other elevated work areas.

In addition to safe work practices, always use the following safety equipment to avoid falls during steel erection.

- Guardrails
- Safety nets
- Personal fall-arrest systems

3.1.1 Guardrails

Guardrails are the most common type of fall protection (*Figure 7*). They protect workers before a fall by providing a barrier between the work area and the ground or lower work areas. They may be made of wood, pipe, steel, or wire rope and must be able to support 200 pounds of force applied to the top rail.

3.1.2 Safety Nets

Safety nets are used for fall protection on bridges, steel buildings, and similar projects. They must be installed as close as possible to, but not more than 30' beneath, the work area. There must be enough clearance under a safety net to prevent a worker who falls into it from hitting the surface below. There must also be no obstructions between the work area and the net.

3.1.3 Personal Fall-Arrest Systems

Personal fall-arrest systems (*Figure 8*) catch workers after they have fallen. They are designed and rigged to prevent a worker from free falling a distance of more than 6' and hitting the ground or a lower work area. Personal fall-arrest systems use specialized equipment, which includes:

- Body harnesses and belts
- Lanyards
- Deceleration devices
- Lifelines
- Anchoring devices and equipment connectors

3.2.0 Falling Objects

Falling objects and flying objects are a common hazard during steel erection. You are at risk from falling objects when you are beneath cranes, scaffolds, forms, or where overhead work is being performed. There is a danger from flying objects when power tools or activities such as pushing, pulling, or prying cause objects to become airborne. Injuries can range from minor abrasions to concussions, blindness, or death.

(A)

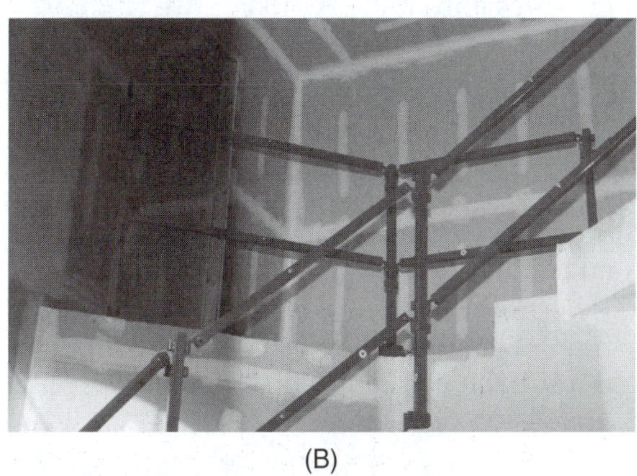

(B)

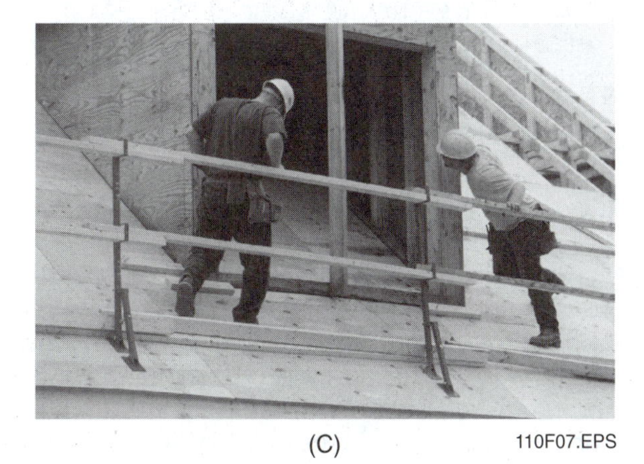

(C)

110F07.EPS

Figure 7 ◆ Typical guardrails.

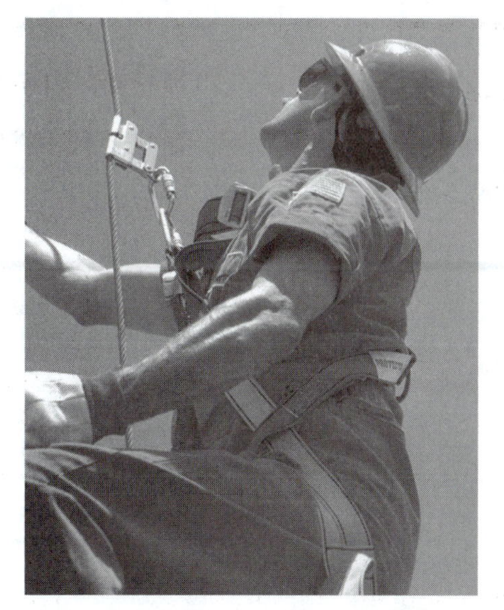

BODY HARNESS AND LIFELINE

LANYARD

ROPE GRAB

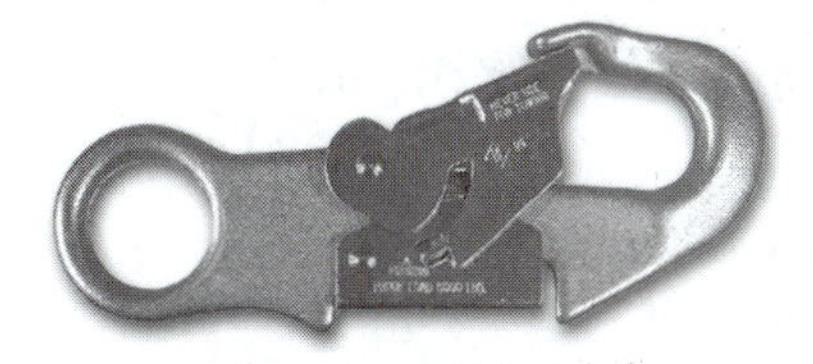

DOUBLE LOCKING SNAP HOOK

110F08.EPS

Figure 8 ◆ Fall-protection gear.

Pay Attention

A rigger was in the process of rigging up three beams in a triangular shape. He had put the choker around the first beam, and tightened it by pulling on the leg of the sling. This pushed up the top end of the sling, so the eye of the sling rode up onto or over the point of the hook, where it caught on the latch. The rigger was unaware that the beam was not properly secured. He then signaled the crane operator to raise the load. When the beam was overhead, the rigger began rigging the second beam. Almost instantly, the eye of the sling slipped off the latch, and the rigged beam dropped onto the victim. Although the rigger survived the accident, he spent several weeks in the hospital.

The Bottom Line: Ensure that your work area is safe before beginning any job, and stay alert for potential safety hazards.

Source: The Occupational Safety and Health Administration (OSHA)

Follow these safeguards to avoid getting struck by falling or flying objects.

- Never walk under a load.
- Always wear a hard hat.
- Use toeboards, screens, and guardrails to prevent falling objects from striking those below.
- Use debris nets, catch platforms, or canopies to catch or deflect falling objects.
- Use safety glasses, goggles, or face shields around machines or tools that can cause flying particles.
- Inspect tools to ensure that protective guards are in good condition.
- Make sure you are trained in the proper operation of powder-actuated tools.
- Avoid working underneath loads being moved.
- Barricade hazardous areas and post warning signs.
- Inspect cranes and hoists to see that all components, such as wire rope, lifting hooks, and chains are in good condition.
- Do not exceed the lifting capacity of cranes and hoists.
- Secure tools and materials to prevent them from falling on anyone below.

3.3.0 Material Handling

Working with steel can be dangerous. Not only is there a risk of falling or being crushed, there is a danger of cuts, burns, eye injuries, and electrocution. Steel, and the equipment used to work with it, must be handled safely to avoid accidents. The types of accidents that can happen include injuries from:

- Sharp edges
- Welding operations
- Cranes engaging electrical power lines
- Moving equipment such as cranes

To prevent these types of accidents, the following safeguards should be used when handling materials.

- Handles and holders must be attached to loads to reduce the chance of getting fingers pinched or crushed.
- Workers must use appropriate fall-arrest equipment.
- Gloves or other hand and forearm protection must be worn for loads with sharp or rough edges.
- Eye protection must be worn to avoid getting particles in the eyes. Welding operations require the use of special lenses to prevent serious vision damage.
- When the loads are heavy or bulky, wear steel-toed safety shoes or boots to prevent foot injuries caused by slips or accidentally dropping a load.
- Never overload equipment when mechanically moving materials.

4.0.0 ◆ ADDITIONAL HAZARDS

Many injuries result from improper or unsafe use of tools and equipment related to steel erection. Becoming familiar with the basic tools and materials of your trade and understanding how to handle them safely are important steps toward working safely. The need to keep tools clean and in good working order cannot be overstressed. Safe and proper use of tools will not only help preserve the tools in good working condition, but can also help to prevent accidents. All tools are capable of inflicting injury, even hand tools.

Make Good Housekeeping a Habit

Two employees of a steel-erection contractor had been bolting and welding at a height of 19' from a mobile elevated platform. After finishing work on a column, they were riding in the lift with the platform fully extended, when the right front wheel rolled over a pile of debris that had been left on the concrete floor. The scissor lift tipped over, causing one of the employees to sustain a serious head injury. He was taken to an area hospital, where he died.

The Bottom Line: Proper inspection of the site could have prevented this accident. You must always be aware of any potential hazards around a work site.

Source: The Occupational Safety and Health Administration (OSHA)

Some rules for the care and safe use of power tools are:

- Do not attempt to operate any power tool without being certified on that particular tool.
- Always wear appropriate safety equipment and protective clothing for the job being done. For example, wear safety glasses and close-fitting clothing that cannot become caught in the moving parts of a tool. Roll up long sleeves, tuck in your shirt tail, and tie back long hair.
- Never leave a power tool running unattended.
- Assume a safe and comfortable position before starting a power tool.
- Do not distract others or let anyone distract you while operating a power tool.
- Be sure that any electric power tool is properly grounded before using it.
- Be sure that power is not connected before performing maintenance or changing accessories.
- Do not use dull or broken accessories.
- Use power tools only for their intended purpose.
- Do not use a power tool with the guards or safety devices removed.
- Use a proper extension cord of sufficient size to service the particular electric tool you are using.
- Do not operate an electric power tool if your hands or feet are wet.
- Become familiar with the correct operation and adjustments of a power tool before attempting to use it.
- Be sure there is proper ventilation before operating gasoline-powered equipment indoors.
- Keep a fire extinguisher nearby when filling and operating gasoline-powered equipment.
- Store tools properly when not in use.

5.0.0 ◆ RESPONDING TO ACCIDENTS

Accidents and injuries can be avoided by working safely and following the rules. When accidents do happen, it's important to stay calm and immediately tell your supervisor. Depending on the type of injury, a worker trained in first aid should evaluate the injury and determine the appropriate treatment plan. For example, if the injury is a minor cut, a bandage can be applied. If the cut is deep and requires stitches, the worker should be taken to the hospital. If the cut is severe and there is an extreme loss of blood or amputation is involved, call 911 or your local emergency service number immediately.

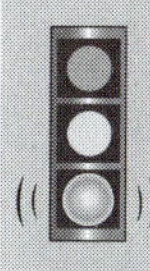

NOTE

Emergency-response plans vary depending on the company. Learn your company's emergency response plan and be prepared.

Summary

Accidents on steel-erection jobs are almost always fatal. Falls are the most common types of accidents that happen on these jobs. Workers can also get injured from falling objects, sharp edges, faulty rigging equipment, and unsafe tools and equipment. Most of these accidents happen because someone is not paying attention or doing their job safely. It is your responsibility to be aware of the hazards and follow all safety procedures; this can save your life and the lives of your co-workers.

Review Questions

1. All _____ are required to have identification tags.
   a. eyebolts
   b. hitches
   c. hoists
   d. slings

2. The maximum load weight that the sling is designed to carry is called the _____.
   a. load control
   b. lift value
   c. rated capacity
   d. rejection value

3. It is safe to lift a load that weighs more than the maximum load weight of the sling as long as you notify your supervisor before lifting the load.
   a. True
   b. False

4. Hitches help control the movement of a load.
   a. True
   b. False

5. All of the following are considered rigging hardware *except* _____.
   a. anchor bolts
   b. eyebolts
   c. lifting clamps
   d. rigging hooks

6. A hoist is a device that uses a pneumatic jack to lift a load.
   a. True
   b. False

7. Construction workers are injured during material handling more than any other type of accident.
   a. True
   b. False

8. Falls from elevations are almost always fatal.
   a. True
   b. False

9. Guardrails are the most common type of fall protection.
   a. True
   b. False

10. Steel-toed boots should not be worn when moving heavy or bulky loads.
    a. True
    b. False

GLOSSARY

Trade Terms Introduced in This Module

Hitch: The way the sling is arranged to hold the load. Some common examples include a bridle hitch, choker hitch, basket hitch, and single vertical hitch. Also called a rigging configuration.

Hoist: A device that mechanically lifts loads using a pulley system. Hoists can be manual or powered by compressed air.

Rigging: A system of attaching a load to a hoist or crane.

Rigging hardware: Shackles, eyebolts, lifting clamps, and hooks that are used to connect the sling load to the crane or hoist.

Sling: A length of synthetic material, chain, wire, or rope that is wrapped around the load and attached to the hoist or crane.

Figure Credits

Liftall Company, Inc.	110F02 (A) and (B)
The Crosby Group	110F04 (A), (B), and (C)
J.C. RenFroe and Sons, Inc.	110F04 (D)
Jet Equipment and Tools	110F06
Construction Safety Association of Ontario	110F07 (A)
Fall Protection Systems, Inc.	110F07 (B) and (C)
Protecta International. Inc.	110F08

NCCER CURRICULA — USER UPDATE

NCCER makes every effort to keep its textbooks up-to-date and free of technical errors. We appreciate your help in this process. If you find an error, a typographical mistake, or an inaccuracy in NCCER's curricula, please fill out this form (or a photocopy), or complete the online form at **www.nccer.org/olf**. Be sure to include the exact module ID number, page number, a detailed description, and your recommended correction. Your input will be brought to the attention of the Authoring Team. Thank you for your assistance.

Instructors – If you have an idea for improving this textbook, or have found that additional materials were necessary to teach this module effectively, please let us know so that we may present your suggestions to the Authoring Team.

NCCER Product Development and Revision

13614 Progress Blvd., Alachua, FL 32615

Email: curriculum@nccer.org
Online: www.nccer.org/olf

❑ Trainee Guide ❑ AIG ❑ Exam ❑ PowerPoints Other ______________________________

Craft / Level: Copyright Date:

Module ID Number / Title:

Section Number(s):

Description:

Recommended Correction:

Your Name:

Address:

Email: Phone:

Module 75111-03

Walking and Working Surfaces

COURSE MAP

This course map shows all of the modules in Field Safety. The suggested training order begins at the bottom and proceeds up. The local Training Program Sponsor may adjust the training order.

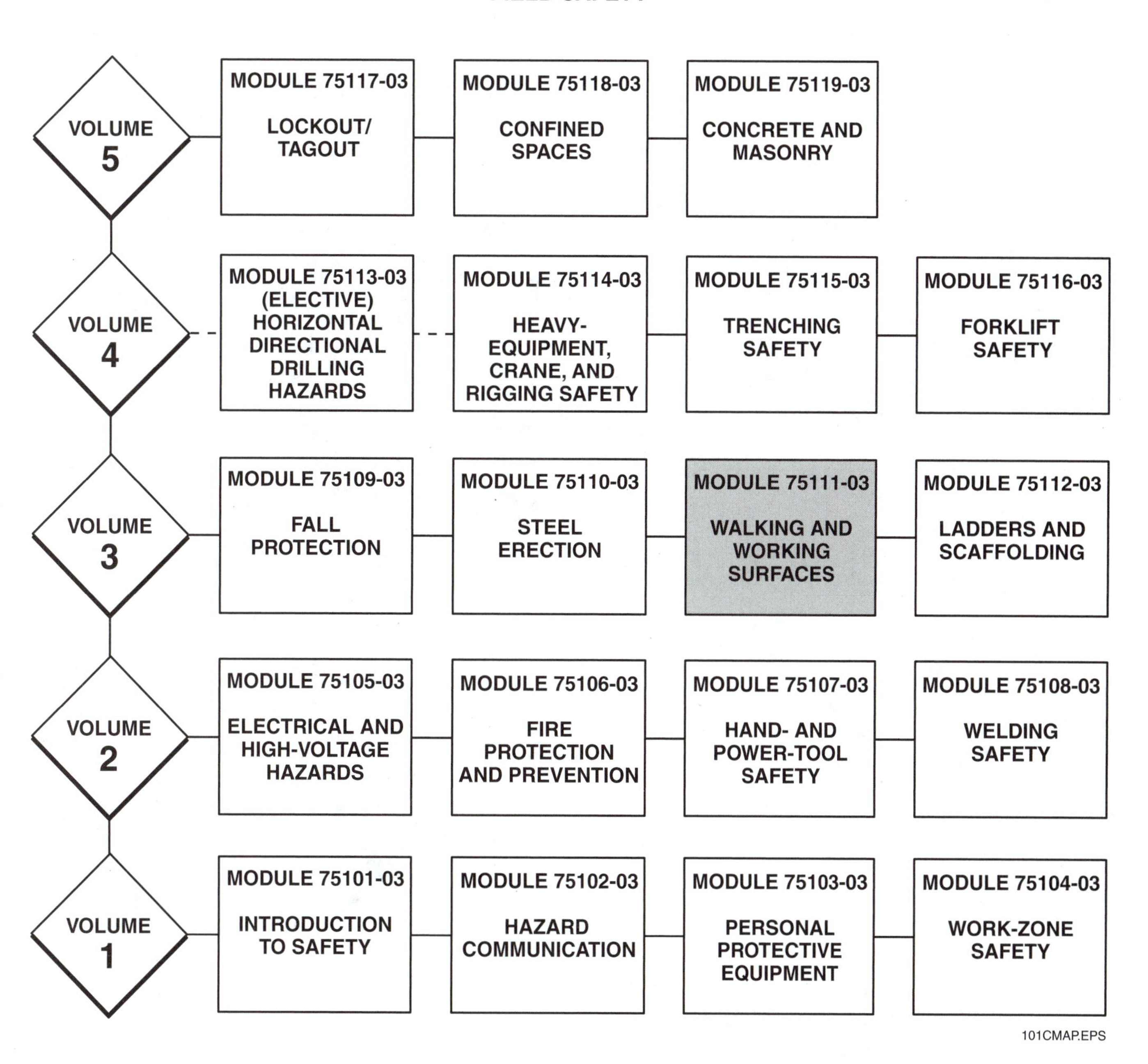

MODULE 75111-03 CONTENTS

Figures

MODULE 75111-03

Walking and Working Surfaces

Objectives

When you have completed this module, you will be able to do the following:

1. Explain the hazards associated with walking and working surfaces.
2. Describe how to avoid accidents and injuries on walking and working surfaces.
3. Explain how to respond to accidents and injuries on walking and working surfaces.

Prerequisites

Before you begin this module, it is recommended that you successfully complete the following: Field Safety, Modules 75101-03 through 75110-03.

Required Materials

1. Pencil and paper
2. Appropriate personal protective equipment

1.0.0 ◆ INTRODUCTION

Slips, trips, and falls on walking and working surfaces cause 15% of all accidental deaths in the construction industry. Some accidents occur due to environmental conditions such as snow, ice, or wet surfaces. Others happen because of poor housekeeping and careless behavior such as leaving tools, materials, and equipment out and unattended.

Accidents from slips, trips, and falls can be avoided if everyone is aware of their surroundings and follows the rules on the site. It is your responsibility to keep all walking and working areas clean and dry.

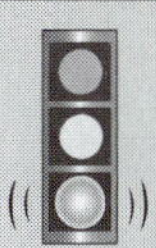

NOTE

Fall protection is covered in detail in the *Fall Protection* module in Volume Three.

2.0.0 ◆ TYPES OF WALKING AND WORKING SURFACES

Walking and working surfaces vary depending on the job. Some workers are exposed to openings in the floors and walls, while others experience icy or wet walking and working surfaces. *Figure 1* shows an example of an icy surface. The danger in this picture is that it's hard to tell that there's any ice on the surface at all.

111F01.EPS

Figure 1 ◆ Ice on a walking surface.

Covering Openings

Carpenters were setting trusses on the second floor of a house they were building. There was no guardrail or floor cover over the floor opening for the stairway. While placing a truss in position, one of the carpenters was killed when he slipped and fell through the floor opening to the concrete basement below.

The Bottom Line: Unguarded openings can be deadly.

Source: The Occupational Safety and Health Administration (OSHA)

Some common examples of the areas where workers may experience hazards include:

- Floors
- Walls
- Platforms
- Ramps and runways
- Stairs
- Ladders and scaffolding
- Roofs

Always use caution when walking or working on or near these surfaces, especially if there are hazards in the area.

2.1.0 Floors

Floors can be a hazard in a number of ways. Ice, grease, oil, or wet processes can make them slippery. Tools, equipment, materials, or litter can clutter them. Unguarded openings in the floor or ground can cause fatal falls. Some of the types of openings to be aware of on a construction site are:

- Stairs
- Hatches
- Chutes
- Trapdoors
- Manholes

To avoid slipping, tripping, and falling on floors, follow these guidelines:

- Check for ice and snow. If the ice cannot be removed, wear shoes with skid-resistant cleats.
- Make sure the floor is clean and dry. If wet processes are used in the work area, make sure there is proper drainage, grating, and mats.
- Keep the floor clear of tools, equipment, materials, or litter that could be tripped over or slipped on.
- Avoid openings in floors unless they are properly guarded or covered.

2.2.0 Walls

Openings in walls such as windows, doors, and chutes are generally at eye level and seem easy to avoid. However, wall openings are dangerous when they are not protected by guardrails or fences. Any rain entering the opening may cause a slipping hazard. For example, if a worker loses his or her balance and falls near an unguarded wall opening, he or she could slip through the opening and fall to the ground or a lower work area. Also, any tools or materials that fall through the opening can seriously injure those below. If you are working near a wall opening, make sure it is barricaded and that the work area is dry and free of clutter.

Looks Can be Deceiving

A 41-year-old ironworker was working as a member of an eight-person crew installing steel decking on the roof of a new six-story building. The worker left the work area to get a small piece of decking material to finish the job. When he did not return, the crew searched for him and found him lying unconscious on the fifth floor of the stairwell. The worker had removed a 3' × 6' piece of decking from the floor. He wasn't aware that the piece of decking was covering a 2' square ventilation opening in the top of the stairwell. After he picked up the decking, he stepped forward and fell through the opening. He died 12 hours later as a result of his injuries.

The Bottom Line: Always use the proper covers and barricades when there is an opening in a walking or working surface.

Source: The National Institute for Occupational Safety and Health (NIOSH)

Follow Instructions

A 21-year-old laborer had been throwing old roofing materials off a roof with six unguarded skylights. During a work break, the victim sat down on one of the skylights, which began to break under his weight. He tried to raise himself from the skylight with his arms, but the plastic dome failed completely. He was killed when he fell 27' to the concrete floor below. The victim had been warned by his supervisor and co-workers not to sit on the skylights.

The Bottom Line: Ignoring safety warnings can be fatal.

Source: The National Institute for Occupational Safety and Health (NIOSH)

2.3.0 Platforms

Platforms are work areas elevated above the floor or ground. They are sometimes located above dangerous equipment such as galvanizing tanks or degreasing units. Many platforms do not have guardrails. These platforms, called open-sided platforms, are hazardous because they do not protect workers from falling over the edge. To help prevent accidents, make sure platforms are dry and clear of materials and debris before stepping onto them.

2.4.0 Ramps and Runways

The hazards of ramps and runways are similar to those of floors. Workers can trip on tools and equipment or slip on wet or icy surfaces (*Figure 2*). Ramps and runways can be more dangerous, however, because they are sloped. When slips, trips, and falls happen at a downhill angle, it increases the speed at which the worker slides or rolls down the ramp or runway. The types of injuries received are often more serious because the worker hits the ground with greater force. Imagine if a worker was carrying a tool with a sharp blade or edge during such an accident. It is likely that the worker would be injured from the fall as well as from the blade or edge of the tool.

111F02.EPS

Figure 2 ◆ Ramp covered with ice and snow.

Put Your Tools Away

An 18-year-old sheet metal helper died after he fell through an unprotected vent opening to a concrete floor 33' below. The victim was part of a crew engaged in replacing corrugated metal roof sheeting and installing chain-link fencing material on top of 3' × 8' fiberglass panels used as skylights. The fencing material was being installed to eliminate the hazard of falls posed by the fiberglass skylights because three months earlier, a worker on the same site had fallen to his death through one of the skylights. When the supervisor ordered the crew to stop work temporarily, the members of the crew moved toward a vent to warm themselves. As they moved, the victim tripped over some tools that were left out and fell through the vent.

The Bottom Line: To avoid a tripping or falling hazard, always make sure that the work site is clean, and that all tools and equipment have been properly put away.

Source: The National Institute for Occupational Safety and Health (NIOSH)

To avoid slipping, tripping, and falling on ramps and runways, follow these guidelines:

- Check the surface before using it. If you find that the ramp or runway is icy or wet, don't use it until is dry and free of ice. If the ice cannot be removed, wear shoes with skid-resistant cleats.
- Make sure the ramp is clear of tools, equipment, materials, or debris.
- Make sure any tools or equipment you are carrying are turned off and secured. This will help prevent injuries if you fall.

2.5.0 Stairs

Stairs are used to travel between levels, in and out of pits, and on and off platforms. Depending on the location of the stairs, they can be wet, icy, or slippery. They can also be damaged or cluttered with tools and equipment. All of these conditions can cause workers to slip, trip, or fall. Don't use stairs if you notice any of these conditions or if the stairs do not have a guardrail.

2.6.0 Ladders

The biggest hazard of using a ladder is falling. Workers can fall from the ladder or the ladder can slip out from under them. This usually happens when the ground or ladder is wet (*Figure 3*) or icy, or if the ladder is not properly secured. In any case, serious injuries or death can occur.

The following safe practices can help prevent slips and falls from ladders:

- Wear safe, strong work boots that are in good condition.
- Watch where you step. Be sure your footing is secure.

111F03.EPS

Figure 3 ◆ Water on a ladder.

- Maintain clean, smooth walking and working surfaces. Fill holes, ruts, and cracks.
- Be sure to clean up slippery material.
- Pick up litter.
- If you must climb to reach something, use a sound ladder that has been safely set up and properly secured at the top and bottom (*Figure 4*).
- When climbing a ladder, always face the ladder and use both hands.
- When using a ladder, don't overreach. Climb down and move the ladder to the desired position.

2.7.0 Scaffolding

Scaffolding is used to support workers and materials on elevated platforms (*Figure 5*). Working on scaffolding can be very dangerous because of weather conditions, poor housekeeping, and carelessness. Always pay attention and follow these guidelines to avoid injury and death:

- Always keep scaffolding planks clear of extra tools and materials. Clean up any slippery substances that get spilled on scaffolding.
- Anchor freestanding scaffolding with guy wires to prevent tipping or sliding.
- Locate and read the posted safety rules and regulations for scaffolding use.
- When removing objects and tools from the work area, lower them down with a rope. Never throw or drop them from the scaffolding.
- Make sure all personnel are off the platform before moving it.
- Keep tools and materials back from the edge of work platforms.
- Use only approved scaffolding.
- Inspect the scaffolding regularly.
- Install all braces and accessories according to the manufacturer's recommendations.
- Ensure that the scaffolding is plumb and level.
- Install all safety rails according to regulations.
- Make sure the wheels are locked before ascending or disassembling rolling scaffolding.
- Do not work on scaffolding during storms or high winds, or when the scaffolding is covered with ice or snow.
- If the scaffolding is wobbly or bouncy, or can be pulled down easily, it is not safe. Scaffolding must have a sound structure and include toeboards and handrails. Wood floor planks must be free of knots. Broken members must be repaired immediately.

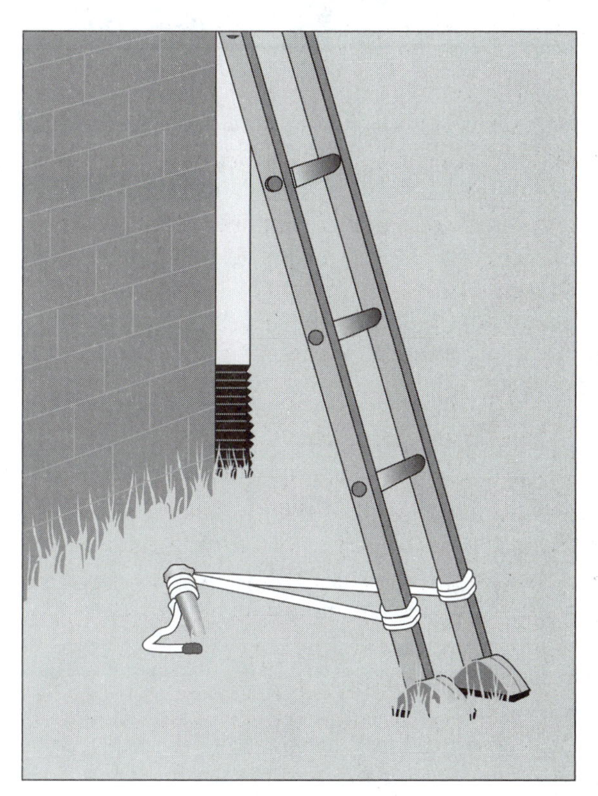

Figure 4 ◆ Properly secured ladder.

Guardrails are Essential

A laborer was working on the third level of tubular welded frame scaffolding that was covered with ice and snow. The planking on the scaffolding wasn't sturdy and the scaffolding didn't have a guardrail. There was also no access ladder for the various scaffolding levels. The worker slipped and fell approximately 20' to the pavement below. He received a fatal head injury.

The Bottom Line: When working on scaffolding, use common sense. Make sure the scaffolding has guardrails, is sturdy, and free of ice, snow, and debris.

Source: The Occupational Health and Safety Administration (OSHA)

Figure 5 ◆ Typical scaffolding.

2.8.0 Roofs

Walking and working on roofs presents two different kinds of risks—falling through an opening such as a skylight or vent, and falling off of the roof. Both can happen because of weather hazards, poor housekeeping, or carelessness. The following guidelines will help ensure your safety when working on a roof:

- Wear boots or shoes with rubber or crepe soles that are in good condition.
- Always wear fall-protection devices, even on shallow-pitch roofs.
- Rain, frost, and snow are all dangerous because they make a roof slippery. If possible, wait until the roof is dry. Otherwise, wear special roof shoes with skid-resistant cleats in addition to fall protection.
- Brush or sweep the roof periodically to remove any accumulated dirt or debris.
- On pitched roofs, install the necessary roof brackets and toeboards (*Figure 6*) as soon as possible. They can be removed and repositioned as shingle-type roofing is installed.
- Remove any unused tools, cords, and other loose items from the roof. They can be a serious hazard.
- Be alert to any other potential hazards such as live power lines.
- Use common sense. Taking chances can lead to injury or death.

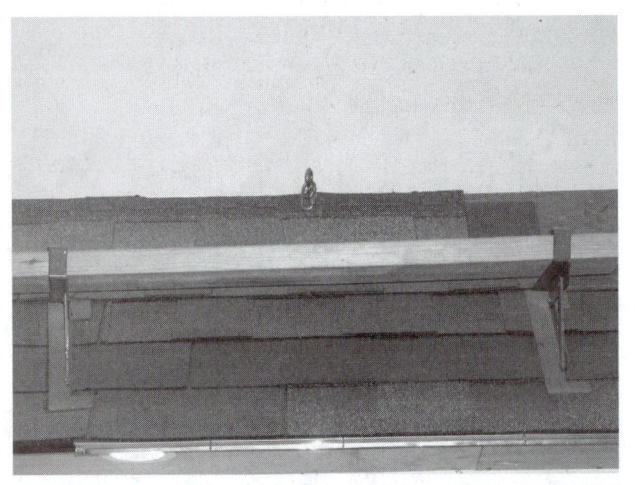

111F06.EPS

Figure 6 ◆ Example of a toeboard.

Through the Roof

A 24-year-old plumber was killed when he fell through an unguarded skylight opening to a concrete floor approximately 22' below. The victim and a co-worker were installing plumbing fixtures on the roof of a new building. The roof had several 4' × 4' openings framed with 2" × 6" wood. These openings were intended for smoke-vent skylights when the structure was complete. Although the victim and others had been working on this project for several days before the incident, no fall protection or guards of any type were in place. The victim was walking away from his co-worker and looking back over his shoulder to talk to him when he stepped through a skylight opening.

The Bottom Line: Always watch what you are doing and use the appropriate fall-protection equipment.

Source: The National Institute for Occupational Safety and Health (NIOSH)

3.0.0 ◆ ACCIDENT AND INJURY RESPONSE

Accidents and injury can be avoided by working safely and following the rules. When accidents do happen, it's important to stay calm and tell your supervisor immediately. Depending on the type of injury, a worker who is trained in first aid should evaluate the injury and decide on the best treatment plan. For example, if the injury is a minor cut, a bandage can be applied. If the cut is deep and requires stitches, the worker should be taken to the hospital. If the cut is severe and there is an extreme loss of blood or amputation is involved, immediately call 911 or your local emergency response number.

NOTE

Emergency response plans vary depending on the company. Learn your company's emergency response plan and be prepared to follow it.

Summary

Slips, trips, and falls from unsafe walking and working surfaces are a common and serious type of accident on a construction site. You have a responsibility to yourself and others to work safely and follow the rules of good housekeeping. Personal injury and damage to equipment is less likely to happen when you do so. It is also important that everyone on a job site knows and follows their company's emergency-response plan.

Review Questions

1. Less than 5% of accidental deaths in the construction industry are from slips, trips, and falls.

a. True
b. False

2. All of the following are considered walking and working surfaces *except* _____.

a. counter tops
b. floors
c. ladders
d. walls

3. It is safe to work in a wet area that does not have drains, grating, or mats.

a. True
b. False

4. Openings in walls such as windows, doors, and chutes present all of the following hazards *except* _____.

a. they allow rain into the work area, which can create a slipping hazard
b. tools and materials can fall through them and injure workers on lower levels
c. they are difficult to see and avoid
d. workers can trip and fall through them

5. Platforms must never be located above dangerous equipment.

a. True
b. False

6. Walking and working surfaces that are located above ground and sometimes over large equipment are _____.

a. ceilings
b. floors
c. platforms
d. ramps

7. Wearing shoes with skid-resistant cleats on wet surfaces helps prevent slips.

a. True
b. False

8. Tools and other objects should be lowered to the ground with a rope when they are removed from scaffolding.

a. True
b. False

9. All of following guidelines must be adhered to when working or walking on a roof *except* _____.

a. use common sense
b. clean off the roof periodically to remove debris
c. wear fall protection only on slanted roofs
d. remove all unused tools, cords, or other loose items

10. When an accident happens, the first thing you should do is stay calm and _____.

a. take care of the injury yourself
b. go to the hospital
c. tell your supervisor
d. call 911 or your local emergency-response number

Figure Credits

Gerald Shannon	111F01, 111F02
Brigid McKenna	111F03
Bil-Jax, Inc.	111F05
Thomas P. Burke	111F06

NCCER CURRICULA — USER UPDATE

NCCER makes every effort to keep its textbooks up-to-date and free of technical errors. We appreciate your help in this process. If you find an error, a typographical mistake, or an inaccuracy in NCCER's curricula, please fill out this form (or a photocopy), or complete the online form at **www.nccer.org/olf**. Be sure to include the exact module ID number, page number, a detailed description, and your recommended correction. Your input will be brought to the attention of the Authoring Team. Thank you for your assistance.

Instructors – If you have an idea for improving this textbook, or have found that additional materials were necessary to teach this module effectively, please let us know so that we may present your suggestions to the Authoring Team.

NCCER Product Development and Revision

13614 Progress Blvd., Alachua, FL 32615

Email: curriculum@nccer.org
Online: www.nccer.org/olf

❑ Trainee Guide ❑ AIG ❑ Exam ❑ PowerPoints Other ______________________

Craft / Level: Copyright Date:

Module ID Number / Title:

Section Number(s):

Description:

Recommended Correction:

Your Name:

Address:

Email: Phone:

Module 75112-03

Ladders and Scaffolding

COURSE MAP

This course map shows all of the modules in Field Safety. The suggested training order begins at the bottom and proceeds up. The local Training Program Sponsor may adjust the training order.

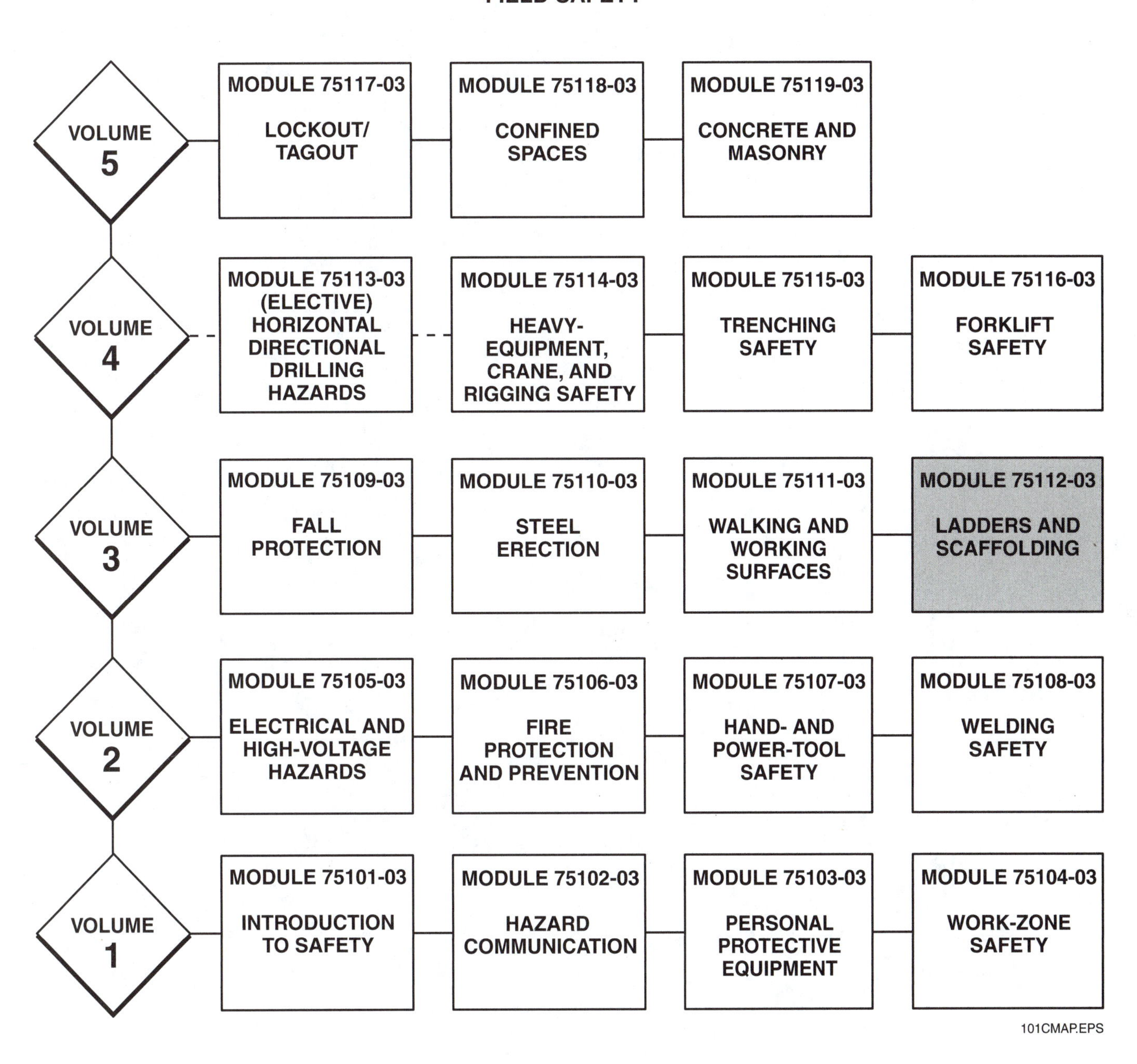

MODULE 75112-03 CONTENTS

Figures

MODULE 75112-03

Ladders and Scaffolding

Objectives

When you have completed this module, you will be able to do the following:

1. Identify the different types of ladders and scaffolding used on a work site.
2. Describe how to safely use ladders and scaffolding.
3. Properly set up and use ladders and scaffolding.

Prerequisites

Before you begin this module, it is recommended that you successfully complete the following: Field Safety, Modules 75101-03 through 75111-03.

Required Materials

1. Pencil and paper
2. Appropriate personal protective equipment

1.0.0 ◆ INTRODUCTION

Ladders and **scaffolding** are a common sight on construction jobs. They are also a frequent source of fatal accidents. The most common accidents associated with ladders and scaffolding are falling, being struck by falling objects, and electrocution. In one instance, a worker died when he slipped from a fixed ladder attached to a water tower and fell 40' to the ground. He died because he was not using the proper safety equipment.

Safety is your top priority on a job. It is your responsibility to learn how to set up, use, and maintain equipment. Not only will this keep you safe, it could save the lives of your co-workers.

2.0.0 ◆ LADDERS

Ladders are some of the most important tools on a job site. Different types of ladders should be used in different situations (*Figure 1*). Aluminum ladders are corrosion-resistant and can be used in situations where they might be exposed to the elements. They are also lightweight and can be used where they need to be frequently lifted and moved. Wooden ladders, which are heavier and sturdier than fiberglass or aluminum ladders, can be used when heavy loads must be moved up and down. Fiberglass ladders are nonconductive and also very durable, so they are useful in situations involving electrical work or where some amount of rough treatment is unavoidable. Both fiberglass and aluminum are easier to clean than wood.

Bad Weather and Unsafe Conditions Can Kill

A laborer was working on the third level of a tubular welded frame scaffolding which was covered with ice and snow. The planking on the scaffolding wasn't sturdy and the scaffolding didn't have a guardrail. There was also no access ladder for the various scaffolding levels. The worker slipped and fell approximately 20' to the pavement below. He died of a head injury.

The Bottom Line: Make sure that all scaffolding has solid planking and guardrails.

Source: The Occupational Safety and Health Administration (OSHA)

Selecting the right ladder for the job at hand is important to completing a job as safely and efficiently as possible. When selecting a ladder, consider its features and how it meets the needs of the job. Always consider the highest **duty rating** and weight limit needed, as well as the height requirements. A ladder that is too long or too short will not allow the work surface to be reached easily, safely, or comfortably.

When selecting a ladder for a job, it is important to choose one that will extend at least 36" above the landing surface you are trying to reach. Always place the base of the ladder so that the distance between the base and the wall is one-quarter of the ladder length from the base elevation to the point where the ladder touches the wall, as shown in *Figure 2*.

Figure 1 ◆ Types of ladders.

Figure 2 ◆ Proper positioning.

Keep the following precautions in mind when setting up and using any type of ladder:

- Do not use a metal ladder when performing any type of electrical work or whenever there is a possibility that you might come into contact with electrical conductors.
- Place the ladder in a stable manner.
- Place the ladder so that it leaves 6" of clearance in back of the ladder and 30" of clearance in front of the ladder.
- Place the ladder so that it leans against a solid and immovable surface. Never place a ladder against a window, door, doorway, sash, loose or movable wall, or box.
- Face the ladder when climbing up or down.
- Climb or descend the ladder one rung at a time. Never run up or slide down a ladder.
- Do not use ladders during high winds. If you must use a ladder in windy conditions, make sure you lash the ladder securely in order to prevent slippage.
- Check to make sure the soles of your shoes are free of oil, mud, and grease.
- Keep both hands free so that you can hold the ladder securely while climbing. Use a rope to raise and lower any tools and materials that you might need.
- Never rest any tools or materials on the top of a ladder.
- Move the ladder in line with the work to be done. Never lean sideways away from the ladder in order to reach the work area.
- Never stand on the top two rungs of a ladder.
- Use ladders only for short periods of elevated work. If you must work from a ladder for extended periods, use a lifeline fastened to a safety belt.
- Lay the ladder on the ground when you have finished using it, unless it is anchored securely at the top and bottom where it is being used.
- Never use makeshift substitutes for ladders.
- Never use stepladders for straight-ladder work.

Each ladder is designed for a specific purpose and climbing conditions, and they generally fall into these four categories.

- Straight
- Extension
- Step
- Fixed

2.1.0 Straight Ladders

Straight ladders consist of two rails, rungs between the rails, and safety feet on the bottom of the rails (*Figure 3*). The straight ladders used in construction are generally made of wood or fiberglass. Metal ladders conduct electricity and should not be used around electrical equipment.

Ladders should always be inspected before use. Be sure to follow these guidelines when inspecting a ladder.

- Check the rails and rungs for cracks or other damage including loose rungs. If you find any damage, do not use the ladder.
- Check the entire ladder for loose nails, screws, brackets, or other hardware. If you find any hardware problems, tighten the loose parts or have the ladder repaired before you use it.

Figure 3 ◆ Straight ladder.

- Make sure the feet are securely attached, that there is no damage, and that they are not worn down (*Figure 4*). Do not use a ladder if its safety feet are not in good working order.

Figure 4 ◆ Ladder safety feet.

Setup is the next step after inspecting a ladder. Use these guidelines to make sure you are setting up the ladder safely.

- Place the straight ladder at the proper angle before using it. A ladder placed at an improper angle will be unstable and could cause you to fall.
- Straight ladders should be used only on stable and level surfaces unless they are secured at both the bottom and the top to prevent any accidental movement (*Figure 5*).
- The distance between the foot of a ladder and the base of the structure it is leaning against must be one-quarter the distance between the ground and the point where the ladder touches the structure.

Once you've inspected and set up the ladder, you can begin working on it. Here are some safeguards to use while working on a ladder.

- Never try to move a ladder while someone is on it. If a ladder must be placed in front of a door that opens toward the ladder, the door must be locked or blocked open so that it cannot strike the ladder.
- Never use a ladder as a work platform by placing it horizontally.
- Make sure the ladder you are about to climb or descend is properly secured.
- Make sure the ladder's feet are solidly positioned on firm, level ground.
- Always check to ensure that the top of the ladder is firmly positioned and in no danger of shifting to the right or left once you begin your climb.
- Keep both hands on the rails when climbing a straight ladder.
- Always keep your body's weight in the center of the ladder between the rails.
- Never go up or down a ladder while facing away from it. Face the ladder at all times.
- Don't carry tools in your hands while you are climbing a ladder. Instead use a hand line or tagline and pull tools up once you've reached the place you will be working.

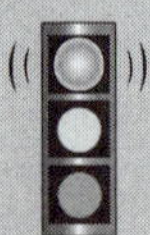

WARNING!

Remember that the addition of your own weight will affect the ladder's steadiness once you mount it. It is important to test the ladder first by applying some of your weight to it without actually beginning to climb. This way, you will be sure that the ladder remains steady as you ascend.

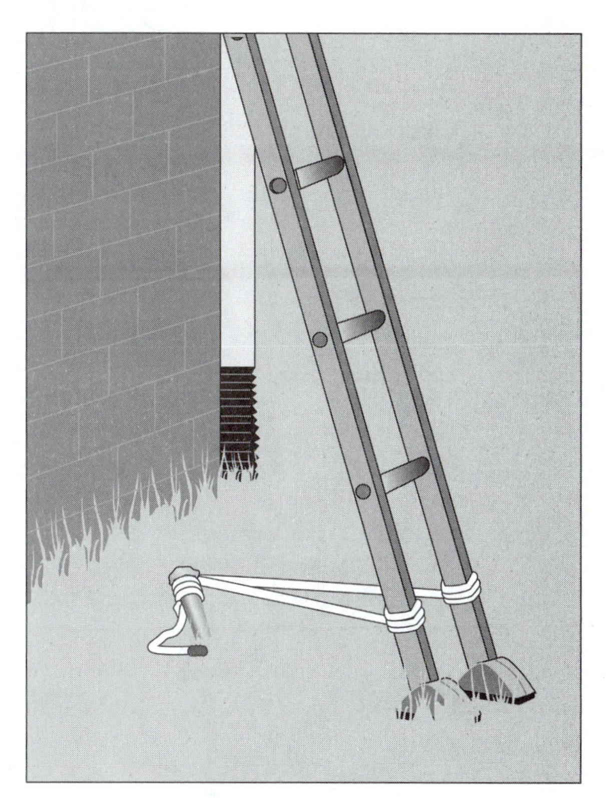

Figure 5 ◆ Securing a ladder.

2.2.0 Extension Ladders

An extension ladder is actually two straight ladders. They are connected so you can adjust the overlap between them and change the length of the ladder as needed (*Figure 6*).

Figure 6 ◆ Extension ladders.

Extension ladders are positioned and secured following the same rules as straight ladders. There are, however, some safety rules that are unique to extension ladders.

- When you adjust the length of an extension ladder, always reposition the movable section from the bottom, not the top, so you can make sure the rung locks are properly engaged after you make the adjustment.
- Make sure the section locking mechanism is fully hooked over the desired rung.
- Make sure that all ropes used for raising and lowering the extension are clear and untangled.
- Make sure the extension ladder overlaps between the two sections (*Figure 7*). For ladders up to 36' long, the overlap must be at least 3'. For ladders 36' to 48' long, the overlap must be at least 4'. For ladders 48' to 60' long, the overlap must be at least 5'.
- Never stand above the highest safe standing level on a ladder. On an extension ladder, this is the fourth rung from the top. If you stand higher, you may lose your balance and fall. Some ladders have colored rungs to show where you should not stand.

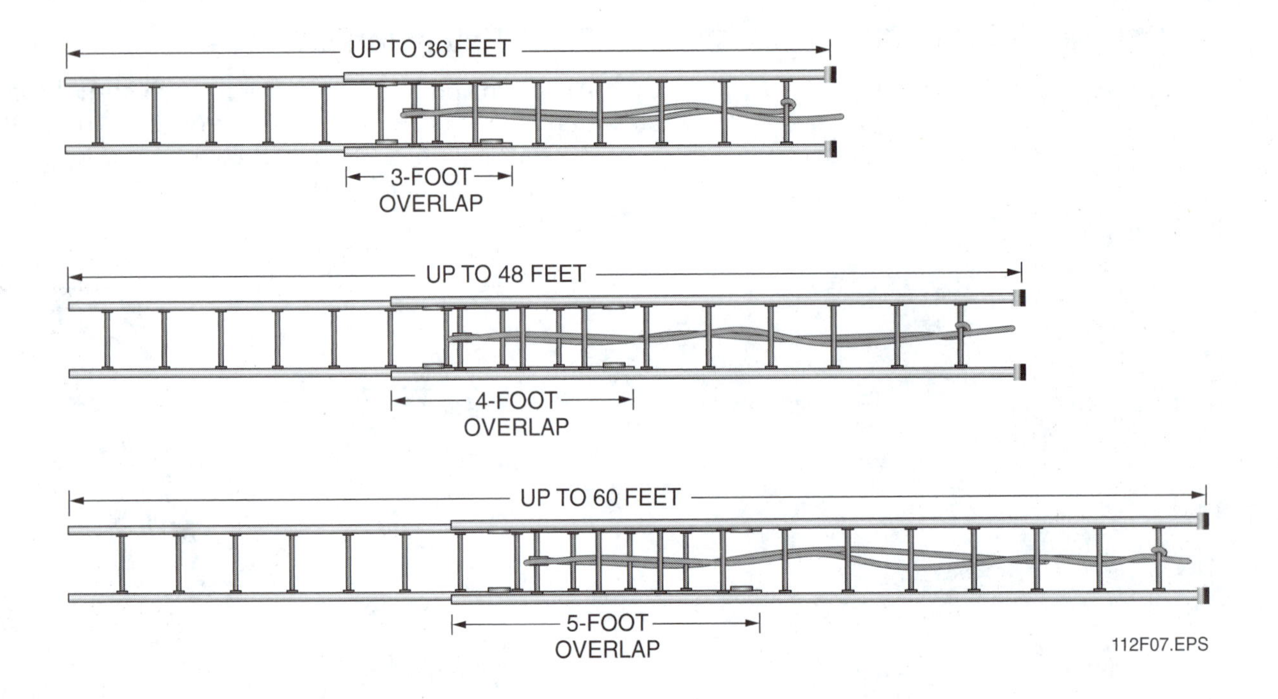

Figure 7 ◆ Overlap lengths for extension ladders.

2.3.0 Stepladders

Stepladders are self-supporting ladders made of two sections hinged at the top *(Figure* 8). The section of a stepladder used for climbing consists of rails and rungs like those on straight ladders. The other section consists of rails and braces. Spreaders are the hinged arms between the sections that keep the ladder stable and prevent it from folding while in use. A stepladder may have a pail shelf to hold paint or tools.

Figure 8 ◆ Typical stepladder.

Inspect stepladders the way you inspect straight and extension ladders. For stepladders, though, pay special attention to the hinges and spreaders to be sure they are in good repair. Also, be sure the rungs are clean. A stepladder's rungs are usually flat, so oil, grease, or dirt can easily build up on them and make them slippery.

Follow these rules when using stepladders.

- Be sure that all four feet are on a hard, even surface when you position a stepladder. If they're not, the ladder can rock from side to side or corner to corner when you climb it.
- Never stand on the top step or the top of a stepladder. Putting your weight this high will make the ladder unstable. The top of the ladder is made to support the hinges, not to be used as a step.
- Make sure the spreaders are locked in the fully open position when the ladder is in position.
- Never use the braces for climbing even though they may look like rungs. They are not designed to support your weight.

Figure 9 shows some common ladder safety precautions.

Do's

- Be sure your ladder has been properly set up and is used in accordance with safety instructions and warnings.
- Wear shoes with non-slip soles.

- Keep your body centered on the ladder. Hold the ladder with one hand while working with the other. Never let your belt buckle pass beyond either ladder rail.

- Move materials with extreme caution. Avoid pushing or pulling anything while on a ladder. You may lose your balance or tip the ladder.

Don'ts

- DON'T stand above the highest safe standing level.
- DON'T stand above the second step from the top of a stepladder and the 4th rung from the top of an extension ladder. You may lose your balance and fall.

- DON'T climb a closed stepladder. It may slip out from under you.
- DON'T climb on the back of a stepladder. It is not designed to hold a person.

- DON'T stand or sit on a stepladder top or pail shelf. They are not designed to carry your weight.
- DON'T climb a ladder if you are not physically and mentally up to the task.

- DON'T exceed the duty rating, which is the maximum load capacity of the ladder.
- DON'T permit more than one person on a single-sided stepladder or on any extension ladder.

112F09.EPS

Figure 9 ◆ Ladder safety.

2.4.0 Fixed Ladders

Fixed or stationary ladders (*Figure 10*) are ladders that cannot be readily moved or carried because they are a permanent part of a building or structure. Fixed ladders must be capable of supporting at least two loads of 200 pounds (91 kilograms) each. Each step or rung must be capable of supporting a single concentrated load of at least 200 pounds applied in the middle of the step or rung.

Think Before You Act

During the construction of a building, a masonry worker was instructed by his foreman to prepare a batch of mortar on the second level and use the stairway to carry it to the third level. This worker decided it would be quicker and easier to use the top section of an extension ladder (without safety feet) instead of the stairway. He set up the ladder by placing one end of the ladder on the wet concrete floor and leaning the other end of the ladder against the wall. He then started to climb. When he was halfway up, the ladder slipped on the wet floor, causing him to fall approximately 12' to his death.

The Bottom Line: Always follow safety instructions.

Source: The National Institute for Occupational Safety and Health (NIOSH)

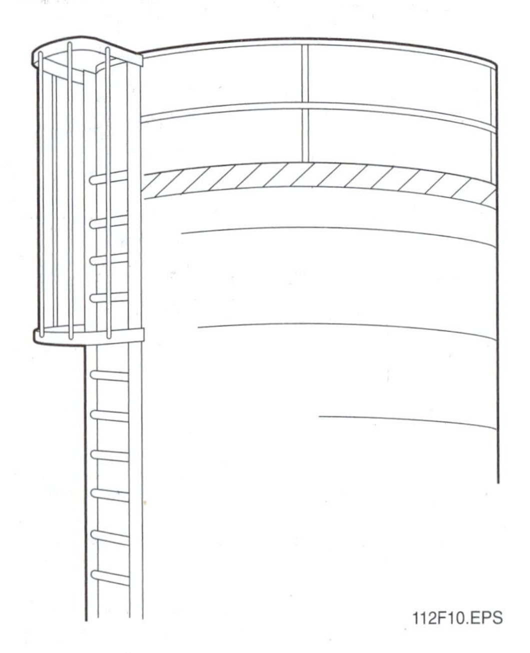

Figure 10 ◆ Typical fixed ladder.

The angle of a fixed ladder must be no greater than 90 degrees from the horizontal, as measured at the back of the ladder. The ladder must be equipped with cages, wells, ladder safety devices, or self-retracting lifelines when the climb is less than 24', but the top of the ladder is a distance greater than 24' above lower levels.

Ladders where the total length of the climb equals or exceeds 24' must be equipped with one of the following:

- Ladder safety devices
- Self-retracting lifelines and rest platforms at intervals not to exceed 150' (15.2 meters)
- A cage or well and multiple ladder sections

3.0.0 ◆ SCAFFOLDING

Scaffolding consists of elevated working platforms that support workers and materials. They are a very common sight on many construction jobs. Typical scaffolding is shown in *Figure 11*. The main part of the scaffolding is the working platform. A working platform should have a guardrail system that includes a top rail, midrail, toeboard, and screening. To be safe and effective, the top rail should be approximately 42" high, the midrail should be located halfway between the toeboard and the top rail, and the toeboard should be a minimum of 4" high.

If people will be passing or working under the scaffolding, the area between the top rail and the toeboard must be screened. Finally, the platform planks must be laid closely together. For safety purposes, the ends of the planks must overlap at least 6" and no more than 12".

Watch Your Step

A maintenance employee was descending from a fixed ladder and fell approximately 5' to the floor. The employee injured his left ankle and right knee, requiring surgery. As a result, he missed several months of work.

The Bottom Line: Even a simple and repetitive task, such as climbing a fixed ladder, requires your maximum attention.

Source: U.S. Department of Energy

Figure 11 ◆ Typical scaffolding.

There are many types of scaffolding used in residential and commercial construction. The three most common are:

- Built-up scaffolding
- Swing scaffolding
- Beam-suspended scaffolding

3.1.0 Scaffolding Hazards

Improper or careless use of scaffolding can result in accidents, injury, and/or death. Those who work on scaffolding can minimize their risks by being aware of the hazards involved and following the proper safety procedures and guidelines to minimize those hazards.

The main hazards involved with the use of scaffolding are:

- Falls
- Workers being struck by falling objects
- Electric shock

Falls can happen because fall protection has not been provided, is not used, or is installed or used improperly. Poorly planked scaffolding causes many falls. Working on scaffolding when conditions are dangerous such as in high winds, ice, rain, and lightning also leads to accidents. Falls also happen when scaffolding collapses because of improper construction.

Fall protection is required on any scaffolding 10' or more above a lower level. Fall-protection devices consist of guardrail systems, personal fall-arrest systems, and/or safety nets. A guardrail system normally serves as adequate fall protection for most scaffolding.

Guardrail systems must extend around all open sides of the scaffolding. The side facing the work surface need not have a guardrail if it is located less than 14" away from the work surface. Any opening on a scaffolding platform must be protected by a guardrail system, including the access opening(s) and platforms that do not extend across the entire width of the scaffolding. Detailed information about fall protection can be found in the *Fall Protection* module in Volume Three.

People who work or pass under scaffolding may be hit by falling objects. Tools, materials, debris, and scaffolding parts may fall to the surface below. Those working on scaffolding may also be injured if there are others working above them, or if the structure or workpiece extends above the work level of the scaffolding. Any worker who is exposed to the danger of falling objects is required to wear a hard hat. Depending on the situation, additional protection may be needed such as debris nets, screens or mesh, canopy structures, and toeboards. Barricades that prevent access under the scaffolding can also be used to protect workers and others.

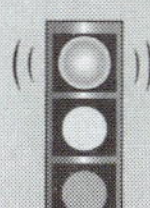

WARNING!
If the scaffolding you are working on shifts or begins to collapse, stop what you are doing and exit the scaffolding immediately.

Inspect All Materials

A crew laying bricks on the upper floor of a three-story building built a 6' platform to connect two scaffolds. The platform was correctly constructed of two 2" × 12" planks with standard guardrails. One of the planks however, was not scaffolding-grade lumber. It also had extensive dry rot in the center. When a bricklayer stepped on the plank, it broke and he fell 30' to his death.

The Bottom Line: Make sure that all planking is sound and secure. Your life depends on it.

Source: The Occupational Safety and Health Administration (OSHA)

Scaffolding

Which scaffolding would you want to work on? Why?

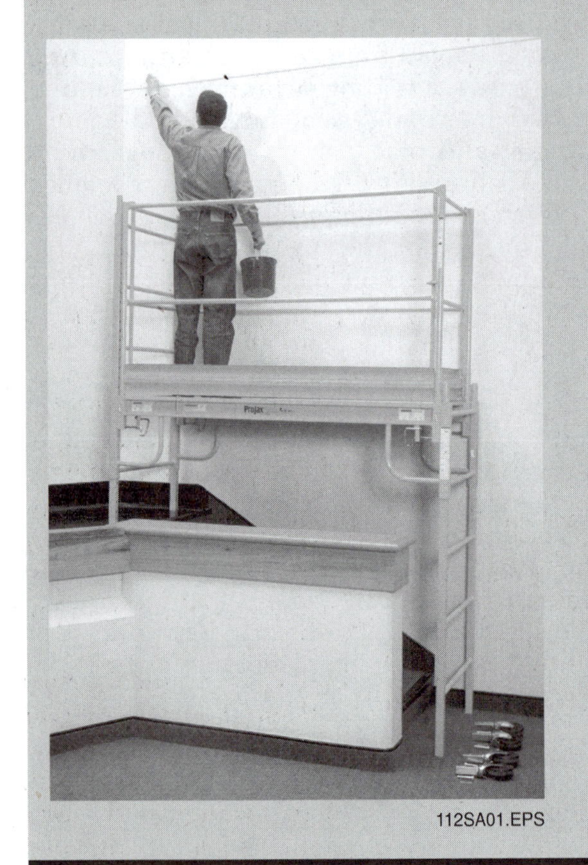

112SA01.EPS

112SA02.EPS

Because most scaffolding is made of metal, the chance of electric shock is always a hazard. Never assume that you can work around high-voltage wires just by avoiding contact. High voltages can arc through the air and cause electrocution without direct contact. When scaffolding must be erected close to power lines, the utility company must be called in to de-energize, move, and/or cover the lines with insulating protective barriers.

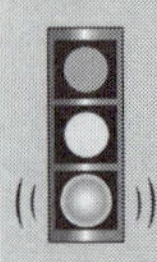

NOTE

Always refer to the competent person on site if you have any questions about the safety of scaffolding.

3.2.0 General Safety Guidelines

Safety begins by getting training in the proper use of scaffolding. It is equally important to always use the right safety equipment, including a hard hat and personal fall-protection systems.

Be Prepared

Two employees were painting the exterior of a three-story building when one of the two outriggers on their two-point suspension scaffolding failed. One painter safely climbed back onto the roof while the other fell approximately 35' to his death. Neither painter was wearing an approved safety harness and lanyard attached to an independent lifeline.

The Bottom Line: Be prepared for the unexpected. When working on scaffolding, always wear approved personal protective equipment.

Source: The Occupational Safety and Health Administration (OSHA)

Never work on scaffolding, in a work cage (*Figure 12*), or on a platform if you:

- Are subject to seizures
- Become dizzy or lightheaded when working at an elevation
- Take medication that might affect your stability and/or performance
- Are under the influence of drugs and/or alcohol

When working on scaffolding, always follow these guidelines.

- Erect and use scaffolding according to the manufacturer's instructions. They must also be erected and used in accordance with all local, state, and federal/OSHA requirements.

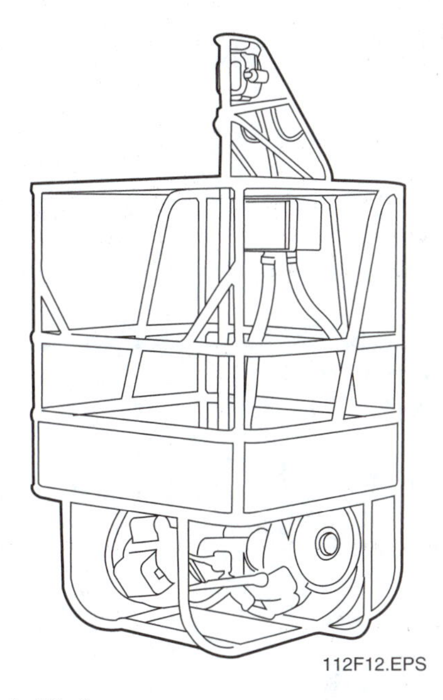

112F12.EPS

Figure 12 ◆ Work cage.

- Never interchange parts of a scaffolding system made by different manufacturers.
- Attach a green, red, or yellow tag (*Figure 13*) as needed to any scaffolding that is assembled and erected to alert users of its current mechanical and/or safety status. Do not rely solely on the tag. Inspect all parts of scaffolding before each use.
- If the scaffolding shifts, exit the scaffolding immediately.

3.2.1 Safety Guidelines for Built-Up Scaffolding

Built-up scaffolding (*Figure 14*) is built from the ground up at a job site. Use the following guidelines when erecting and using tubular built-up scaffolding:

- Inspect all scaffolding parts before assembly.
- Never use parts that are broken, damaged, or deteriorated. Be cautious of rusted materials.
- Follow the manufacturer's recommendations for the proper methods of erecting and using scaffolding.
- Do not interchange parts from different manufacturers.
- Do not force braces or other parts to fit. Adjust the level of the scaffolding until the connections can be made easily.
- Provide adequate sills or underpinnings for all scaffolding built on filled or soft ground. Compensate for uneven ground by using adjusting screws or leveling jacks.
- Do not use boxes, concrete blocks, bricks, or other similar objects to support scaffolding.
- Keep scaffolding free of clutter and slippery material.

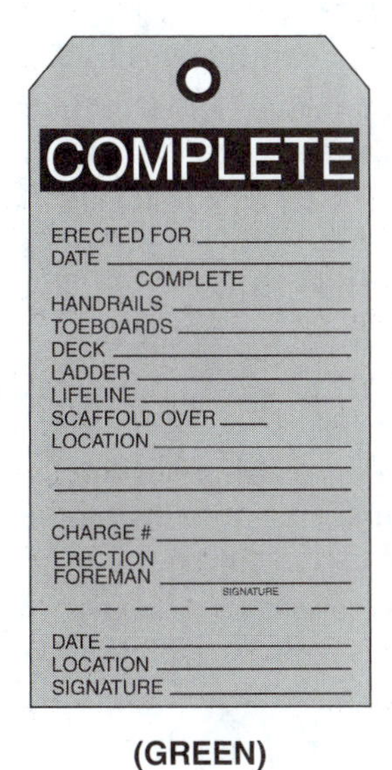

(GREEN)

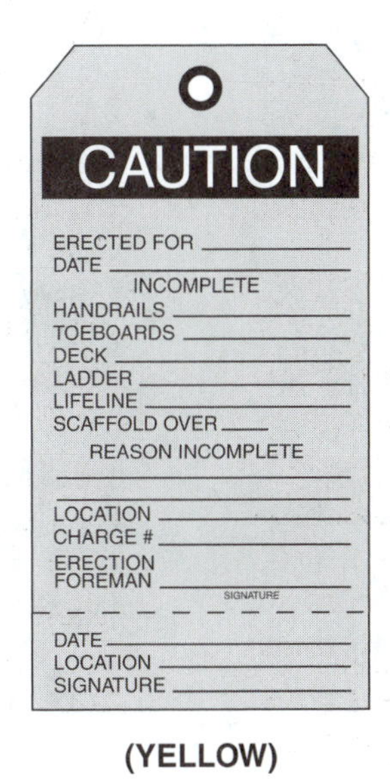

(YELLOW)

(RED)

112F13.EPS

Figure 13 ◆ Typical scaffolding tags.

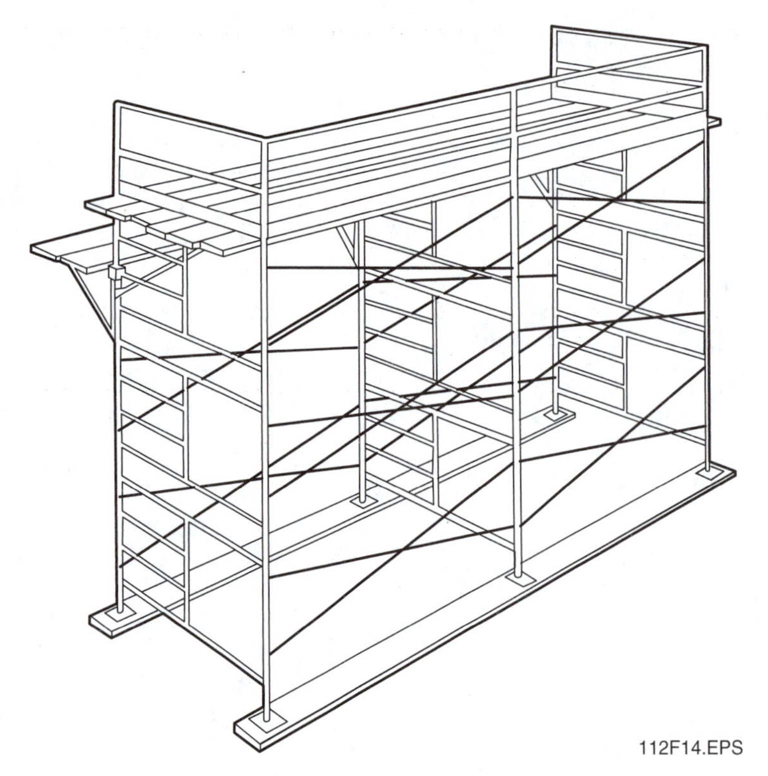

Figure 14 ◆ Built-up scaffolding.

- Be sure scaffolding is plumb and level at all times. Follow the prescribed spacing and positioning requirements for the parts of the scaffolding. Anchor or tie-in scaffolding to the building at prescribed intervals.
- Use ladders rather than cross braces to climb the scaffolding. Position ladders with caution to prevent the scaffolding tower from tipping.
- Do not work on scaffolding that is more than 10' high without guardrails, midrails, and toeboards on open sides and ends.
- Lock the casters of mobile scaffolding when it is positioned for use.
- Do not ride on mobile scaffolding.
- Avoid building scaffolding near power lines.

3.2.2 Safety Guidelines for Swing and Other Suspended Scaffolding

Swing scaffolding (*Figure 15*) is suspended by ropes or cables in a manner that allows it to be raised or lowered as needed. Another type of suspended scaffolding is a work cage. A work cage is typically suspended with rigging devices that attach to I-beams with various sizes of clamps and rollers. It's important to following these guidelines when erecting and using swing and other suspended scaffolding:

- Follow the manufacturer's recommendations for installation, use, and maintenance of the equipment. Before installation, inspect all parts of a structure to which rigging and tieback lines will be secured to ensure that they can support the load.
- Be sure rigging devices are of the proper size and design to support the scaffolding and that they are installed properly. If counterweights are used to secure the inner end of outrigger beams, they should be fastened to the outrigger. Roofing materials and sand bags are not appropriate counterweights.
- Check that tieback lines are installed perpendicular to the face of the structure and are secured to a solid support.
- Check for power lines or electric service wires on the job site. If they pose a hazard, contact the utility company to have them temporarily deactivated.
- Observe the scaffolding's load capacity; never overload the equipment.
- Inspect all scaffolding equipment each day. Check ropes and cables thoroughly for wear, fraying, corrosion, brittleness, damage, or other conditions that may weaken them. Have them replaced as necessary by qualified personnel.
- Keep suspension ropes and cables straight and perpendicular to the platform during use. Do not affix them to anything to change the line of travel.
- Lash the scaffolding to the building or structure to prevent it from swaying.

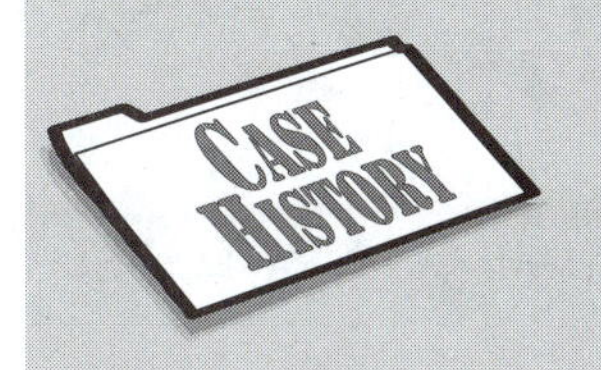

Work Independently

Two workers were sandblasting a 110'-high water tank while working on a two-point suspension scaffolding 60' to 70' above the ground. The scaffolding attachment point failed, releasing the scaffolding cables, and the scaffolding fell to the ground. Both workers died instantly because they were not tied off independently, nor was the scaffolding equipped with an independent attachment system.

The Bottom Line: Make sure that all safety equipment is in place before using any scaffolding.

Source: The Occupational Safety and Health Administration (OSHA)

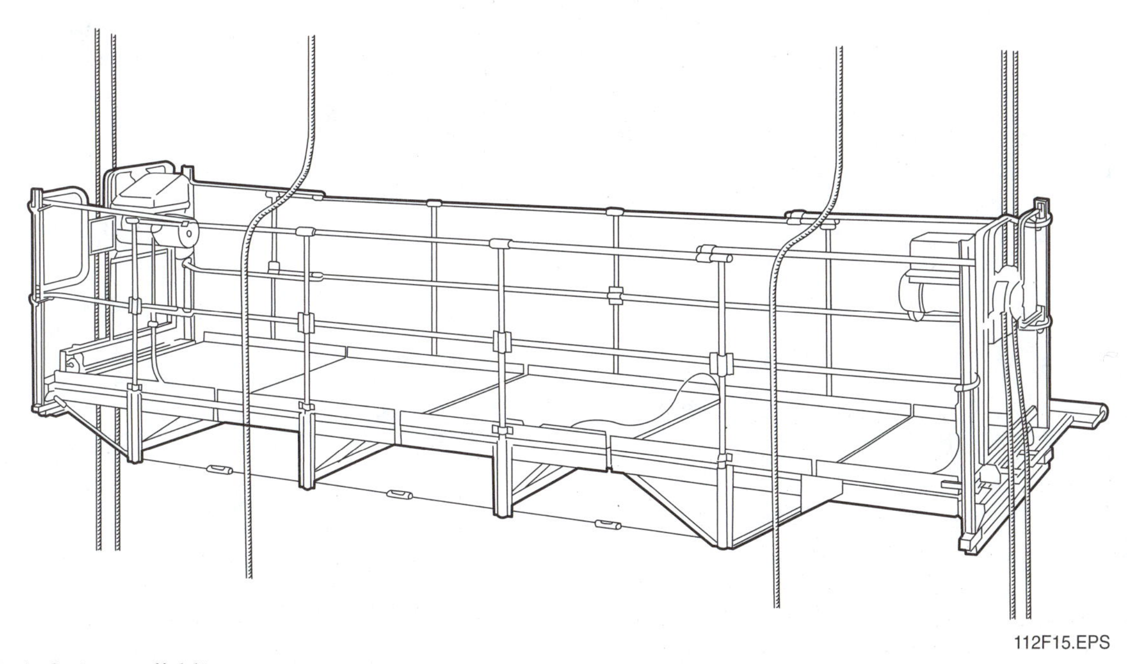

Figure 15 ◆ Swing scaffolding.

- Stay off scaffolding during storms or high winds; watch for icy or slippery platforms.
- Guarantee safe access to the swing stage at all times.
- Use two-way radios for communication between workers on the scaffolding and on the ground.
- Do not combine two or more **two-point swing scaffolding** units to form one unit.
- Raise and lash the scaffolding in a safe position when not in use.

4.0.0 ◆ RESCUE AFTER A FALL

Every elevated job site should have an established rescue and retrieval plan. Planning is especially important in remote areas that are not readily accessible to a telephone. Before beginning work, make sure that you know what your employer's rescue plan calls for you to do in the event of a fall. Find out what rescue equipment is available and where it is located. Learn how to use equipment for self-rescue and the rescue of others.

If a fall occurs, any employee hanging from the fall-arrest system must be rescued safely and quickly. Your employer should have previously determined the method of rescue for fall victims, which may include equipment that lets the victim rescue himself or herself, a system of rescue by co-workers, or a way to alert a trained rescue squad. If a rescue depends on calling for outside help such as the fire department or rescue squad, all the needed phone numbers must be posted in plain view at the work site. In the event a co-worker falls, follow your employer's rescue plan. Call any special rescue service needed. Communicate with the victim and monitor him or her constantly during the rescue.

Know the Dangers of Your Job

A worker was standing on a 6"-wide plank laid between two adjacent I-beams 14' above a concrete floor. He was using a jackhammer to chip away an old concrete and brick floor from a horizontal I-beam. He lost his balance and fell to the floor below, sustaining fatal head injuries. He was not wearing any fall-protection gear.

The Bottom Line: Always use appropriate fall-arrest equipment. Do not use scaffolding unless it is constructed properly.

Source: The Occupational Safety and Health Administration (OSHA)

Summary

Ladders and scaffolding are used in a variety of different construction jobs. The type of ladders or scaffolding will vary with each job, but the dangers won't. There is always a risk of falling and being stuck by falling objects, each of which can result in serious injury or death. It is your responsibility to be aware of and follow the safety procedures associated with ladders and scaffolding.

Review Questions

For Questions 1 through 3, match the type of ladder to the corresponding description.

1. _____ Aluminum

2. _____ Fiberglass

3. _____ Wooden

a. Corrosion-resistant and can be used in situations where it might be exposed to the elements
b. Heavier than other ladders and can be used when heavy loads must be moved up and down
c. Durable and useful in situations where some amount of rough treatment is unavoidable

4. If you lean a straight ladder against the top of a 16' wall, the base of the ladder should be _____ from the base of the wall.

a. 1½'
b. 3'
c. 4'
d. 6'

5. Straight ladders can be used on unstable surfaces if safety feet are used.

a. True
b. False

6. It is safe to stand on the top step of a stepladder as long as you are holding onto a solid component of the building.

a. True
b. False

7. A fixed ladder must be able to support at least _____ pounds.

a. 150
b. 200
c. 250
d. 300

8. Fall protection is required on any scaffolding that is _____ or more above a lower level.

a. 10'
b. 12'
c. 14'
d. 16'

9. Electric shock is one of the hazards of working on scaffolding.

a. True
b. False

10. Built-up scaffolding is suspended by ropes or cables so that it can be lowered and raised as needed.

a. True
b. False

GLOSSARY

Trade Terms Introduced in This Module

Duty rating: American National Standards Institute (ANSI) rating assigned to ladders. It indicates the type of use the ladder is designed for (industrial, commercial, or household) and the maximum working load limit (weight capacity) of the ladder. The working load limit is the maximum combined weight of the user, tools, and any materials bearing down on the rungs of a ladder.

Ladder: A wood, metal, or fiberglass framework consisting of two parallel side pieces (rails) connected by rungs on which a person steps when climbing up or down. Ladders may either be of a fixed length that is permanently attached to a building or structure, or portable. Portable ladders have either fixed or adjustable lengths and are either self-supporting or not self-supporting.

Scaffolding: A temporary built-up framework or suspended platform or work area designed to support workers, materials, and equipment at elevated or otherwise inaccessible job sites.

Two-point swing scaffolding: A manual or power-operated platform supported by hangers at two points suspended from overhead supports in a way that allows it to be raised or lowered to the working position.

Figure Credits

Louisville Ladder Group	112F01
Ridge Tool Company	112F06, 112F08
Werner Ladder Company	112F09
Bil-Jax, Inc.	112SA01
Gary Wilson	112SA02

NCCER CURRICULA — USER UPDATE

NCCER makes every effort to keep its textbooks up-to-date and free of technical errors. We appreciate your help in this process. If you find an error, a typographical mistake, or an inaccuracy in NCCER's curricula, please fill out this form (or a photocopy), or complete the online form at **www.nccer.org/olf**. Be sure to include the exact module ID number, page number, a detailed description, and your recommended correction. Your input will be brought to the attention of the Authoring Team. Thank you for your assistance.

Instructors – If you have an idea for improving this textbook, or have found that additional materials were necessary to teach this module effectively, please let us know so that we may present your suggestions to the Authoring Team.

NCCER Product Development and Revision

13614 Progress Blvd., Alachua, FL 32615

Email: curriculum@nccer.org
Online: www.nccer.org/olf

❑ Trainee Guide ❑ AIG ❑ Exam ❑ PowerPoints Other ______________________

Craft / Level: Copyright Date:

Module ID Number / Title:

Section Number(s):

Description:

Recommended Correction:

Your Name:

Address:

Email: Phone:

Module 75113-03

Horizontal Directional Drilling Hazards

COURSE MAP

This course map shows all of the modules in Field Safety. The suggested training order begins at the bottom and proceeds up. The local Training Program Sponsor may adjust the training order.

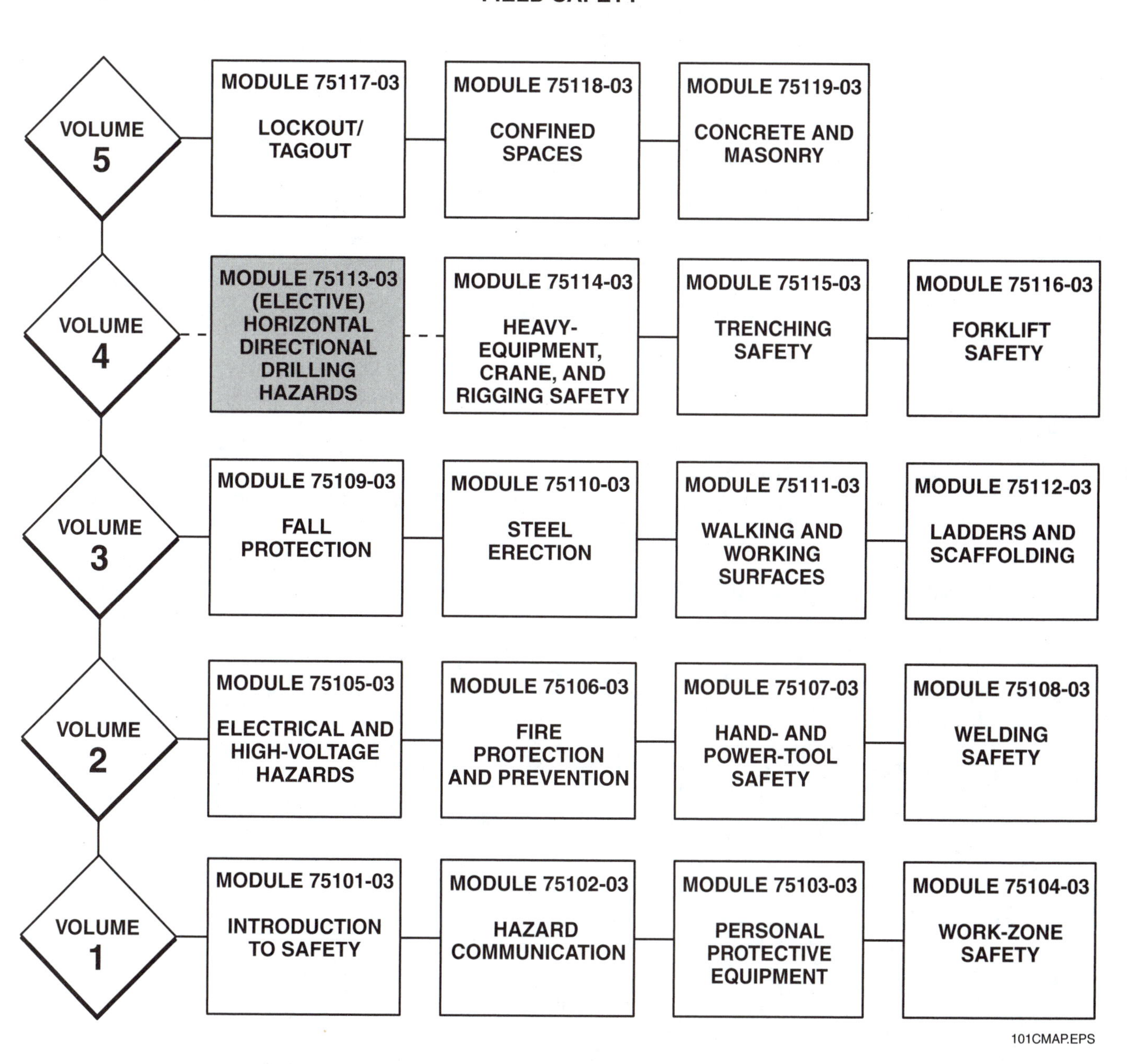

MODULE 75113-03 CONTENTS

Figures

Tables

MODULE 75113-03

Horizontal Directional Drilling Hazards

Objectives

When you have completed this module, you will be able to do the following:

1. Describe the horizontal directional drilling (HDD) process.
2. Identify common safety hazards associated with horizontal directional drilling jobs.
3. Describe how to avoid the hazards associated with horizontal directional drilling jobs.
4. Respond to utility strikes that may cause personal injury, equipment damage, and property damage.
5. Identify safety alert signs and symbols.

Prerequisites

Before you begin this module, it is recommended that you successfully complete the following: Field Safety, Modules 75101-03 through 75112-03. This *Horizontal Directional Drilling Hazards* module (75113-03) is an elective and is not required for successful completion of this course.

Required Materials

1. Pencil and paper
2. Appropriate personal protective equipment

1.0.0 ◆ INTRODUCTION

Safety is an important part of working in the construction industry. Safety is equally important for horizontal directional drilling (HDD) jobs where specialized equipment is used to bore holes for installing utility lines (*Figure 1*). Workers on any construction project must have training to avoid common hazards such as slips, falls, pinched fingers and toes, vehicle accidents, and back injuries. In addition to having the knowledge and ability to recognize and avoid common job-site hazards on HDD projects, craft professionals must also recognize and avoid the hazards specific to HDD.

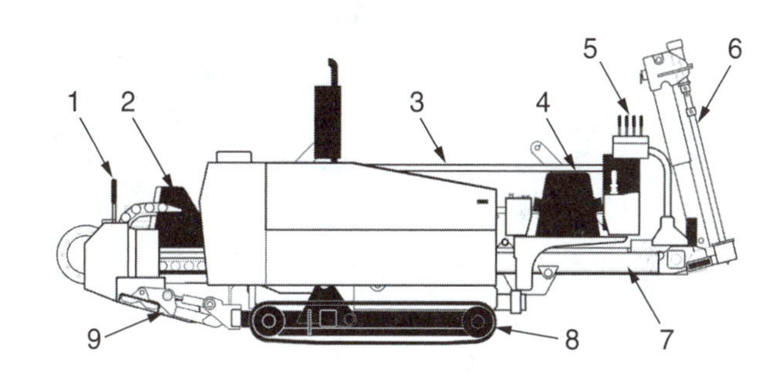

1. Setup/drive controls and tethered ground drive controller
2. Carriage
3. Pipeloader
4. Operator's station
5. Anchoring system controls
6. Anchoring system
7. Drill frame
8. Tracks
9. Stabilizers

113F01.EPS

Figure 1 ◆ Typical horizontal directional drilling rig.

The hazards on an HDD job include a rotating **drill string**; making up and breaking out of pipe joints (*Figure 2*); and the potential of striking **utilities** in the job-site area. In this module, you will learn how to recognize and avoid these hazards.

Drill Rod Dangers

A three-man crew was installing an underground telephone cable in a residential area. They had just completed a bore hole under a driveway using a horizontal boring machine. The bore hole rod had been removed from the hole. While the rod was still rotating, the operator straddled it and stooped over to pick it up. His trouser leg became entangled in the rotating rod and he was flipped over, striking tools and materials. He died as a result of his injuries.

The Bottom Line: Never step over a rotating drill rod, no matter how safe it looks.

Source: The Occupational Safety and Health Administration

113F02.EPS

Figure 2 ◆ Drill pipe.

HDD jobs can vary from installing pipes that are as small as 2" up to those with diameters as large as 48". The size of the drill rigs used will also vary. Rigs can be small, medium, or large (*Figure 3*). Small rigs are used for installing utility cable and smaller pipes in crowded city areas. Medium-sized rigs are used to install municipal pipelines. Large rigs are used to install large diameter pipes.

During a typical **boring** operation, sections of pipe are joined together at each end and pushed through the ground. The drill head (*Figure 4*), which is placed on the front of the drill pipe, rotates and pushes through the ground to make the bore hole. Once the drill head reaches the end of the bore hole, a back-reamer (*Figure 5*) is attached and the drill pipe is pulled back through the bore hole. Product pipe, such as cable or utility lines, may be pulled through the bore at the same time.

113F03.EPS

Figure 3 ◆ Drill rigs.

Figure 4 ◆ Drill heads.

Figure 5 ◆ Back-reamer.

The entry and exit points of a bore hole are called the *entry pit* and the *receiving pit* (*Figure 6*). These pits are used to collect drilling fluids. Drilling fluids are drained from the pits at the end of the operation. These pits vary in size, depending on the job. For small operations, the pits are typically 1' to 3' deep, 1' to 2' wide, and 2' to 6' long. For larger operations, the entry and receiving pits may require trenching and shoring.

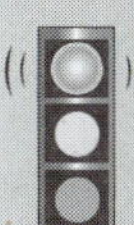

WARNING!

Horizontal directional drilling involves trenching hazards as well as electrical and fire safety hazards related to utility strikes. Before entering any HDD job site, be sure you have completed the *Trenching Hazards*, *Electrical and High-Voltage Hazards*, and *Fire Prevention and Protection* modules.

2.0.0 ◆ HORIZONTAL DIRECTIONAL DRILLING PROCESS

There are four basic steps in the HDD process:

- Job-site planning
- Site setup
- Boring and **reaming** operations
- Cleanup

Established safety guidelines must be followed during each of these steps to ensure the safety of everyone on the job site.

2.1.0 Job-Site Planning

Job-site planning is the first step in the HDD process. During job-site planning, the site supervisor walks the site and comes up with a plan for drilling. This plan includes identifying the locations of signs and barricades, as well as determining where the trailer that carries the boring machine will be placed for unloading and loading.

During this step, the site supervisor also chooses the entry point, path, and exit point of the boring operation. Locating and marking existing utility lines such as gas, phone, and electricity are also part of the planning process (*Figure 7*). Subsite locators (*Figure 8*) are portable devices that are used at the site to identify utility lines and obstructions. *Table 1* shows standard surface utility marking colors.

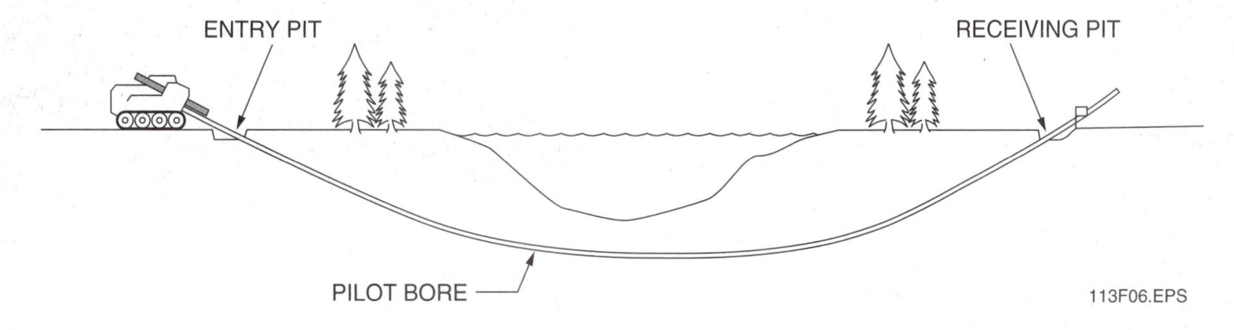

Figure 6 ◆ Entry and receiving pits.

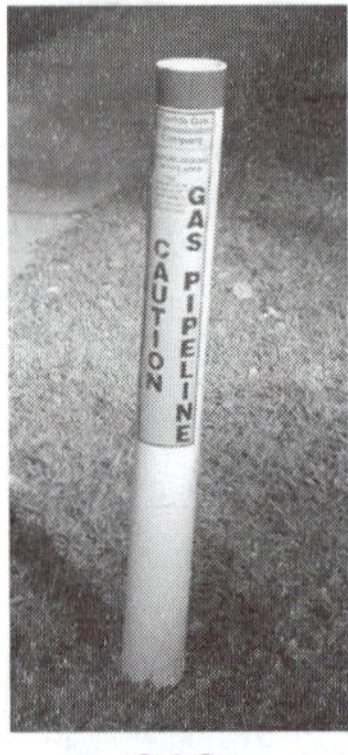

GAS PIPELINE

FIBER-OPTIC CABLE

TELEPHONE CABLE

113F07.EPS

Figure 7 ◆ Examples of marked utilities.

113F08.EPS

Figure 8 ◆ Sub-site locator.

Table 1 Standard Surface Utility Marking Colors

Color	Utility
Red	Electric power
Yellow	Gas, oil, steam, and petroleum
Blue	Water and irrigation
Green	Sewer and storm drains
Orange	Fiber-optic, telephone, and cable TV
Pink	Temporary survey marking
White	Proposed construction area or work limits
Purple	Reclaimed water

The most common way of identifying existing utility hazards before drilling is to use established services such as One-Call. This service provides information on buried utilities in the area. Contractors must call this service at least 72 hours before doing any drilling. If this information is known in advance, utility strikes are less likely to happen. If this type of service does not exist in your area, local utility companies must be contacted to identify utility lines in the area.

After the utilities have been located, perform a physical inspection of the site to determine any other potential hazards or problems. When inspecting the perimeter of a job site, look for the following:

- Notices for buried utilities
- Storm drains and sewers
- Overhead power lines
- Gas or water meters
- Junction boxes
- Mailboxes
- Light poles
- Manhole covers and catch basins
- Sunken ground

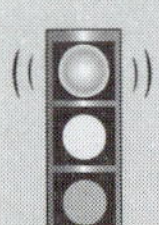

WARNING!

If there is any doubt about the location of utility lines, the area in question should be thoroughly searched and uncovered by hand, if necessary, to verify that it is safe.

Unsafe Conditions

A drill operator was injured while working on a 60-degree slope without adequate secure footing. The drill operator stood on the slope, which made reaching the controls awkward. He lost his footing, pinching his thumb between the stabilizer and the auger. The resulting cut required 10 stitches. No permanent damage to his thumb was anticipated, but the injury could have been considerably worse.

The Bottom Line: Never work under unsafe conditions.

Source: U.S. Department of Energy

2.2.0 Site Setup

After the area has been inspected, it is time to set up the drilling site and equipment. Before the boring machine can be unloaded from the trailer, make sure the following safeguards are in place:

- Properly placed barricades (*Figure 9*)
- Visible warning signs
- Pedestrian and traffic controls (*Figure 10*)

113F09.EPS

Figure 9 ◆ Properly placed barricades.

113F10.EPS

Figure 10 ◆ Typical traffic control/warning sign.

Be On the Lookout

On August 8, 2001, a contractor employee was injured when a drill stem disengaged. The discharge swivel, located approximately 25' above the ground, unexpectedly released from its supporting chains and fell. The employee received a slight blow to the head, pushing him off his work platform onto a pile of soft sand approximately 1' below. He suffered a broken right wrist and a dislocated left wrist from a back-out wrench he placed on the tool slot of the drill stem. This accident was caused by equipment failure.

The Bottom Line: Make sure you inspect all equipment before doing a job.

Source: U.S. Department of Energy

2.2.1 Unloading the Boring Machine

The next step in site setup is to unload the boring machine from the trailer (*Figure 11*). There are several hazards to recognize and avoid when unloading the boring machine.

Being struck by moving equipment – Workers are at risk of being stuck by moving equipment when a rig is unloaded from the trailer. Rigs can overturn or slide rapidly down the trailer unloading ramps. This is dangerous because HDD rigs are heavy. Even a small rig can weigh up to 3,000 pounds.

113F11.EPS

Figure 11 ◆ Unloading the boring machine.

What's Wrong With This Picture?

113SA01.EPS

Amputation Caused by Lack of Training

On Friday, November 16, 2001, a hollow-stem auger drill rig operator lost a finger in a pinch-point injury. The middle and index fingers of the same hand were broken at the knuckle. He also suffered a large cut on the back of his hand beginning at the knuckle of the middle finger, crossing the index finger, and extending toward the thumb. This cut severed the tendon to his index finger.

An investigation of the accident found that the manual alignment of extension rod sections to the threaded rod cap was a practice that the operator considered a routine and accepted risk. It isn't! This action regularly introduced the operator's hand into the pinch-point hazard. That's why he lost his finger.

During the investigation, it was discovered that the drilling sub-contractor never provided written procedures or training to this worker.

The Bottom Line: Make sure you know how to do your job safely and that you get the training you need.

Source: U.S. Department of Energy

Get out of the way if you see that a rig is about to turn off or slide down the ramps. It's your responsibility to be safe whether you are the worker unloading the rig or you are simply on site. You must be aware of everything that is happening on the site. This includes making sure that the unloading ramps are safely secured and that the path from the trailer to the entry site is clear of pedestrians and other equipment. Also, always follow the manufacturer's safeguards and rules for the equipment being used.

Being pinched between the drill rig and another object – The weight of objects and equipment will shift as the drill rig is unloaded. It's important to be aware of how a weight shift will affect any pieces of equipment you are touching. For example, if you are controlling the chain that is being used to lower the drill rig off the trailer, and there is a shift in weight, the chain can instantly go from slack to tight. When this happens, your hands are at risk of being severely pinched by the chain. Always be aware of equipment you are touching or handling and move quickly when you notice a shift in weight. You can also get pinched if the drill rig is in front of you and there is an object directly behind you that limits the amount of space you have to move.

Getting crushed by the drill rig if it flips or rolls off the trailer – Because drill rigs are so heavy, there is an added risk of being crushed if a rig flips or rolls while being unloaded. To avoid being crushed and killed by a flipping or rolling rig, make sure that the unloading ramps are properly secured before moving the rig onto the ramps. Move the rig as slowly as possible and be aware of any quick shifts in weight.

Strains and sprains from moving loading ramps – Loading ramps were designed to hold heavy equipment like HDD rigs. Because of this, they are often heavy themselves. They are also oddly shaped and can be awkward to move. Always get help when moving loading ramps. Also, be aware of your own limits. If you find that a loading ramp is too heavy and you are having trouble moving it, ask for help. If you don't, you could strain or sprain your back, arms, or legs.

2.2.2 Positioning and Setting Up the Equipment

After the boring machine has been unloaded, it must be properly positioned. Support equipment must also be set up and tested. This equipment is used to protect workers from electrical and mechanical hazards. It includes the:

- Ground strike system
- **Grounding mat** for both the operator and the machine
- **Operator presence sensing system (seat switch)**

Positioning and setting up the equipment has many hazards. Being aware of these hazards will help you to be more cautious when doing your job. The hazards involved in positioning and setting up the equipment are:

- Strains and sprains when moving equipment
- Tripping and falling on tools or equipment
- Pinch points when handling the drill head, drill pipe, and hand tools
- Electrocution by exposed, energized utilities in the area
- Leaks and injection hazards from the hydraulic lines
- Bore and receiving pit cave-ins
- Contact with the rotating head and drill rod
- Tripping over equipment on the ground
- Falling into the bore or receiving pit

Follow these safeguards to avoid setup/positioning hazards:

- Make sure the ground strike system, which protects workers from being electrocuted when a power line is hit, has been set up and tested. Ground strike systems include an electrical sensing stake that is driven into the ground, a monitor (*Figure 12*), a nonconductive cable that connects the stake and the monitor, and an alarm system. When a power line is struck, audible and visual warning signals go off. Ground strike systems vary depending on the equipment used. The setup and testing of the ground strike system also varies depending on the equipment. Always follow the equipment manufacturer's instructions when setting up and testing a ground strike system.

Figure 12 ◆ Typical ground strike system monitor.

- Set up the grounding mat for both the machine and the operator. Grounding mats are mesh mats that are placed underneath the drill rig. They protect workers from electrocution by providing a single electrical path to ground should the equipment come in contact with an energized conductor during a utility strike. They can be considered part of the ground strike system.
- Keep all workers and equipment away from power lines. Power lines can be overhead or underground. Sub-site locators can help to locate underground power lines. Overhead power lines are easier to see and must be avoided at all times.
- Stay away from hydraulic and pneumatic lines. Pressure can build up in these lines and cause leaks. The pressure is strong enough to cut the skin.
- Pin all air hose connections to prevent hose whipping. Hose whipping can cause serious cuts and bruises. Immediately step away from any runaway hose and shut down the machine.
- Use caution when entering and exiting bore and receiving pits. The soil in pits can be unstable. This can cause the worker to fall or the pit to cave in. There is also a danger of getting stuck in the drill string. If applicable, make sure a shoring system is set up to protect bore and receiving pits from cave-ins.
- Test the operator presence sensing system (seat switch). The operator presence sensing system can tell when a worker is in the seat. If the worker leaves the seat while the drill is operating, the rig will shut off.
- Test the operation of all controls to make sure they are working correctly. This includes the ground strike system as well as the controls on the rig.
- Make sure the handheld locator and the transmitter, which is installed in the bore head, have been calibrated. This will help to prevent dangerous underground strikes.
- Turn the boring machine off while workers are attaching the bore head. Workers can be struck by the bore head if the boring machine is powered up.

2.3.0 Boring Operations

Boring operations (*Figure 13*) can begin once the equipment is in position and set up. Potential hazards of the boring process include:

- Electric shock
- Rotating equipment that can strike workers, such as the drill string

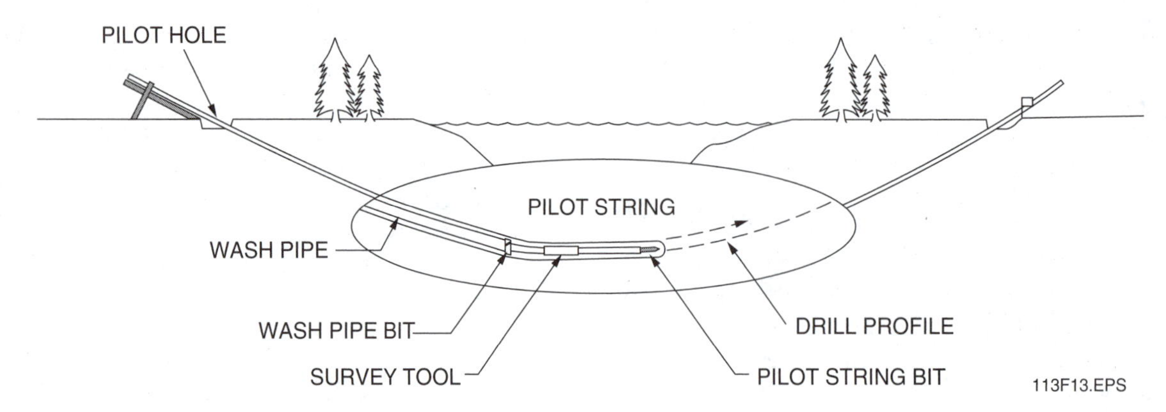

Figure 13 ◆ Pilot bore.

- Equipment such as wrenching that is on the drill string when it starts rotating
- Rotating equipment that workers or their clothes can get caught on
- Pinch points near the drilling rig
- Trips and falls near rotating drill rods

Following these safeguards can prevent you and others from getting hurt during the boring process:

- Wear nonconductive gloves when operating the equipment. Nonconductive gloves help to protect against electric shock in case you accidentally come in contact with an energized conductor or equipment.
- Inspect gloves for damage (*Figure 14*). Gloves and sleeves can be inspected by rolling the outside and inside of the protective equipment between your hands. Squeeze the inside of the gloves or sleeves to bend the outside area and create enough stress to expose any cracks, cuts, or other defects. When the entire surface has been checked in this manner, the equipment is then turned inside out, and the procedure is repeated. It is very important to not leave the rubber protective equipment inside out.
- Keep all body parts clear of the drill rig while drilling is taking place. This will eliminate the possibility of being caught in or struck by the rotating drill string.
- Remove all tools that are used to loosen or tighten joints. If a wrench is left in place while the drill string is turning, it could hit the worker or fly off and hit a co-worker.
- Turn the rig off and bleed down pneumatic and hydraulic power lines before service or repair work is done. Make sure this is done properly. Equipment that becomes energized during repair can cause serious injury. For more information about energized equipment, refer to the *Lockout/Tagout* module.

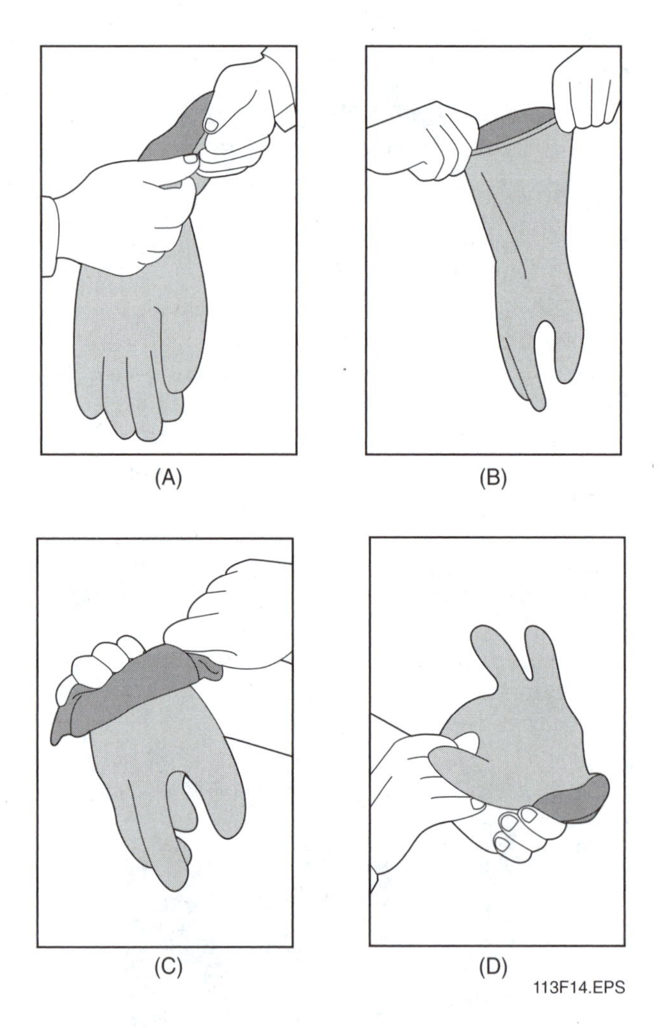

Figure 14 ◆ Glove inspection.

- Disconnect the battery when servicing or repairing the electrical systems on the drill. This will ensure that equipment does not become energized and start operating.
- Wear nonconductive boots. Nonconductive boots will provide extra protection against electrocution in the event of a power line strike.

- Turn the drill rig off when work is being done on the drill string. This helps to ensure that no one is caught or stuck on the rotating drill string.
- Make sure the operator communicates with all personnel before beginning to rotate the drill rod. It is important that everyone on the site knows what's going on at all times. Workers can easily be injured or killed if they approach equipment or an area of the site that is not safe because drilling is taking place.

2.3.1 Potholing

If there is any risk of a utility strike, a pothole must be dug at or near the likely location of the utility to physically identify gas lines, electrical utilities, fiber-optic lines, and communication lines. Potholing is also used to find out if the drill is where it is supposed to be according to locating equipment. This is helpful because there are times when the shifting of the soil caused by the boring operation moves the drill string out of its intended path.

Potholing can be done using mechanical equipment such as air blasters and excavation vacuums or digging by hand. The hazard of using mechanical equipment comes from contacting or cutting the utility line you are trying to locate. Digging the pothole by hand is a safe alternative.

2.3.2 Reaming and Product Pulls

Reaming and product pulls are the second part of boring operations. During the reaming process, the drill head is replaced with a back-reaming tool and the product pipe is pulled back through the bore, as shown in *Figure 15*. Many of the hazards linked to reaming and product pulls are similar to those that occur during boring. Keep the following hazards and safeguards in mind when reaming and performing product pulls.

- Workers may be struck by or caught on rotating equipment such as the drill pipes or rotating drill string. This commonly occurs when workers are not paying attention. It also happens when they enter an area of the site where they are not supposed to be, such as receiving pits. Be aware of moving parts when working near the drilling rod and attachments, or when the product is being pulled back. Also, turn the drilling rig off while workers are in the pit and/or working with the drill rod, reamer, or attachments. This will help workers avoid being struck by or caught in rotating equipment.
- Strains, sprains, crushed fingers, or crushed hands may result from contact with equipment. Drill rigs and the pipes that are used to do the drilling are heavy. It's very easy to injure yourself when working with this equipment. Make sure to use mechanical lifting devices or get help when handling heavy equipment such as pipes, reamers, and attachments.
- Falls and slips can happen when workers are in receiving pits. Make sure to shut off the rig and water supply when the reamer enters the receiving pit. Also, make sure the receiving pit has been vacuumed.

2.4.0 Cleanup

Cleanup is the last step in the HDD process. The hazards of site cleanup are similar to the hazards of unloading and setting up the drill rig, including:

- Getting struck by moving equipment
- Being trapped between the boring machine and another object
- Getting crushed by the boring machine if it flips or rolls off the trailer
- Strains and sprains from moving loading ramps
- Pinch points created by the tilting of the trailer

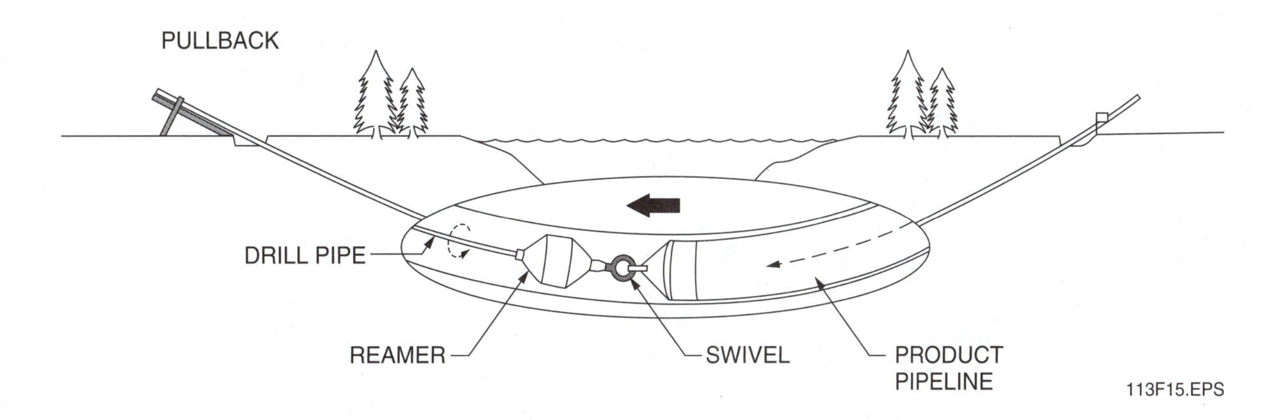

Figure 15 ◆ Back-reaming.

During cleanup, perform the following:

- Return the site to its original condition.
- Mark the location of the installed pipe.
- Make sure all of the signs, barricades, tools, and equipment are put away.
- Load the boring machine.

3.0.0 ◆ RESPONDING TO UTILITY STRIKES

Drills can strike electrical lines, gas pipes, communication lines, and sewage pipes during the boring operation. Striking utilities can cause personal injury and equipment damage. The hazards related to striking utilities include electrocution, gas line explosions, and exposure to contaminated water or sewage. This section explains what to do when a strike happens. *Figure 16* shows industry-standard strike safety symbols.

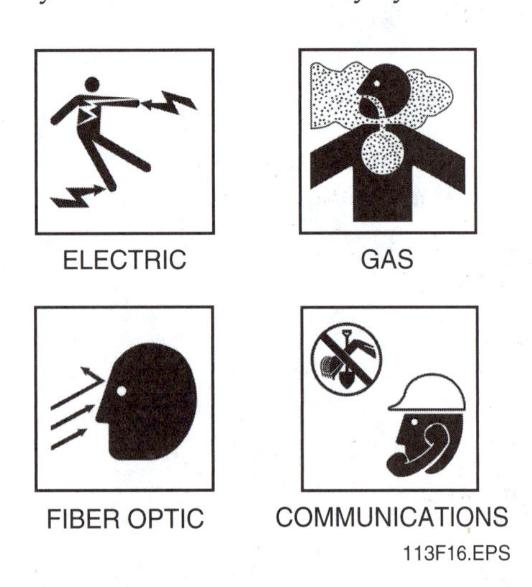

Figure 16 ◆ Symbols indicating a strike.

3.1.0 Electrical Strike

Follow these guidelines if an electrical strike occurs:

- Do not move. There can be a higher amount of electricity in the ground around you or in equipment near you. Coming in contact with this electricity will cause electrocution.
- Remain calm.
- Warn all workers at the site.
- Do not touch the machine, drill pipe, water system, mud-mixing system, or anything connected to the drill.
- If you are the operator, follow the drill manufacturer's procedure for reversing the drill. This will break contact with the electrical line.
- Notify the electric company.

3.2.0 Gas Strike

Follow these guidelines if a gas strike occurs:

- Shut down all motorized equipment.
- Evacuate the area.
- Do not try to back the drill out.
- Notify emergency response personnel.
- Notify the gas company.

3.3.0 Fiber-Optic Strike

Follow these guidelines if a fiber-optic strike occurs:

- Stop all drilling.
- Do not look into the cut ends of the cable. It could burn your eyes.
- Notify the fiber-optic utility owner.

3.4.0 Communication Line Strike

Follow these guidelines if a communication line strike occurs:

- Stop all drilling.
- Notify the communication line utility owner.

3.5.0 Sanitary/Storm Sewer and Water Strike

Follow these guidelines if a sanitary/storm sewer and water strike occurs:

- Stop all drilling.
- Warn all workers at the site.
- Notify the utility owner.
- Follow your company's procedures for decontaminating all affected workers.
- Seek medical attention.

4.0.0 ◆ SAFETY ALERT LEVELS

Safety alert levels on a job site help to ensure that no one gets hurt and that equipment is not damaged. There are three basic safety levels: DANGER, WARNING, and CAUTION. *Figure 17* defines the hazards associated with each level.

Also, watch for the words *NOTICE* and *IMPORTANT*. NOTICE is used to alert and stop you from doing something that could damage equipment or personal property. It is also used to warn you against unsafe actions. IMPORTANT is used to remind you to do your job safely.

Loose Connection

While conducting drilling on a site, the drill string was being pulled from the hole. The driller loosened the top connection on an 8' drill rod where it was connected to the drive head. He then proceeded to raise the drive head and drill string to break the bottom connection to the drill rod. After unscrewing the bottom connection, the driller started to raise the drive head and attached drill rod in order to swing the drill rod out so that the driller's helper could remove it and place it on the pipe rack. As the drive head was raised, the drill rod (weighing approximately 70 pounds) came loose from the top connection. The helper, who was standing by to receive the drill rod, saw it come loose from the drive head and was able to step out of the way and let it fall to the drill deck. That was quick thinking on his part.

During an investigation of the incident it was determined that, although the threads on the drive head subassembly showed signs of wear, they were not worn enough to be entirely responsible for the rod falling. It was determined that the driller had loosened the top connection between the drill rod and the top drive subassembly too much. This compromised the holding power of the tapered threads, allowing the drill rod to come loose and fall. Fortunately, no one was hurt when this happened.

The Bottom Line: It is important to remember that your carelessness not only affects you, it affects those around you. Make sure all connections are tightened properly.

Source: U.S. Department of Energy

5.0.0 ◆ SAFETY ALERT SYMBOLS

Safety alert symbols can save your life. You will find these universal symbols on equipment and machinery around the job site. It is important to recognize these symbols and to know their meanings so that you can do your job safely.

Safety symbols are organized by the three basic safety levels:

- *Danger* – The DANGER symbols in *Figure 18* represent hazards that will result in death or serious injury if safety procedures are not followed.

The DANGER symbol points out **extremely hazardous** situations that **must be avoided** in order to prevent **death or serious injury.**

The WARNING symbol points out **potentially hazardous** situations that **should be avoided** in order to prevent **death or serious injury.**

The CAUTION symbol points out **potentially hazardous** situations that **should be avoided** in order to prevent **minor or moderate injury.**

113F17.EPS

Figure 17 ◆ Safety alert levels.

DANGER

Turning shaft will kill you or crush arm or leg. Stay away.

DANGER

Electric shock. Contacting electric lines will cause death or serious injury. Know location of lines and stay away.

DANGER

Deadly gases. Lack of oxygen or presence of gas will cause sickness or death. Provide ventilation.

DANGER

Moving parts. Being struck by wrench will kill or injure. Do not use drilling unit to turn or move drill string when wrench is used.

113F18.EPS

Figure 18 ◆ Danger symbols.

- *Warning* – The WARNING symbols in *Figure 19* represent potential hazards that may result in death or serious injury if safety procedures are not followed.
- *Caution* – The CAUTION symbols in *Figure 20* represent potential hazards that may result in minor or moderate injury if safety procedures are not followed.

WARNING

Explosion possible. Serious injury or equipment damage could occur. Follow directions carefully.

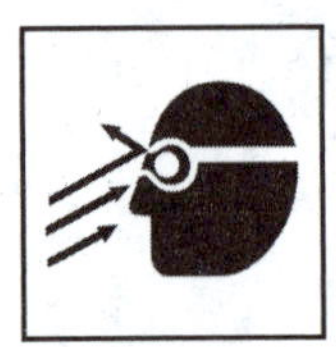

WARNING

Fluid or air pressure could pierce skin and cause injury or death. Stay away.

WARNING

Job-site hazards could cause death or serious injury. Use correct equipment and work methods. Use and maintain proper safety equipment.

WARNING

Crushing weight could cause death or serious injury. Use proper procedures and equipment or stay away.

WARNING

Moving parts could cut off hand or foot. Stay away.

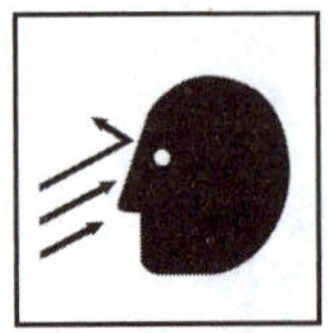

WARNING

Looking into fiber-optic cable could result in permanent vision damage. Do not look into ends of fiber-optic or unidentified cable.

WARNING

Fire or explosion possible. Fumes could ignite and cause burns. No smoking, no flame, no spark.

WARNING

Moving traffic - hazardous situation. Death or serious injury could result. Avoid moving vehicles, wear high visibility clothing, post appropriate warning signs.

WARNING

Hot pressurized cooling system fluid could cause serious burns. Allow to cool before servicing.

WARNING

Improper control function could cause death or serious injury. If control does not work as described in instructions, stop machine and have it serviced.

113F19.EPS

Figure 19 ◆ Warning symbols.

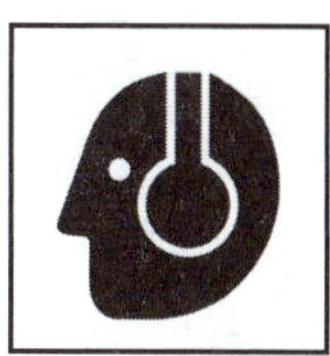

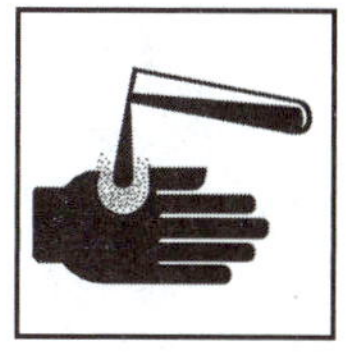

Figure 20 ◆ Caution symbols.

Summary

Safety on an HDD job site is everyone's responsibility. Safety should be considered in every part of the process including job-site planning, job-site inspection, site and equipment setup, boring operations, and cleanup. Some of the common hazards on an HDD site are slips, falls, pinched fingers and toes, vehicle accidents, back injuries, and getting struck by or caught in a rotating drill string. The best way to avoid these injures is to be alert and follow established safety procedures.

Utility strikes are another common hazard on HDD jobs. When utility lines are struck, workers are at risk of being electrocuted, suffocated by poisonous gas, killed by explosions, and contaminated by sewage. Good job-site planning and the use of sub-site locators can help to prevent these types of accidents.

It's also important to know how to use all equipment properly. Accidents are less likely to occur when everyone knows the hazards and safeguards associated with the equipment.

Review Questions

Refer to the following figure to answer Question 1.

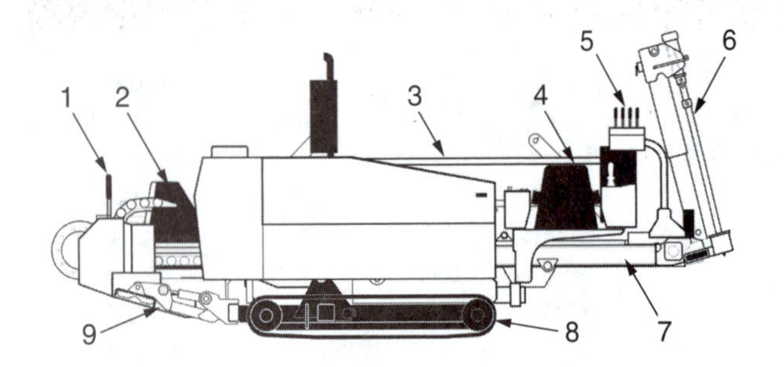

113E01.EPS

1. *Number 9* on this diagram is pointing to the _____.
   a. anchoring system
   b. carriage
   c. drill frame
   d. stabilizers

2. Entry pits for small HDD operations are typically _____ deep.
   a. ½' to 2'
   b. 1' to 3'
   c. 3' to 6'
   b. 7' to 9'

3. The largest pipe size that can be installed with an HDD rig is _____.
   a. 8"
   b. 16"
   c. 24"
   d. 48"

4. One-Call is an example of a service that may be contacted to help locate underground utility lines.
   a. True
   b. False

5. One way to make sure all underground utilities and obstructions have been located is to use a(n) _____.
   a. ground strike system
   b. grounding mat
   c. sub-site tracker
   d. sub-site locator

6. Reclaimed water lines are normally color coded _____.
   a. red
   b. yellow
   c. blue
   d. purple

7. All of the following hazards are related to unloading the boring machine *except* _____.
   a. being struck by moving equipment
   b. being pinched between the drill and another object
   c. getting electrocuted by power lines
   d. getting crushed by the drill rig

8. When positioning the boring machine, a grounding mat should be set up for both the machine and the operator.
   a. True
   b. False

9. Electric shock is considered a hazard during an HDD job.
   a. True
   b. False

10. If you are the operator of a rig when an underground power line is struck, you should _____.
    a. try backing the rig away to break contact
    b. immediately get off the machine
    c. quickly drive forward
    d. call the fire department

PROFILE IN SUCCESS

Frank McDaniel, Casey Industrial, Inc.

Corporate Safety & Training Manager

What accomplishment are you particularly proud of?
I developed and implemented a performance-management process. This is a process that effectively combines traditional and behavior-based safety in a practical format with the goal of sustained continuous improvement.

How did you choose a career in the Safety Industry?
I was at a crossroads in my career path. I had the option of continuing in project management or striking out on a new course. I liked the idea of helping to keep people safe and took a new direction.

What types of training have you been through?
Over the years, I have been involved in numerous training programs including: various behavior-based safety courses, Loss-Control Management courses, Safety management courses, Construction management courses, and Risk management courses.

What kinds of work have you done in your career?
I have directed the safety efforts on major convention center construction, theme park expansion, hospital, major airport, and high-rise condominium projects. I've also done in-home service, high-voltage, and heavy highway work, as well as railroad demolition & re-work, shopping mall construction, and oil refinery and co-generation plant work.

What factors have contributed most to your success?
There are multiple factors that I can point out that have contributed to my success in safety. First, my background as a journeyman carpenter and superintendent. I know how to talk to the work force to gain maximum buy-in to safety improvement processes. Second, my mentor Robert Debner has been a great influence. Bob introduced me to the behavioral processes that are now the cornerstone in my continuous improvement efforts. Third, the interaction with my peers. Leaders in the industry like Gus Leysens and Sherwood Kelly of Willis, Mike Powers of Tri-City Electrical, and Dave Langton have been instrumental as sounding boards. Groups like the NCCER Safety Council are a great resource for sharing ideas.

What advice would you give to those new to Safety industry?
Place value in your workforce. By soliciting their improvement ideas you involve them in the process, gain their buy-in, and increase your odds of success. Seek out those who have achieved success in safety. Learn as much as you can from their successes as well as their failures. Maintain your integrity; Safety is a small community. Sacrifice your integrity and you will limit your professional growth. Do not fear trying something new. Remember, safety improvement is an adventure. You will find it a wild, exciting, and rewarding ride.

GLOSSARY

Trade Terms Introduced in This Module

Boring: The process of drilling a hole into the ground.

Drill string: The length of pipe that connects the boring machine to the drill head.

Grounding mat: A mesh mat that is placed underneath the drill rig to provide a single electrical path to ground should the equipment come in contact with an energized conductor during a utility strike.

Operator presence sensing system (seat switch): A device that is used to stop the drill rig when the operator leaves the seat with the drill turned on.

Reaming: The process of widening or enlarging a bore hole.

Utilities: Services, such as electricity or water, that are provided by a utility company.

Figure Credits

Ditch Witch®	113F01, 113F05, 113F16, 113F17, 113F18, 113F19, 113F20
Gerald Shannon	113F02, 113F12
Vermeer Manufacturing Co.	113F03 (top and middle), 113F04, 113F08
The Robbins Company	113F03 (bottom)
Brigid McKenna	113F07, 113F09, 113F10
Steven P. Pereira	113F11, 113SA01

NCCER CURRICULA — USER UPDATE

NCCER makes every effort to keep its textbooks up-to-date and free of technical errors. We appreciate your help in this process. If you find an error, a typographical mistake, or an inaccuracy in NCCER's curricula, please fill out this form (or a photocopy), or complete the online form at **www.nccer.org/olf**. Be sure to include the exact module ID number, page number, a detailed description, and your recommended correction. Your input will be brought to the attention of the Authoring Team. Thank you for your assistance.

Instructors – If you have an idea for improving this textbook, or have found that additional materials were necessary to teach this module effectively, please let us know so that we may present your suggestions to the Authoring Team.

NCCER Product Development and Revision

13614 Progress Blvd., Alachua, FL 32615

Email: curriculum@nccer.org
Online: www.nccer.org/olf

❑ Trainee Guide ❑ AIG ❑ Exam ❑ PowerPoints Other ______________________________

Craft / Level: Copyright Date:

Module ID Number / Title:

Section Number(s):

Description:

Recommended Correction:

Your Name:

Address:

Email: Phone:

Module 75114-03

Heavy-Equipment, Crane, and Rigging Safety

COURSE MAP

This course map shows all of the modules in Field Safety. The suggested training order begins at the bottom and proceeds up. The local Training Program Sponsor may adjust the training order.

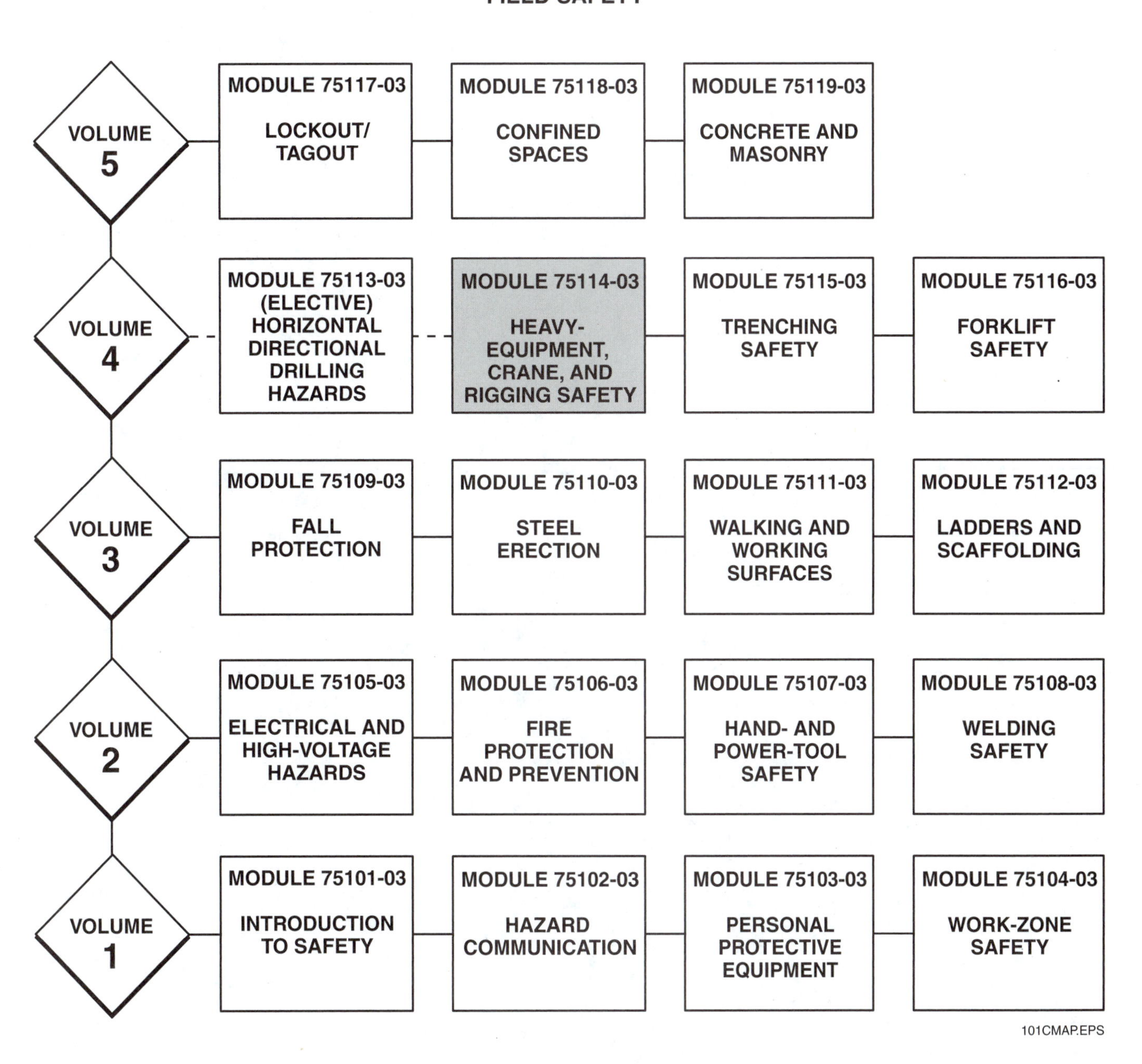

MODULE 75114-03 CONTENTS

Figures

MODULE 75114-03

Heavy-Equipment, Crane, and Rigging Safety

Objectives

When you have completed this module, you will be able to do the following:

1. Describe the types and uses of heavy equipment.
2. Identify the hazards associated with the operation of heavy equipment, including cranes and rigging.
3. Describe the safeguards and safety procedures used when working with heavy equipment.

Prerequisites

Before you begin this module, it is recommended that you successfully complete the following: Field Safety, Modules 75101-03 through 75112-03. *Horizontal Directional Drilling Hazards* (module (75113-03) is an elective and is not required for successful completion of this course.

Required Materials

1. Pencil and paper
2. Appropriate personal protective equipment

1.0.0 ◆ INTRODUCTION

Heavy equipment is used in many different jobs including construction, mining, road maintenance, equipment transportation, and snow removal. Working with heavy equipment can be extremely dangerous. Workers can be struck or crushed by falling loads, fall from equipment, be electrocuted by power lines, or get run over or trapped by vehicles. The swing radius of heavy equipment can also be a hazard if the job is not carefully planned and properly barricaded. *Figure 1* shows an example of what can happen when this is not done.

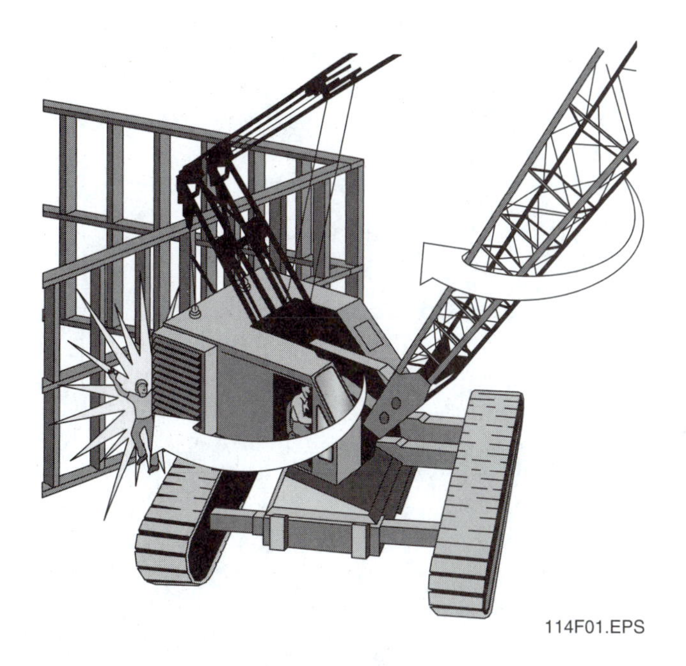

Figure 1 ◆ Worker caught in the swing radius of a crane.

Dangers exist for both **equipment operators** and other workers on the site. In one example, a contractor was operating a backhoe when another employee attempted to walk between the swinging back end of the backhoe and a concrete wall. As the employee approached the backhoe from the operator's blind side, the back end hit the victim, crushing him against the wall.

Because working with or near heavy equipment is dangerous, you must understand that your first responsibility on a job is safety. This includes your own safety, the safety of others on the site, and the safe use of equipment on the site. You must know the hazards and safety procedures of every job you are on, regardless of the work you are doing.

Consider all of the Dangers

Two employees were attempting to adjust the brakes on a backhoe. The victim told the backhoe operator to raise the wheels off the ground with the front bucket and the outriggers so that he could get to the brakes. The victim then crawled under the machine and began to adjust the brakes. He did this without considering that there was only a 36" space from the ground to the drive shaft. While adjusting the brakes, the hood of his rain jacket wrapped around the drive shaft and broke his neck. He died instantly.

The Bottom Line: Loose clothing can be caught in moving parts of machinery. You must consider all possible dangers when working on equipment.

Source: The Occupational Safety and Health Administration (OSHA)

2.0.0 ◆ HEAVY EQUIPMENT

Those who work around heavy equipment need a basic knowledge of all types of heavy equipment, as well as the hazards and safeguards of this equipment. The types of jobs performed by heavy equipment operators vary. The following are just a few of the jobs that use heavy equipment:

- Civil, residential, and industrial construction work
- Mining
- Snow and ice control operations
- Maintaining road surfaces, such as patching potholes and cracks (surface failures), raising depressed concrete, resurfacing asphalt, painting stripes, sweeping highways, inspecting, and repairing
- Loading, lashing, and unloading equipment and materials
- Minor **equipment maintenance** and repair
- Transporting equipment over public highways

Construction workers may be expected to operate many different types of heavy equipment (*Figure 2*). This equipment can vary greatly in size and weight. Some of the equipment used in heavy construction includes:

- Trucks
- Cranes
- Compacting equipment
- Backhoes
- Backhoe loaders
- Scrapers
- Bulldozers
- Excavators
- Motor graders
- Skid steer loaders
- Forklifts
- Concrete paving equipment
- Concrete plant equipment
- Asphalt plant equipment
- Pug mill mixers
- Fine grade trimmers
- Cold milling machines
- Stabilizers
- Trenchers and rock saws
- Breaking equipment

It's important to know how to use this equipment safely and to know the hazards and safeguards on your job site.

Always Wear Your Seat Belt

An employee was driving a front-end loader up a dirt ramp onto a lowboy trailer. The tractor tread began to slide off the trailer. As the tractor began to tip, the operator, who was not wearing a seat belt, jumped from the cab. As he hit the ground, the tractor's rollover cage fell on top of him, crushing him.

The Bottom Line: Always wear your seat belt when operating heavy equipment.

Source: The Occupational Safety and Health Administration (OSHA)

(A) BULLDOZER

(B) ARTICULATED TRUCK

(C) EXCAVATOR

(D) BACKHOE LOADER

(E) SKID STEER LOADER

(F) MOTOR GRADER

114F02.EPS

Figure 2 ◆ Types of heavy equipment.

2.1.0 Job-Site Safety

Safety on the job is everyone's responsibility. The equipment operator has control over equipment application, operation, inspection, lubrication, and maintenance. Therefore, it is primarily the operator's responsibility to use good safety practices in these areas. It's important to understand, however, that even though the operator has much of this responsibility, everyone working on the site is responsible for doing their work safely. They are also responsible for being aware of their co-workers and their actions.

Follow these general job-safety rules to make sure everyone on the site is safe.

- Be alert. Watch out for moving equipment. Assume the operator cannot see you.
- Use retaining guards and safety devices on all equipment.
- Report all defective tools, machines, or other equipment to your supervisor and report all accidents and near-accidents.
- Disconnect power and lock and tagout machines before performing maintenance.
- Use properly fitting wrenches on nuts and bolts.
- Keep your work area clear of scraps and litter.
- Dispose of combustible materials properly.
- Clean up any spilled liquids immediately.
- Store oily rags in self-closing metal containers.
- Never use compressed air to clean yourself or your clothing.
- Make sure all others on the site are at a safe distance from the equipment.
- Use traffic-control devices where required (*Figure 3*).
- Follow manufacturer's safety rules and limitations for all equipment.

114F03.EPS

Figure 3 ◆ Traffic-control devices.

2.2.0 Equipment Safety

The identification of safety hazards begins with a daily machine-safety inspection. Before daily operations begin, the operator should inspect around and under the heavy equipment for visible evidence of potential safety hazards or operational problems. These unsafe conditions must either be corrected immediately or reported, and heavy equipment operations should not proceed until the conditions are safe.

The following is a list of basic items that should be checked daily:

- Engine performance and gauges
- Oil pressure and levels
- Housekeeping tasks
- Fire extinguishers
- Audible warning devices
- **Hydraulic** lines and fluid

Know the Dangers of Your Work Site

Two employees were spreading concrete as it was being delivered by a concrete pumper truck boom. The truck was parked across the street from the work site. Overhead power lines ran above the boom on the pumper truck.

One employee was moving the hose to pour the concrete when the boom of the pumper truck came in contact with the overhead power lines. These lines carry 7,200 volts. The employee was electrocuted and died immediately. He then fell on the employee who was assisting him. The second employee received a massive electrical shock and burns.

No one on the site had received the proper safety training. Otherwise, they would have known how dangerous it is to work under power lines.

The Bottom Line: Always check the job site for hazards before you begin work. Overhead and underground power lines can be deadly.

Source: The Occupational Safety and Health Administration (OSHA)

- Windshield wipers, mirrors, and lights
- Batteries and charging system
- Visible hand signal and load charts
- Guards and clutches
- Outriggers, tracks, wheels, and tires
- Brakes, locks, and safety devices
- Drive chains, steering, and all rollers
- Sheaves, drums, and all cables
- Blocks and hooks
- Boom, gantry, jib, and extension
- Carrier assembly
- All controls, including hoisting, swing, travel, and boom

2.3.0 Personal Safety

Personal safety is as important as site safety. Your actions affect everyone on the site. Follow these guidelines to prevent accidents and injury:

- Wear close-fitting clothing that is appropriate for the activity being performed.
- Tie back long hair before operating equipment.
- Wear safety glasses, a suitable hard hat, goggles, hearing protection, and respiratory equipment where required.
- Fasten loose sleeves when working around machine tools or rotating equipment.
- Remove rings and other jewelry when working.
- Be alert.
- Do not jump on or off of moving equipment.
- Know the location of first-aid equipment, fire extinguishers, and emergency telephone numbers.

2.4.0 Weather Hazards

Heavy equipment operators usually work outdoors. Under certain environmental conditions, such as extreme hot or cold weather, work can become uncomfortable and possibly dangerous. There are specific things to be aware of when working under these adverse conditions.

2.4.1 Cold Weather

The amount of injury caused by exposure to abnormally cold temperatures depends on wind speed, length of exposure, temperature, and humidity. Freezing is increased by wind and humidity or a combination of the two factors. Follow these guidelines to prevent injuries such as **frostbite** during extremely cold weather:

- Always wear the proper clothing.
- Limit your exposure as much as possible.
- Take frequent, short rest periods.
- Keep moving. Exercise fingers and toes if necessary, but do not overexert.
- Do not drink alcohol before exposure to cold. Alcohol can dull your sensitivity to cold and make you less aware of over-exposure.
- Do not expose yourself to extremely cold weather if any part of your clothing or body is wet.
- Do not smoke before exposure to cold. Breathing can be difficult in extremely cold air. Smoking can worsen the effect.
- Learn how to recognize the symptoms of over-exposure and frostbite.
- Place cold hands under dry clothing against the body, such as in the armpits.

Use the Correct Procedure

An operating engineer operating a 65-ton rubber tire hydraulic crane had completed a pick and was demobilizing the crane from the pick area. He backed the crane from a concrete slab pick-up area using an exit ramp. The boom of the crane was retracted and a lifting beam and load block secured. He intended to lower the lifting beam to a storage area on the ground located to his right rear. Upon exiting the slab, the operator noticed that the crew bus was parked parallel to the slab to his right and a control point was barricaded directly behind the operator's right rear. He stopped the crane, boomed out approximately 55', and swung the boom over the crew bus to his right so the load (lifting beam) would go around the back of the bus and not over it. Once the boom was perpendicular to the bus, the crane tipped over because the operator had failed to extend the outriggers. Fortunately, the crane did not strike the bus and the worker was unhurt. The driver was at fault for the accident because he failed to use the correct procedures.

The Bottom Line: Use outriggers and other safety controls to prevent costly accidents.

Source: U.S. Department of Energy

Cold Weather Clothing Tips

Use the following tips to prevent injury due to cold weather:

- Dress in layers.
- Wear thermal-type woolen underwear.
- Wear outer clothing that will repel wind and moisture.
- Wear a face helmet and head and ear coverings.
- Carry an extra pair of dry socks when working in snowy or wet conditions.
- Wear warm boots and make sure that they are not so tight that circulation becomes restricted.
- Wear wool-lined mittens or gloves covered with wind- and water-repellent material.

2.4.2 Cold Exposure Symptoms and Treatment

If you live in a place with cold weather, you will most likely be exposed to it when working. Spending long periods of time in the cold can be dangerous. It's important to know the symptoms of cold weather exposure and how to treat them. Symptoms of cold exposure include:

- Shivering
- Numbness
- Low body temperature
- Drowsiness
- Weak muscles

Follow these steps to treat cold exposure:

Step 1 Get to a warm inside area as quickly as possible.

Step 2 Remove wet or frozen clothing and anything that is binding such as necklaces, watches, rings, and belts.

Step 3 Rewarm by adding clothing or wrapping in a blanket.

Step 4 Drink hot liquids, but do not drink alcohol.

Step 5 Check for frostbite. If frostbite is found, seek medical help immediately.

2.4.3 Symptoms and Treatment of Frostbite

Frostbite is an injury resulting from exposure to cold elements. It happens when crystals form in the fluids and underlying soft tissues of the skin. The frozen area is generally small. The nose, cheeks, ears, fingers, and toes are usually affected. Affected skin may be slightly flushed just before frostbite sets in. Symptoms of frostbite include:

- Skin that becomes white, gray, or waxy yellow. Color indicates deep tissue damage. Victims are often not aware of frostbite until someone else recognizes the pale, glossy skin.
- Skin tingles and then becomes numb.
- Pain in the affected area starts and stops.
- Blisters show up on the area.
- The area of frostbite swells and feels hard.

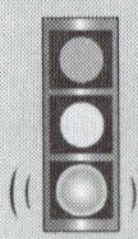

NOTE

In advanced cases of frostbite, mental confusion and poor judgment occur, the victim staggers, eyesight fails, the victim falls, and may pass out. Shock is evident, and breathing may cease. Death, if it occurs, is usually due to heart failure.

Use the following steps to treat frostbite.

Step 1 Protect the frozen area from refreezing.

Step 2 Warm the frostbitten part as soon as possible.

Step 3 Get medical attention immediately.

2.4.4 Hot Weather

Hot weather can be as dangerous as cold weather. When someone is exposed to excessive amounts of heat, they run the risk of overheating. Some conditions associated with overheating include:

- Heat exhaustion
- Heat cramps
- Heat stroke

Heat exhaustion is characterized by pale, clammy skin; heavy sweating with nausea and possible vomiting; a fast, weak pulse; and possible fainting.

Heat cramps can occur after an attack of heat exhaustion. Cramps are characterized by abdominal pain, nausea, and dizziness. The skin becomes pale with heavy sweating, muscular twitching, and severe muscle cramps.

Heat stroke is an immediate, life-threatening emergency that requires urgent medical attention. It is characterized by headache, nausea, and visual problems. Body temperature can reach as

Be Aware of Those Around You

A miner was adding water to the batteries on a scissor lift being used for drilling rock bolts. He was on both knees, leaning inside the battery compartment when a forklift backed into and struck the compartment door. The forklift moved backwards approximately 3', closing the door and causing the miner's left and right upper arms and outer back areas to be squeezed inside the compartment. When the miner yelled out, the forklift operator stopped backing up. He then moved forward so the miner could stand. The miner was severely bruised on his left and right upper arms and upper back. The driver of the forklift had not looked around the work site to see if anyone else was there. Fortunately, the miner recovered from his injuries.

The Bottom Line: Always check your work site before operating equipment. Know where your co-workers are at all times.

Source: U.S. Department of Energy

high as 106°F. This will be accompanied by hot, flushed, dry skin; slow, deep breathing; possible convulsions; and loss of consciousness.

Follow these guidelines when working in hot weather in order to prevent heat exhaustion, cramps, or heat stroke.

- Drink plenty of water.
- Do not overexert yourself.
- Wear lightweight clothing.
- Keep your head covered and face shaded.
- Take frequent, short work breaks.
- Rest in the shade whenever possible.

2.5.0 Driving Equipment Safely

All heavy equipment is moved during the course of a job. It is moved to the site, on the site, and away from the site. Many injuries and deaths happen during the movement of heavy equipment. It is important to be especially safety conscious whenever equipment is moving. Follow these guidelines when transporting equipment on public roads or on the job site.

Always observe these guidelines when driving equipment on public roads:

- Drive slowly and never speed.
- Know the distance it takes to stop safely.
- Allow extra time to enter traffic.
- Stay in the extreme right lane or on the shoulder.
- Travel with your lights on.
- Use proper warning signs and flags: twelve-inch square red flags for day travel, or warning signals visible for 500' for night travel.
- Secure all attachments and loose gear.
- Turn cautiously; allow for extensions or attachments and for structural clearances. Some equipment is top-heavy and will tip over if a turn is made too fast.
- Know the turning radius of your equipment.
- Know and obey all state and local laws.

Double Check Your Equipment

A 49-year old bulldozer operator was crushed to death while compacting earthen fill during the construction of an oil exploration island. The victim was operating the bulldozer on level ground, alternately in forward and reverse at half throttle. With the blade completely down, the victim shifted the transmission control lever toward neutral. The transmission was partially disengaged, but not fully in neutral.

He assumed, without checking, that the bulldozer transmission was in a stable neutral position. He exited the cab on the right side. When he did this, the transmission slipped back into the first gear of reverse, causing the bulldozer to suddenly move. As a result, the victim was pulled between the underside of the fender and the top of the track cleats. As the bulldozer continued in reverse, the victim was fatally crushed beneath the track cleats.

The Bottom Line: Always use parking brakes and double check settings before exiting the equipment. Never leave equipment when it is still running.

Source: The National Institute for Occupational Safety and Health (NIOSH)

Always follow these guidelines for driving equipment on the job site:

- Never drive a machine on a job site, in a congested area, or around people without a spotter or flagger to guide you. The spotter or flagger is responsible for determining and controlling the driver's speed.
- Be sure everyone is in the clear while backing up, hooking up, or moving attachments.
- Never move pipe or similar items over the heads of other workers.
- If you cannot see your area clearly from the operator's seat and have no spotter, dismount and examine the site for possible hazards.
- Wait for an all-clear signal before moving.
- Signal a forward move with two blasts of the horn.
- Signal a reverse move with three blasts of the horn.
- Yield the right-of-way to loaded equipment on haul roads and in pits.
- Maintain a safe distance from all other vehicles.
- When moving, keep the equipment in gear at all times; never coast.
- Maintain a ground speed consistent with ground conditions.
- Pass only when necessary; use caution.
- Watch for overhead electrical power lines.
- Watch for flags indicating buried pipes and power and gas lines. Telephone and utility companies flag the work site when needed with colored flags; gas and chemicals, yellow; phone, orange; electric, red; water, blue; and sanitary and storm sewers, green.

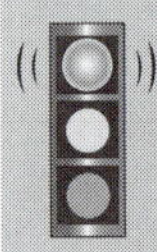

WARNING!

Equipment does not have to come in direct contact with wires to act as a conductor. If operating near power lines, power to these lines should be shut off. Do not operate within 10' of power lines.

2.6.0 Communications and Hand Signals

Operation of heavy equipment requires concentration by the operator. Any distractions may have disastrous results. At times, the operator will not have a clear view of the site or other equipment. This means that a signal person will have to act as the operator's eyes. To accomplish this, there must be a standard form of communication between the operator and the signal person.

Communicate With Co-Workers

A radiological control technician was providing support for maintenance employees performing work in a facility. While other workers prepared the work site, the manlift operator began traveling toward the site. The control technician was walking along next to the manlift. The route used to approach the job site involved navigating through a narrow contamination control zone (CCZ) corridor. This required the operator's full attention. As they approached the job site, the technician moved ahead of the manlift and out of the direct view of the operator. The technician knelt down to pick up some surveying equipment that was in the path. He lost eye contact with the operator. Therefore, the operator did not realize that the technician was in the pathway of the manlift. As a result, the front wheel of the manlift ran over the technician's left foot. When the operator heard the technician screaming, he immediately stopped and backed up. Luckily, the technician did not have any broken bones. He did, however, miss 14 days of work. Using hand signals and maintaining eye contact could have spared this technician his injured foot.

The Bottom Line: Equipment operators have limited visibility. Pedestrians must be aware of and stay away from moving equipment.

Source: U.S. Department of Energy

There are many ways for the operator and the signal person to communicate. One is a radio. This involves the signal person using a radio to tell the operator how to control the load. Disadvantages of this type of communication are interference from other radio or electronic equipment and the inability to hear the radio over background machinery noise.

Hardwired communication circuits are another way to set up communication. This involves a dedicated circuit. With a dedicated circuit, the interference from other radios is removed. The disadvantage of this type of communication is not being able to hear because of loud noise on the site.

To get rid of the problem of noise in voice and radio communications, a standard set of hand signals can be used (*Figure 4*). They are not affected by noise or interference. The only disadvantage is that when using hand signals, the signal person and the operator must be in sight of each other. This can be difficult if there is a lot of heavy equipment and workers on the site. Still, it is generally the safest and most effective way to communicate.

114F04.EPS

Figure 4 ◆ Hand signals.

3.0.0 ◆ CRANE SAFETY

Heavy construction workers routinely use a variety of lifting devices. Cranes vary in size and carry different types of loads (*Figure 5*). Cranes are among the most commonly used pieces of heavy equipment and can also be the most dangerous. Therefore, it is important to understand and follow all safety requirements when working around cranes.

3.1.0 Site Hazards and Restrictions

It takes a combined effort by everyone involved in crane operations to make sure no one is injured or killed. There are many site hazards and restrictions related to crane operations. These hazards include:

- Underground utilities such as gas, oil, electrical, and telephone lines; sewage and drainage piping; and underground tanks
- Electrical lines or high-frequency transmitters
- Structures such as buildings, excavations, bridges, and abutments

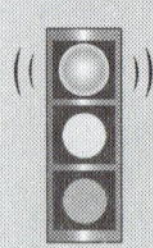

WARNING!

Power lines and environmental issues such as weather are common causes of injury and death during crane operations.

The operator should inspect the work area and identify hazards or restrictions that may affect the safe operation of the crane. This includes:

- Ensuring the ground can support the crane and the load
- Checking that there is a safe path to move the crane around on site
- Making sure that the crane can rotate in the required quadrants for the planned lift

The operator is required to follow the manufacturer's recommendations and any locally established restrictions placed on crane operations, such as time restrictions for noise abatement or traffic considerations.

3.2.0 Manufacturer's Requirements and Restrictions

To operate a mobile crane safely, the operator must use the manufacturer's data and documentation provided with the crane. These manuals provide information on required startup checks and periodic inspections, as well as inspection

114F05.EPS

Figure 5 ◆ Typical cranes.

Know Equipment Limits

A journeyman glazier died after attempting to lift a 1,000-pound case of glass with the 60'-long articulated boom of a manlift. As the victim attempted to raise the case of glass, which weighed substantially above the 600-pound lift capacity of the manlift, the off-side wheels on the base of the manlift lifted 4½' off the ground. When the victim realized he was in danger of turning the manlift over, he immediately reversed the controls to lower the boom. When the controls were reversed, the boom dropped approximately 2'. This caused slack in the sling, which allowed the sling to slip free from the manlift. The case of glass dropped to the ground.

Because the boom was no longer supporting the 1,000-pound case of glass, the manlift hurled skyward and the victim was catapulted from the bucket. He fell more than 30' from the bucket to his death.

The Bottom Line: Do not exceed equipment capacity. Equipment failure can be deadly.

Source: The National Institute for Occupational Safety and Health (NIOSH)

guidelines. Also, these manuals provide information on configurations and capacities, in addition to many safety precautions and restrictions of use. Violation of any of these requirements or precautions is hazardous to the safe operation of the crane and could make the operator liable if an accident should occur. Operators should always follow the manufacturer's instructions.

3.3.0 Emergency Response

Operators must react quickly and correctly to any crane malfunction or emergency situation that might arise. Proper responses to emergency situations must be learned. The operator's first priority is to prevent injury and loss of life. The operator's second priority is to prevent damage to equipment or surrounding structures.

3.3.1 Fire

The operator's judgment is crucial in determining the correct response to fire. The first response is to cease crane operation and, if time permits, lower the load and secure the crane. In all cases of fire, evacuate the area even if the load cannot be lowered or the crane secured. After emergency services have been notified, a qualified individual may judge if the fire can be combated with a fire extinguisher. A fire extinguisher can be successful at fighting a small fire in its beginning stage, but a fire can get out of control very quickly. The operator must keep in mind that priority number one is preventing loss of life or injury to anyone. Do not be overconfident. Even trained firefighters using the best equipment can be overwhelmed and injured by fires.

3.3.2 Contact With Energized Electrical Power Sources

If a crane or load comes in contact with or becomes entangled in power lines, the operator must assume that the power lines are energized. Any other assumption could be fatal. The following guidelines should be followed if the crane comes in contact with an electrical power source.

- Stay in the cab of the crane.
- Do not allow anyone to touch the crane or the load.
- Reverse the movement of the crane to break contact with the energized power line.
- If staying in the cab is not possible, the operator should jump clear of the crane, landing with both feet together on the ground.
- Call the local power authority or owner of the power line.
- Have the lines verified secure and properly grounded before allowing anyone to approach the crane or load.

3.3.3 Malfunctions During Lifting Operations

Mechanical malfunctions during a lift can be very serious. If a failure causes the radius to increase unexpectedly, the crane can tip or the structure could collapse. Loads can also be dropped during a mechanical malfunction. A sudden loss of load on the crane can cause a whiplash effect that can tip the crane or cause the boom to fail.

The chance of these types of failures occurring in modern cranes is greatly reduced because of system redundancies and safety backups. However, failures do happen, so stay alert at all times.

Call Power Companies Before Beginning Work

A three-man crew consisting of a foreman and two boring machine operators was working at an excavation pit that was 7' deep, 5' wide, and 34' long. The pit had been dug to allow a boring machine to drill under an existing roadway. A 12,000V power line, suspended 20' above grade, was directly adjacent and parallel to the pit. The employer had made no attempt to contact the power company to have the power line de-energized.

The foreman was operating the truck-mounted crane that was to feed pipe to the boring machine. He parked it perpendicular to the pit, facing down a slight incline. He extended the boom to about half its length (30') to pick up a 10'-long, 16"-diameter steel pipe that weighed approximately 1,000 pounds. The crane lifted the pipe with a sling. The two boring machine operators standing in the pit attempted to guide the pipe into position in the pit. The pipe became lodged due to misalignment in the pit. The victim attempted to free it by pulling the pipe toward him with one hand. When he did this, the boom moved slightly, causing it to make momentary contact with the overhead power line. The boom, cable, and pipe became energized. The victim was electrocuted when he formed a path to ground as current passed from the pipe through his hand, arm, chest, and legs. The victim was pronounced dead on arrival at the hospital emergency room.

The Bottom Line: Overhead and underground power lines can be deadly. Contact the power company prior to working near these lines.

Source: The National Institute for Occupational Safety and Health (NIOSH)

If a mechanical problem occurs, the operator should lower the load immediately. Next, the operator should secure the crane, tag the controls out of service, and report the problem to the supervisor. The crane should not be operated until it is repaired by a qualified technician.

3.3.4 Hazardous Weather

High winds and lightning may cause severe problems on the job site (*Figure 6*). The crane operator needs to be prepared for these types of situations to avoid accidents. It is rare that high winds and lightning arrive without some warning. So, when a warning is given, it is important for the operator to secure crane operations as soon as practical. To do this, the operator should place the boom in the lowest position possible and secure the crane. Once the crane is secured, leave it immediately and seek shelter in a building.

Figure 6 ◆ Impact of wind on cranes.

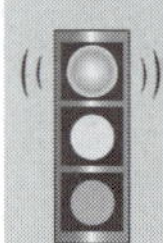

WARNING!

Even with the boom in the lowest position, it may be taller than surrounding structures and could still be a target for lightning strikes.

4.0.0 ◆ RIGGING SAFETY

Rigging operations involve the planned movement of material and equipment from one location to another. Rigging operations are complex procedures that are potentially dangerous, even for fully trained and qualified rigging workers. Ropes, chains, hoists, slings, and other types of equipment are used for rigging (*Figure 7*). Some rigging operations, like tilt-up panel construction, use cranes. Others may use a loader to move materials around on a job site.

All slings, hardware, and rigging devices must be clearly marked with their rated capacity. Paper tags cannot be used. Rated capacity means the same thing as the working load limit or a safe working load. Determining the rated capacity of slings and rigging hardware is necessary to make sure that loads are safely and effectively lifted and transported.

When using slings, follow these safety guidelines:

- Never try to shorten a sling by wrapping it around the hoist hook before attaching the eye to the hook.
- Never try to shorten the legs of a sling by knotting, twisting, or wrapping the slings around one another.
- Never try to shorten a chain sling by bolting or wiring the links together.
- Make sure all personnel are clear of the load before you take full load strain on the slings.
- Never try to adjust the slings while a strain is being taken on the load.
- Make sure all personnel keep their hands away from the slings and the load during hoisting.
- Use sling softeners whenever possible.

Sling softeners protect from abrasion, cuts, heat, and chemical damage. Whether the softeners or pads are manufactured or made in the field, using them regularly will considerably extend the life of the sling. If no standard sleeves are available, use rubber belting or sections of old slings. Softeners prevent kinking in wire rope. Corner buffers are specially made for this purpose, but if they are not available, the rigger can use whatever is handy to protect the rope. Wood, old web slings, and factory-made buffers will provide some level of protection. Softeners for chain slings protect the links contacting the corners of the load and keep the chain links from scraping or crushing the load itself.

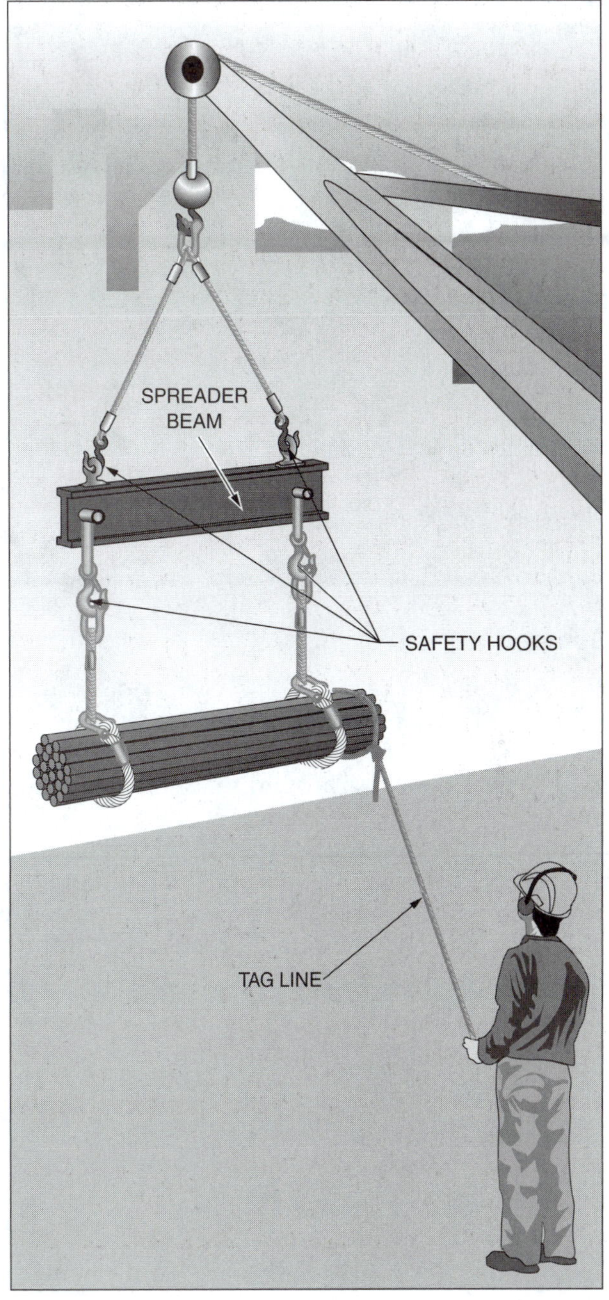

Figure 7 ◆ Example of safe rigging.

Operate Within Safe Lifting Capacity

A 50-year old carpenter at a municipal construction site died after he was struck by a loaded concrete bucket during a crane tip-over. The victim was removing forms from a newly constructed concrete wall while a concrete finishing crew was filling empty forms about 15' to 20' away. Concrete was being hoisted from street level with a crawler-mounted mobile crane and landed under the direction of a rooftop spotter. As the crane operator hoisted a bucket load of concrete, swung it over the roof, and boomed out toward the empty forms, the crane lost stability, tipping toward the victim. The concrete bucket was filled beyond capacity. When the crane operator realized what was occurring, he radioed a warning to the spotter who relayed the warning to rooftop workers. The victim had started to move when the uncontrolled concrete bucket swung toward him, striking his head and shoulder. The victim was pronounced dead at the scene.

The Bottom Line: Do not exceed the equipment's operating capacity. Stay aware of the movement of heavy equipment.

Source: The National Institute for Occupational Safety and Health (NIOSH)

Summary

Heavy equipment is used in a variety of different construction jobs. The size and use of heavy equipment will vary with each job, but the dangers won't. Working around or on heavy equipment can be extremely dangerous. Serious injury or death can occur if the equipment is used improperly or unsafely, or is poorly maintained. It is your responsibility to be aware of the hazards on the job site and the safety procedures required to keep you and your co-workers safe.

Review Questions

1. Heavy equipment is only used in highway construction work.

a. True
b. False

2. Forklifts are considered heavy equipment.

a. True
b. False

3. It is safe to smoke before being exposed to cold weather.

a. True
b. False

4. Which of the following is *not* a sign of cold exposure?

a. Vomiting
b. Weak muscles
c. Drowsiness
d. Numbness

5. When removing wet or frozen clothing, you should also remove necklaces, watches, rings, and belts.

a. True
b. False

6. A condition in which crystals form in the fluids and underlying soft tissue of skin is called _____.

a. cold exposure
b. frostbite
c. hypothermia
d. hyperthermia

7. When someone has heat stoke, his/her body temperature can reach _____ .

a. 99°F
b. 101°F
c. 104°F
d. 106°F

8. Equipment must come in direct contact with power lines in order to cause electrocution.

a. True
b. False

9. If a crane or load contacts a power line, the operator should carefully climb out of the cab.

a. True
b. False

10. Paper tags can be used to mark the rated capacity of slings, hardware, and rigging devices.

a. True
b. False

GLOSSARY

Trade Terms Introduced in This Module

Equipment maintenance: The care, cleaning, inspection, and proper use of machinery and equipment.

Equipment operator: A person skilled in operating certain equipment.

Frostbite: Skin and tissue damage due to extreme cold or freezing.

Hydraulic: Tools and equipment that are powered or moved by liquid under pressure.

Figure Credits

Volvo Construction Equipment North America, Inc.	114F02 (B)
Courtesy of John Deere	114F02 (A), (C), (D), (E), (F)
Brigid McKenna	114F03
Grove	114F05

NCCER CURRICULA — USER UPDATE

NCCER makes every effort to keep its textbooks up-to-date and free of technical errors. We appreciate your help in this process. If you find an error, a typographical mistake, or an inaccuracy in NCCER's curricula, please fill out this form (or a photocopy), or complete the online form at **www.nccer.org/olf**. Be sure to include the exact module ID number, page number, a detailed description, and your recommended correction. Your input will be brought to the attention of the Authoring Team. Thank you for your assistance.

Instructors – If you have an idea for improving this textbook, or have found that additional materials were necessary to teach this module effectively, please let us know so that we may present your suggestions to the Authoring Team.

NCCER Product Development and Revision

13614 Progress Blvd., Alachua, FL 32615

Email: curriculum@nccer.org
Online: www.nccer.org/olf

❑ Trainee Guide ❑ AIG ❑ Exam ❑ PowerPoints Other ____________________

Craft / Level: Copyright Date:

Module ID Number / Title:

Section Number(s):

Description:

Recommended Correction:

Your Name:

Address:

Email: Phone:

Module 75115-03

Trenching Safety

COURSE MAP

This course map shows all of the modules in Field Safety. The suggested training order begins at the bottom and proceeds up. The local Training Program Sponsor may adjust the training order.

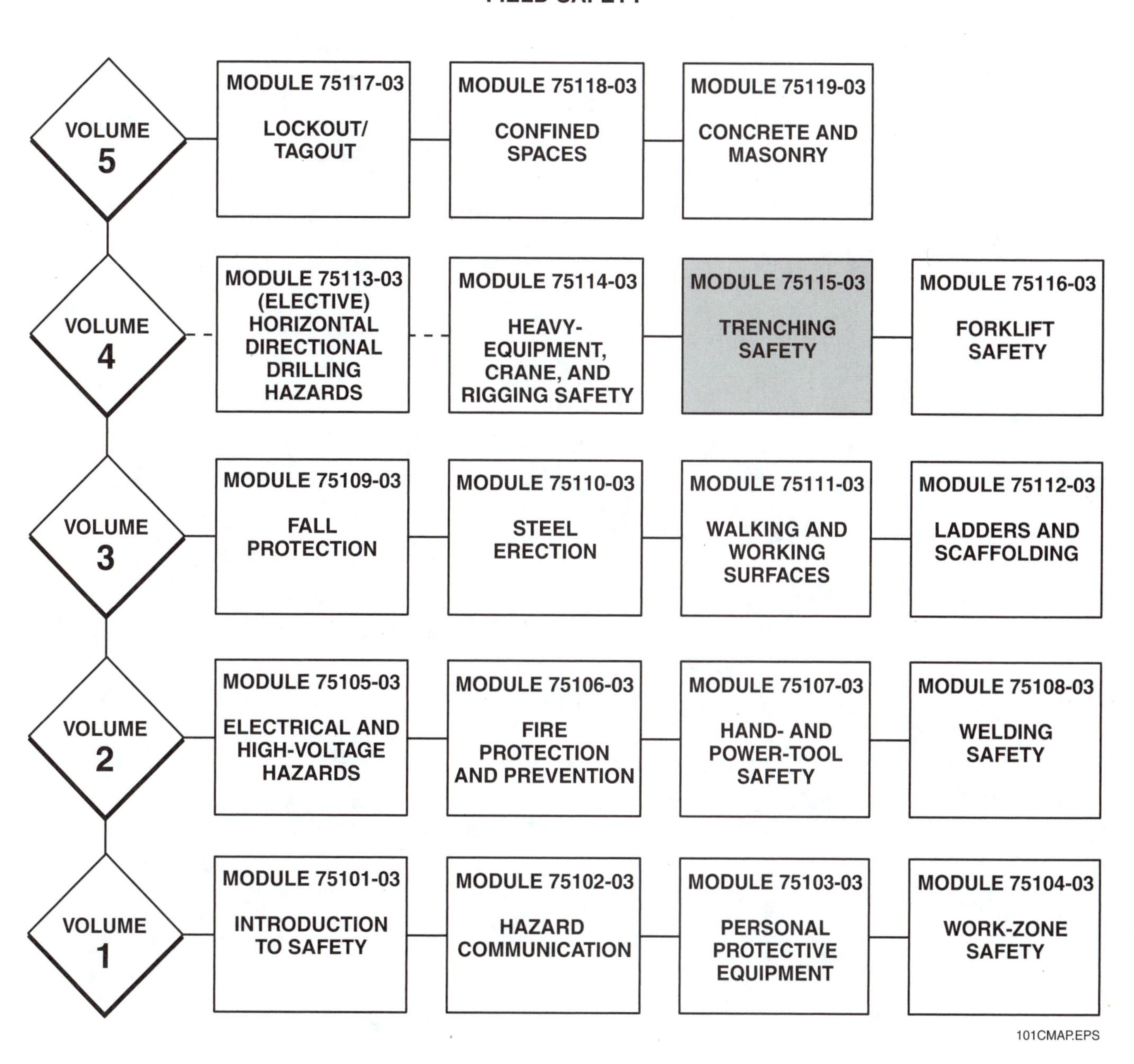

MODULE 75115-03 CONTENTS

Figure

Tables

MODULE 75115-03

Trenching Safety

Objectives

When you have completed this module, you will be able to do the following:

1. Describe the process and purpose of excavation.
2. Explain and identify safety hazards associated with excavation.
3. Demonstrate and explain proper on-site safety and emergency-response procedures.
4. Identify the indications and explain the causes of an unstable trench.
5. Explain the importance of recognizing soil types with regard to excavation.
6. Describe the procedures used in shoring, sloping, and shielding safety methods.

Prerequisites

Before you begin this module, it is recommended that you successfully complete the following: Field Safety, Modules 75101-03 through 75112-03 and Module 75114-03. *Horizontal Directional Drilling Hazards* (module 75113-03) is an elective and is not required for successful completion of this course.

Required Materials

1. Pencil and paper
2. Appropriate personal protective equipment

1.0.0 ◆ INTRODUCTION

Safety is crucial during any excavation job. Excavations are done for a number of reasons, including building cellars and during highway construction (*Figure 1*). During an excavation, earth is removed from the ground, creating a trench. A trench is a narrow excavation made below the surface of the ground in which the depth is greater than the width and the width does not exceed 15'. The soil that is removed from the ground is called spoil. When earth is removed from the ground, extreme pressures may be generated on the trench walls. If the walls are not properly secured by **shoring**, sloping, or **shielding**, they will collapse.

The collapse of unsupported trench walls can instantly crush and bury workers. This type of collapse happens because not enough material is available to support the walls of an excavation.

Worker Injured Attempting Rescue

Employees were laying sewer pipe in a 15'-deep trench. The sides of the trench, 4' wide at the bottom and 15' wide at the top, were not shored or protected to prevent a cave-in. Soil in the lower portion of the trench was mostly sand and gravel and the upper portion was clay and loam. The trench was not protected from vibration caused by heavy vehicle traffic on the road nearby. To leave the trench, employees had to exit by climbing over the backfill. As they attempted to leave the trench, there was a small cave-in covering one employee to his ankles. When the other employee went to his co-worker's aid, another cave-in occurred covering the first worker to his waist. The first employee died of a rupture of the right ventricle of his heart at the scene of the cave-in. The other employee suffered a hip injury.

The Bottom Line: Always make sure you have a safe route out of a trench.

Source: The Occupational Safety and Health Administration (OSHA)

www.SHORING.com

www.SHORING.com

www.SHORING.com

www.SHORING.com

www.SHORING.com

115F01.EPS

Figure 1 ◆ Excavation sites.

One cubic yard of earth weighs approximately 3,000 pounds; that's the weight of a small car. That's more than enough weight to seriously injure or kill a worker. In fact, each year in the United States, more than 100 people are killed and many more are seriously injured in cave-in accidents.

2.0.0 ◆ TRENCHING HAZARDS

Working in and around excavations is one of the most hazardous jobs you will ever do. Safety precautions must be exercised at all times to prevent injury to yourself and others.

Some of the hazards you may encounter during an excavation include:

- Cave-ins due to trench failure
- Falls from workers too close to the trench edge
- Flooding from broken water or sewer mains
- Electrical shock from striking electrical cable in the trench or striking overhead lines
- Toxic liquid or gas leaks from nearby facilities or pipes
- Auto traffic if the excavation site is near a highway
- Falling dirt or rocks from an excavator bucket

2.1.0 Soil Hazards

The type of soil in and around a trench is also a factor that contributes to the collapse of trench walls. *Table 1* shows a sample list of 84 recorded trench failures broken down by the type of soil in which they occurred. Soil type is a major factor to consider in trenching operations. Only a company-assigned competent person has enough experience on the job, training, and education to determine if the soil in and around a trench is safe and stable. However, it is still your responsibility to know the basics about soil and its associated hazards.

Table 1 Failures by Soil Type

Type of Soil	Number of Failures
Clay and/or mud	32
Sand	21
Wet dirt (probably silty clay)	10
Sand, gravel, and clay	8
Rock	7
Gravel	4
Sand and gravel	2

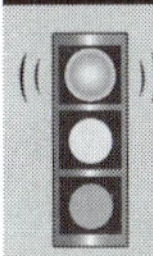

WARNING!
Never enter a trench unless you have approval from your company-assigned competent person.

2.1.1 *Properties of Soil*

Soil is comprised of soil particles, air, and water in varying quantities. Soil particles, or grains, consist of chunks, pieces, fragments, and tiny bits of rock that are released by the weathering of parent rocks. Weathering is a natural process of erosion that can be physical or chemical. Physical processes include freezing and thawing, gravity, and erosion by rivers and rainfall. Chemical processes include **oxidation**, **hydration**, and **carbonation** in which minerals are chemically broken down by the elements and removed via ground and rain water. Properties of a given soil include grain size, soil gradation, and grain shape.

2.1.2 *Types of Soils*

The soil that is found on most construction sites is a mixture of many mineral grains coming from several kinds of rocks. Average soils are usually a mixture of two or three materials, such as sand and silt or silt and clay. The type of mixture determines the soil characteristics. For example, sand with small amounts of silt and clay may compact well and provide a very good excavation soil. In addition to the mineral grains, soil contains water, air, gases, chemicals, and other organic material.

What's Wrong With This Picture?

115SA01.EPS

2.1.3 *Soil Behavior*

Each of the various soil types, depending on the condition of the soil at the time of the excavation, will behave differently (*Table 2*). Sandy soil tends to collapse straight down. Wet clays and loams tend to slab off the side of the trench. These two conditions are shown in *Figure 2*.

Firm, dry clays and loams tend to crack. Wet sand and gravel tend to slide. These conditions are shown in *Figure 3*. You should be aware of the type of soil you are working in and know how it behaves. When you are working in or near a trench, stay alert to changes in the trench. Watch for developing cracks, moisture, or small movements in the trench material. Alert your supervisor and co-workers to any changes you notice in the trench walls. Changes in trench walls may be an early indication of a more severe condition.

Table 2 Field Method for Identification of Soil Texture

Soil Texture	Visual Detection of Particle Size and General Appearance of Soil	Squeezed in Hand	Soil Ribboned Between Thumb and Finger
Sand	Soil has a granular appearance in which the individual grain sizes can be detected. It is free-flowing when in a dry condition.	*When air dry:* Will not form a cast and will fall apart when pressure is released. *When moist:* Forms a cast which will crumble when lightly touched.	Cannot be ribboned.
Sandy Loam	Essentially a granular soil with sufficient silt and clay to make it somewhat coherent. Sand characteristics dominate.	*When air dry:* Forms a cast which readily falls apart when lightly touched. *When moist:* Forms a cast which will bear careful handling without breaking.	Cannot be ribboned.
Loam	A uniform mixture of sand, silt, and clay. Grading of sand fraction quite uniform from coarse to fine. It is mellow, has a somewhat gritty feel, yet is fairly smooth and slightly plastic.	*When air dry:* Forms a cast which will bear careful handling without breaking. *When moist:* Forms a cast which can be handled freely without breaking.	Cannot be ribboned.
Silt Loam	Contains a moderate amount of the finer grades of sand and only a small amount of clay. Over half of the particles are silt. When dry it may appear quite cloddy, which can be readily broken and pulverized to a powder.	*When air dry:* Forms a cast which can be freely handled. Pulverized, it has a soft, flourlike feel. *When moist:* Forms a cast which can be freely handled. When wet, soil runs together and puddles.	It will not ribbon, but has a broken appearance, feels smooth, and may be slightly plastic.
Silt	Contains over 80% silt particles, with very little fine sand and and clay. When dry, it may be cloddy, and readily pulverizes to powder with a soft flourlike feel.	*When air dry:* Forms a cast which can be handled without breaking. *When moist:* Forms a cast which can be freely handled. When wet, it readily puddles.	It has a tendency to ribbon with a broken appearance, feels smooth.
Clay Loam	Fine textured soil breaks into very hard lumps when dry. Contains more clay than silt loam. Resembles clay in a dry condition; identification is made on the physical behavior of moist soil.	*When air dry:* Forms a cast which can be handled freely without breaking. *When moist:* Forms a cast which can be freely handled without breaking. It can be worked into a dense mass.	Forms a thin ribbon which readily breaks, barely sustaining its own weight.
Clay	Fine textured soil breaks into very hard lumps when dry. Difficult to pulverize into a soft flourlike powder when dry. Identification is based on cohesive properties of the moist soil.	*When air dry:* Forms a cast which can be freely handled without breaking. *When moist:* Forms a cast which can be freely handled without breaking.	Forms long, thin, flexible ribbons. Can be worked into a dense, compact mass. Considerable plasticity.
Organic Soils	Identification based on the high organic content. Muck consists of thoroughly decomposed organic material with considerable amount of mineral soil finely divided with some fibrous remains. When considerable fibrous material is present, it may be classified as peat. The plant remains or sometimes the woody structure can easily be recognized. Soil color ranges from brown to black. They occur in lowlands, swamps, or swales. They have high shrinkage upon drying.		

No Way Out

Two employees were laying pipe in a 12'-deep trench when one of the employees saw the bottom face of the trench move. He jumped out of the way along the length of the trench. The other employee was fatally injured as the wall caved in.

The Bottom Line: Always ensure that the walls of a trench are sloped and an emergency exit is provided.

Source: The Occupational Safety and Health Administration (OSHA)

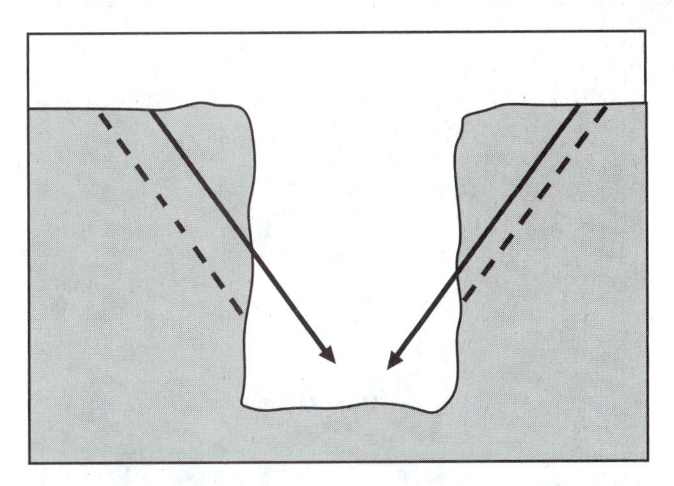

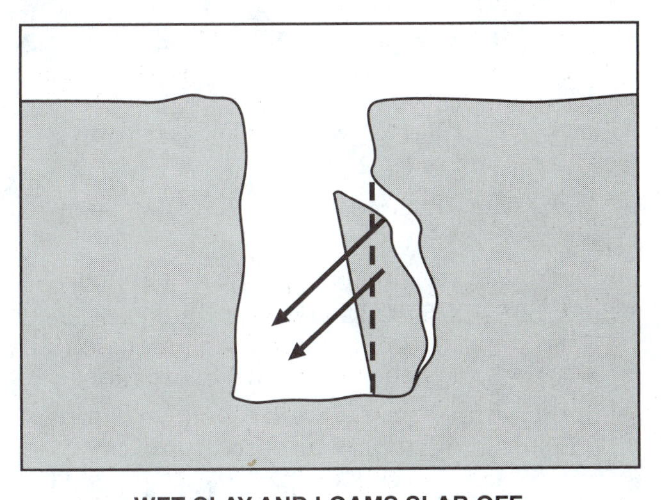

Figure 2 ◆ Behavior of sandy soil and wet clay and loams.

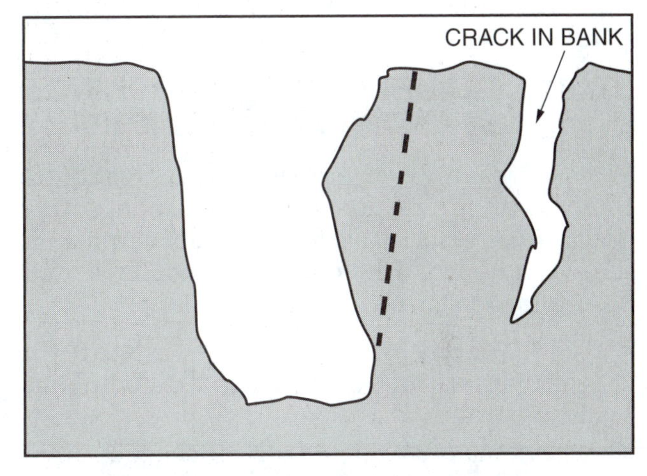

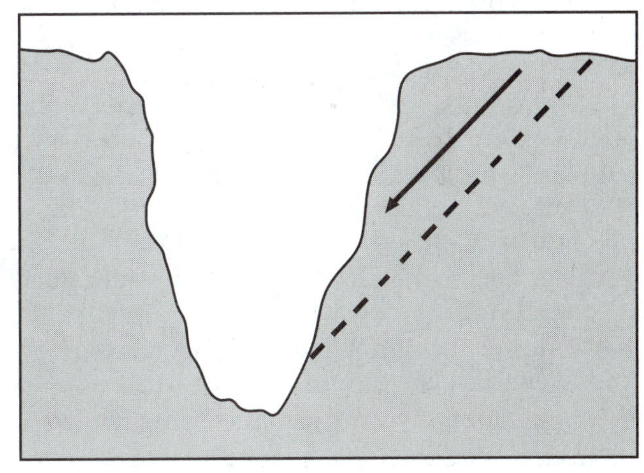

Figure 3 ◆ Behavior of dry clay and loams and wet sand and gravel.

Overhead Power Line Severed by Dump Truck

A subcontractor delivered a load of backfill to a construction area. After dumping the load, the driver failed to lower the dump bed before proceeding forward. The dump bed contacted and cut a 2,400V overhead power line. A spotter was in place while the truck was dumping, but turned away as the truck started to drive away. The line fell clear of the truck as it was cut, and no one was in the area where the end of the severed line fell. Luckily, there were no injuries.

The Bottom Line: Always be aware of overhead and underground power lines when excavating.

Source: The Occupational Safety and Health Administration (OSHA)

3.0.0 ◆ GUIDELINES FOR WORKING IN AND NEAR A TRENCH

When working in or around any excavation or trench, you are responsible for personal safety. You are also responsible for the safety of others in the work trench. The following guidelines must be enforced to ensure everyone's safety.

- Never enter an excavation without the approval of the OSHA-approved competent person on site.
- Inspect the excavation daily for changes in the excavation environment, such as rain, frost, or severe vibration from nearby heavy equipment.
- Never enter an excavation until the excavation has been inspected.
- Wear protective clothing and equipment, such as hard hats, safety glasses, work boots, and gloves. Use respirator equipment if necessary.
- Wear traffic warning vests that are marked with or made of reflective or highly visible material if you are exposed to vehicle traffic.
- Get out of the trench immediately if water starts to accumulate in the trench.
- Do not walk under loads being handled by power shovels, derricks, or hoists.
- Stay clear of any vehicle that is being loaded.
- Be alert. Watch and listen for possible dangers.
- Do not work above or below a co-worker on a sloped or benched excavation wall.
- Barricade access to excavations to protect pedestrians and vehicles (*Figure 4*).
- Check with your supervisor to see if workers entering the excavation need excavation entry permits.
- Ladders and ramps used as exits must be located every 25' in any trench that is over 4' deep.

www.SHORING.com 115F04.EPS

Figure 4 ◆ Barricaded trench.

- Make sure someone is on top to watch the walls when you enter a trench.
- Make sure you are never alone in a trench. Two people can cover each other's blind spots.
- Keep tools, equipment, and the excavated dirt at least 2' from the edge of the excavation.
- Make sure shoring, trench boxes (*Figure 5*), **benching**, or sloping are used for excavations and trenches over 5' deep.
- Stop work immediately if there is any potential for a cave-in. Make sure any problems are corrected before starting work again.

3.1.0 Ladders

There must be at least one method of entering and exiting all excavations over 4' deep. Ladders are generally used for this purpose (*Figure 6*). Ladders must be placed within 25' of each worker.

Trenching Equipment

An operator working on a trenching job tried to position a backhoe to backfill an electrical trench on the top outer edge. As the operator steered into softer, excavated soil, the equipment tilted onto its side and slid downward, stopping only when it engaged the chain link fence 10' below the trench. The operator was wearing a seatbelt and no injuries resulted from this incident.

The Bottom Line: If the employee had not been wearing his seatbelt and the fence had not been in place, this incident could have easily resulted in a very serious injury.

Source: The Occupational Safety and Health Administration (OSHA)

www.SHORING.com 115F05.EPS

Figure 5 ◆ Trench box.

When ladders are used, there are a number of requirements that must be met.

- Ladder side rails must extend a minimum of 3' above the landing.
- Ladders must have nonconductive side rails if work will be performed near equipment or systems using electricity.
- Two or more ladders must be used where 25 or more workers are working in an excavation in which ladders serve as the primary means of entry and exit or where ladders are used for two-way traffic in and out of the trench.
- All ladders must be inspected before each use for signs of damage or defects.

www.SHORING.com 115F06.EPS

Figure 6 ◆ Ladder in a trench.

- Damaged ladders should be labeled *DO NOT USE* and removed from service until repaired.
- Use ladders only on stable or level surfaces.
- Secure ladders when they are used in any location where they can be displaced by excavation activities or traffic.
- While on a ladder, do not carry any object or load that could cause you to lose your balance.
- Exercise caution whenever using a trench ladder.

4.0.0 ◆ INDICATIONS OF AN UNSTABLE TRENCH

A number of stresses and weaknesses can occur in an open trench or excavation. For example, increases or decreases in moisture content can affect the stability of a trench or excavation. The following sections discuss some of the more frequently identified causes of trench failure. These conditions are illustrated in *Figure 7*.

Tension cracks usually form one-quarter to one-half of the way down from the top of a trench. Sliding or slipping may occur as a result of tension cracks. In addition to sliding, tension cracks can cause toppling. Toppling occurs when the trench's vertical face shears along the tension crack line and topples into the excavation. An unsupported excavation can create an unbalanced stress in the soil, which in turn causes **subsidence** at the surface and bulging of the vertical face of the trench. If uncorrected, this condition can cause wall failure and trap workers in the trench, or greatly stress the **protective system**. Bottom heaving is caused by downward pressure created by the weight of adjoining soil. This pressure causes a bulge in the bottom of the cut. Heaving and squeezing can occur even when shoring and shielding are properly installed.

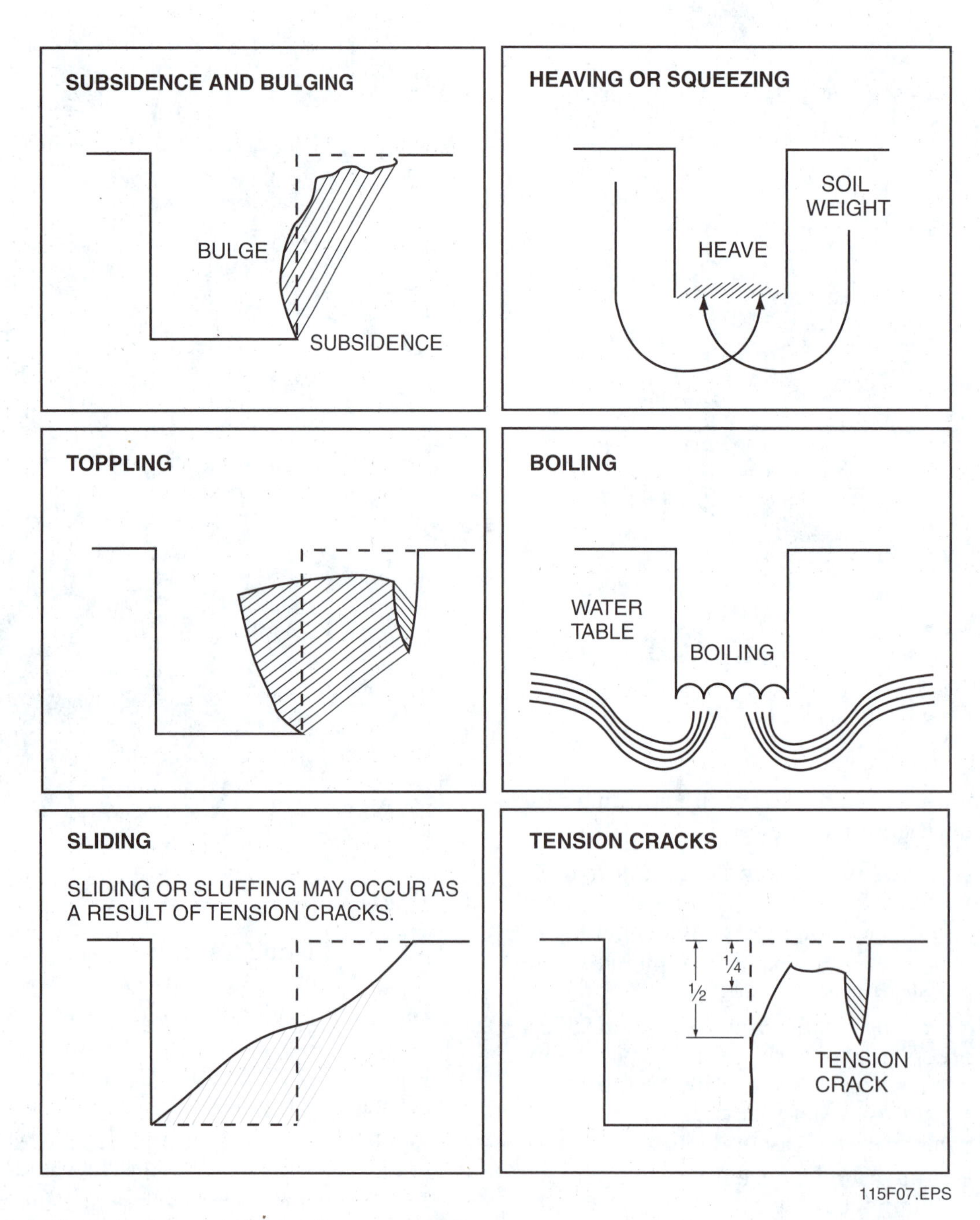

Figure 7 ◆ Indications of an unstable trench.

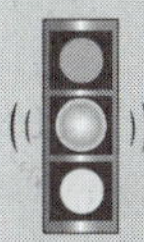

CAUTION

Protective systems are designed for even loads of earth. Heaving and squeezing can place uneven loads on the shielding system and may stress particular parts of the protective system.

Another indication of an unstable trench is boiling. Boiling is when water flows upward into the bottom of the cut. A high water table is one of the causes of boiling. Boiling can happen quickly and can occur even when shoring or trench boxes are used. If boiling starts, stop what you are doing and leave the trench immediately.

5.0.0 ◆ TRENCH FAILURE

The most common hazard during an excavation is trench failure or cave-in. Using common sense and following all applicable safety precautions will make the trench a much safer place to work.

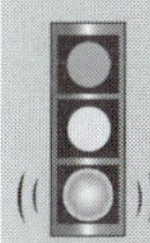

NOTE

Excavations 5' or deeper require manufactured and engineered trench shielding and shoring devices. If shielding or shoring devices are not used, trench walls may be sloped to eliminate the risk of cave-ins. Each of these protective systems should be used in accordance with OSHA regulations.

To understand the seriousness of trench failure, consider what can happen when there is a shift in the earth that surrounds an unsupported trench. Workers could be buried when:

- One or both edges of the trench cave in.
- One or both walls slide in.
- One or both walls shear away and collapse.

Failure of unsupported trench walls is not the only cause of burial. Tons of dirt can be dumped on the workers if the spoil pipe or excavated earth slides into the trench. Such slides occur when the pile is placed too close to the edge of the trench or when the ground beneath the pile gives way. There must be a minimum of 2' between the trench wall and the spoil pile. This area must also be kept free of any tools and materials.

The following conditions will likely lead to a trench cave-in. If you notice any of these conditions, immediately inform your supervisor. These conditions are listed in order of seriousness.

- Disturbed soil from previously excavated ground
- Trench intersections where large corners of earth can break away
- A narrow right-of-way causing heavy equipment to be too close to the edge of the trench
- Vibrations from construction equipment, nearby traffic, or trains
- Increased subsurface water that causes soil to become saturated, and therefore unstable
- Drying of exposed trench walls that cause the natural moisture that binds together soil particles to be lost
- Inclined layers of soil dipping into the trench, causing layers of different types of soil to slide one upon the other and cause the trench walls to collapse

6.0.0 ◆ MAKING THE TRENCH SAFE

There are several ways to make the trench a safer place to work (*Figure 8*). Trench shoring, shielding, and sloping are different methods used to protect workers and equipment. It is important that you recognize the differences between them.

- *Shoring* – Shoring a trench supports the walls of the excavation and prevents their movement and collapse. Shoring does more than provide a safe environment for workers in a trench. Because it restrains the movement of trench walls, shoring also stops the shifting of adjacent soil formations containing buried utilities or on which sidewalks, streets, building foundations, or other structures are built.
- *Trench shields* – Trench shields, also called trench boxes, are placed in unshored excavations to protect personnel from excavation wall collapse. They provide no support to trench walls or surrounding soil, but for specific depths and soil conditions, will withstand the side weight of a collapsing trench wall.
- *Sloping* – Sloping an excavation means cutting the walls of the excavation back at an angle to its floor. This angle must be cut at least to the **angle of repose** for the type of soil being used. The angle of repose is the greatest angle above the horizontal plane at which a material will rest without sliding.

6.1.0 Shoring Systems

Shoring systems are metal, hydraulic, mechanical, or timber structures that provide a framework to support excavation walls. Shoring uses **uprights**, **wales**, and **cross braces** to support walls. *Figure 9* shows a shoring system in place.

www.SHORING.com

www.SHORING.com

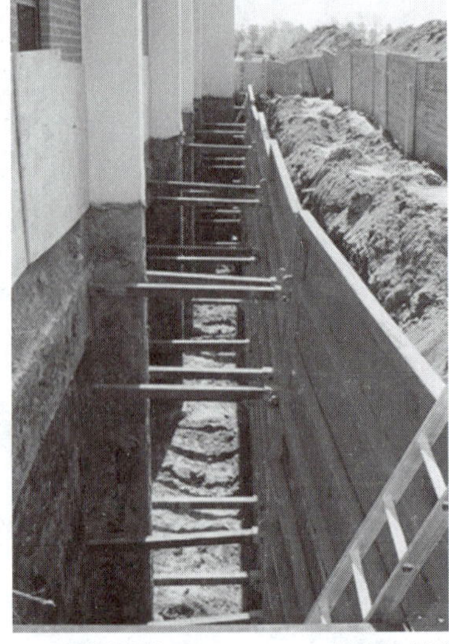

www.SHORING.com 115F08.EPS

Figure 8 ◆ Shoring methods.

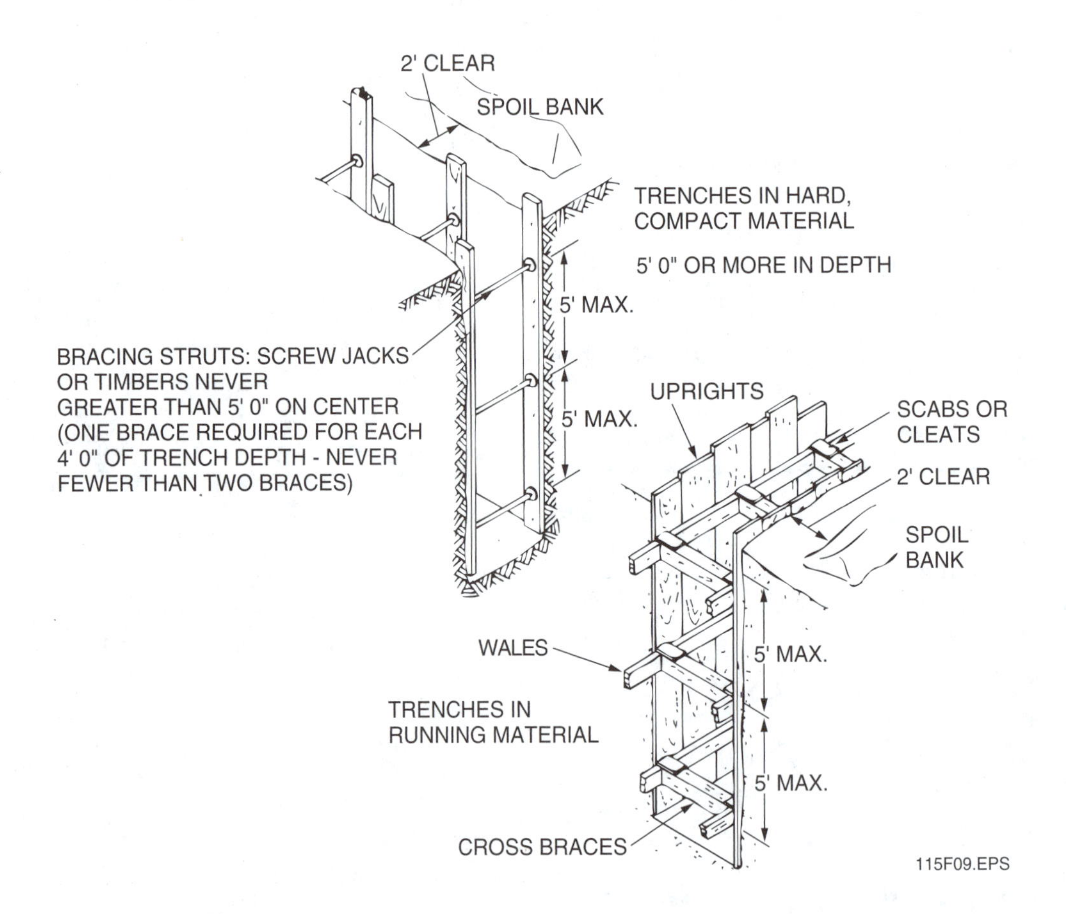

Figure 9 ◆ Shoring system in place.

6.1.1 Hydraulic Shores and Spreaders

Hydraulic shores, shown in *Figure 10*, can be installed and removed quickly. Each shore consists of two vertical rails connected by a hydraulic cylinder. The shores are placed in the trench and the hydraulic cylinders are pumped up to push the vertical rails against the wall. Note that this system uses the hydraulic cylinders as cross bracing. This arrangement is commonly referred to as a **skeleton** shoring system. Hydraulic spreaders, shown in *Figure 11*, can be added to a skeleton shoring system to provide additional cross bracing.

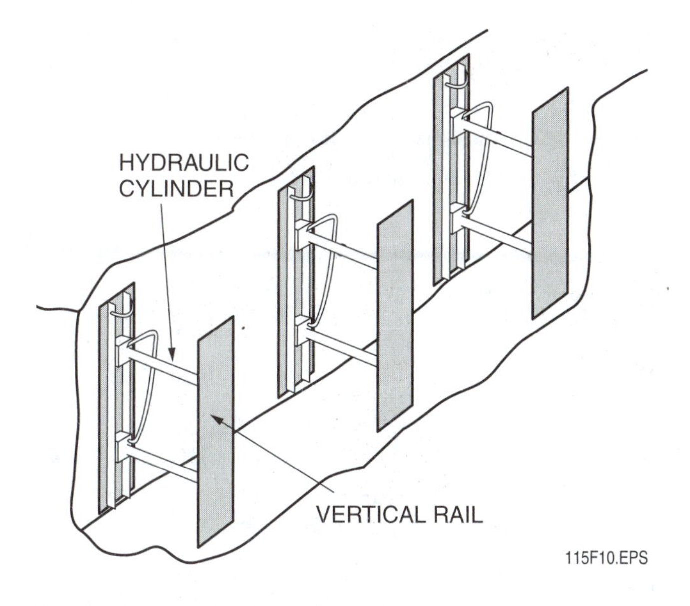

Figure 10 ◆ Hydraulic skeleton shoring system.

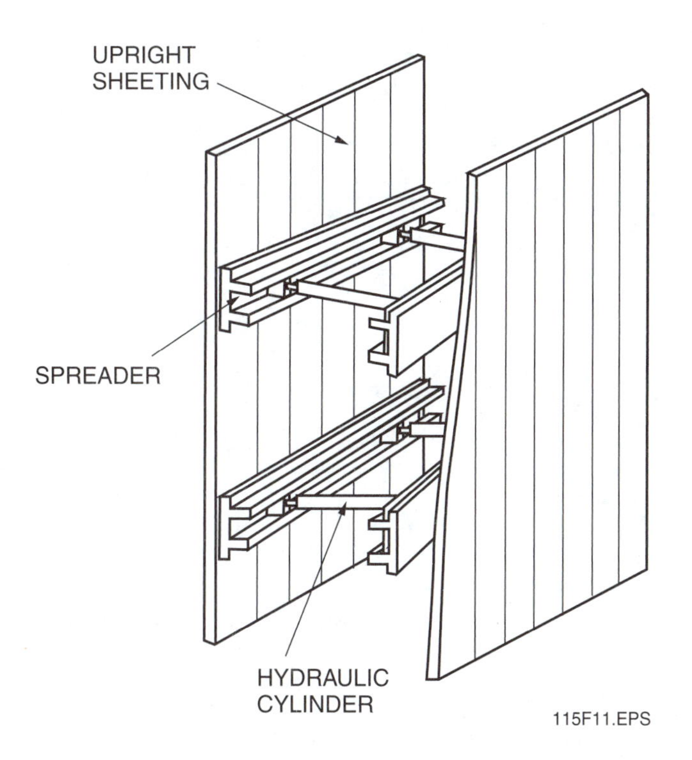

Figure 11 ◆ Hydraulic spreaders.

6.1.2 Vertical Sheeting

When excavating near existing structures or performing long-term excavations, vertical sheeting may be used and supported with hydraulic wales, as shown in *Figure 12*. This method is referred to as tight sheet shoring. Other types of spreaders, including screw jacks, trench jacks, and hydraulic spreaders, are also available.

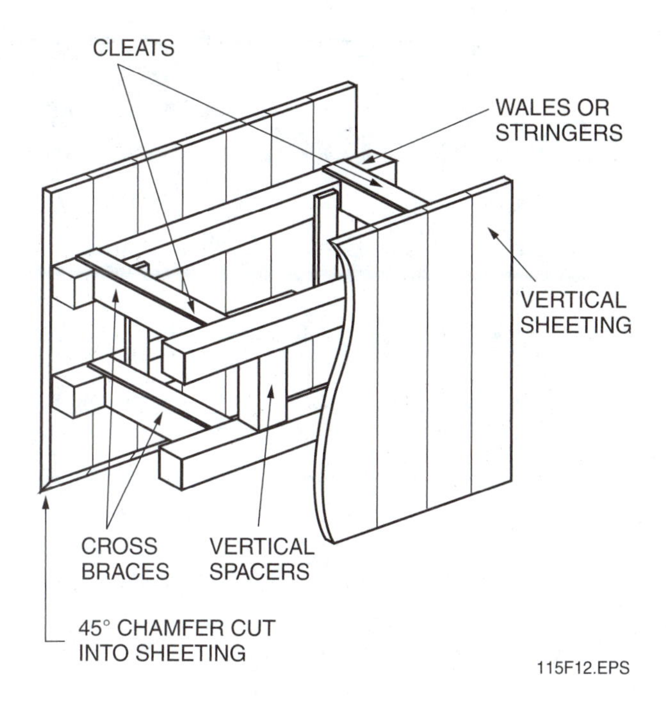

Figure 12 ◆ Tight sheet shoring.

6.1.3 Interlocking Steel Sheeting

Interlocking steel sheeting (*Figure 13*) may be specified under certain conditions such as deep excavations and excavations near buildings or building foundations. Interlocking steel sheeting is commonly used on Department of Transportation (DOT) right-of-ways. It will prevent damage to subbase pavement caused by vibration from vehicle traffic. Interlocking steel sheeting is required when working in waterways. Steel sheeting consists of interlocking panels of steel reinforced with cross members. It is similar in design to **tight sheeting**. Steel sheeting is engineered for a particular application. It must be installed precisely in accordance with the engineer's specifications. Steel sheeting is commonly installed by driving it into the ground using a vibrating hydraulic hammer. It can also be driven with a drop hammer or backhoe bucket.

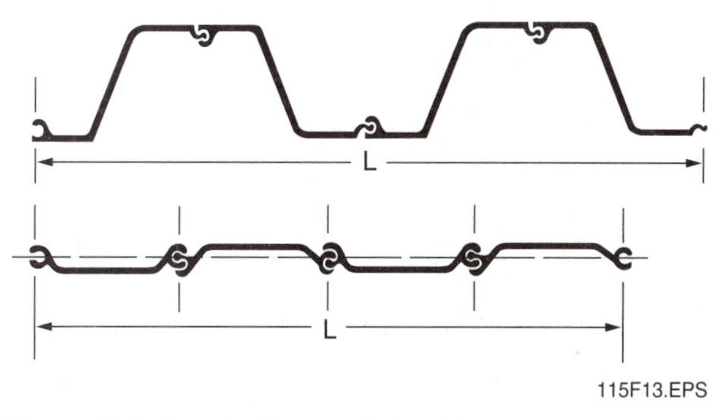

Figure 13 ◆ Interlocking steel sheeting.

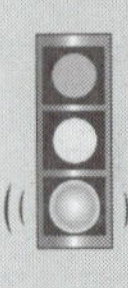

NOTE

The system just described identifies the common components used in shoring systems. You may encounter job-built systems, aluminum or wooden sheeting, or other approved methods.

6.1.4 Shoring Safety Rules

To avoid accidents and injury when shoring an excavation, special safety rules must be followed.

- Never enter an excavation before the shoring is in place.
- Do not install the shoring while you are inside the trench. All shoring must be installed from the top of the trench.
- The cross braces must be level across the trench.
- The cross braces should exert the same amount of pressure on each side of the trench.
- The vertical uprights must be drawn flat against the excavation wall.
- All materials used for shoring must be thoroughly inspected before use and must be in good condition.
- Shoring is removed by starting at the bottom of the excavation and going up.
- The vertical supports are pulled out of the trench from above.
- Every excavation must be backfilled immediately after the support system is removed.

6.2.0 Shielding Systems

A shielding system is a structure that is able to withstand the forces imposed on it by a cave-in, and protect employees within the structure. Shields can be permanent structures or can be designed to be portable and moved along as work progresses. The shielding system is also known as a trench box or trench shield. A trench box is shown in *Figure 14*. If the trench will not stand long enough to excavate for the shield, the shield can be placed high and pushed down as material is excavated.

The excavated area between the outside of the trench box and the face of the trench should be as small as possible. The space between the trench box and the excavation side must be backfilled to prevent side-to-side movement of the trench box. Remember that job site and soil conditions, as well as trench depths and widths, determine what type

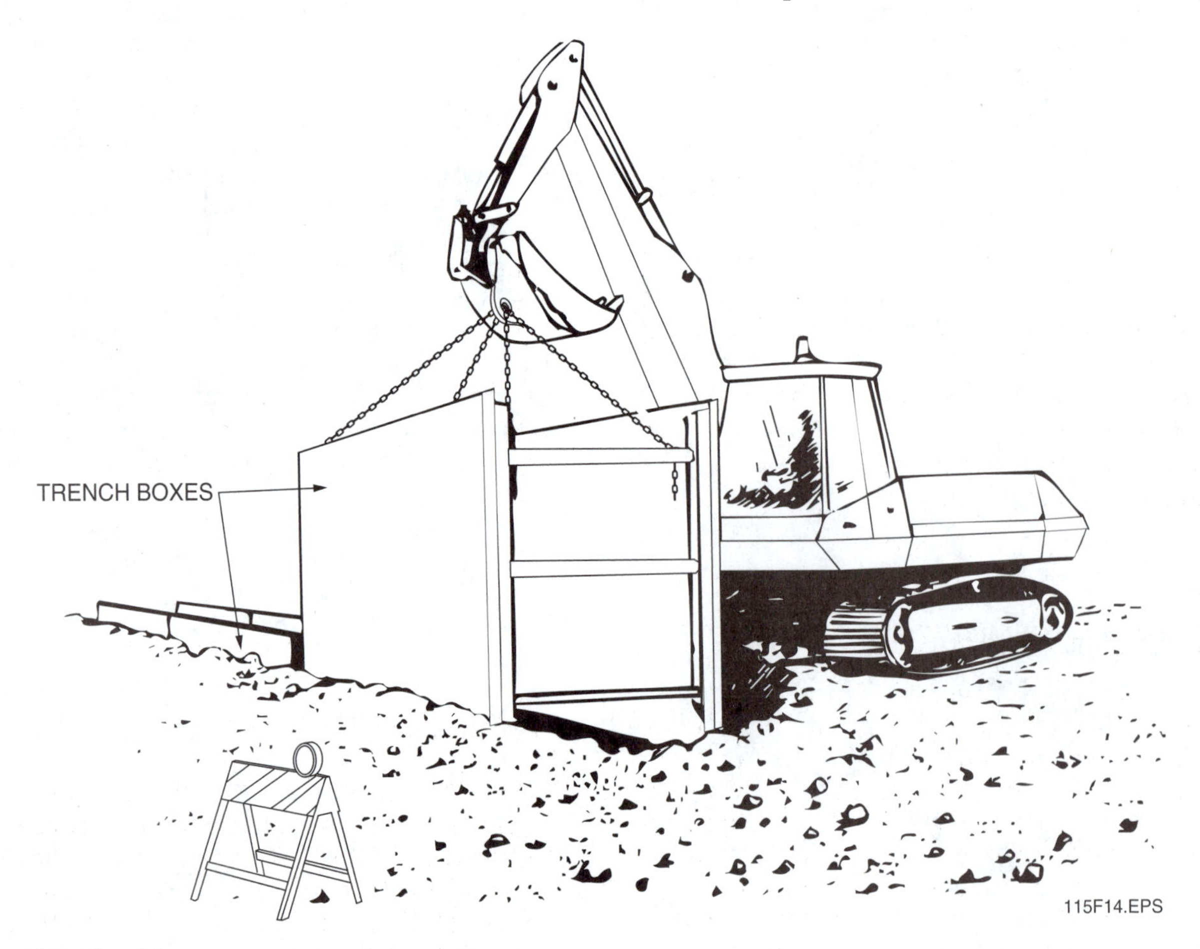

Figure 14 ◆ Trench boxes.

Cave-In Caused by Water

Four employees were laying a lateral sewer line at a building site. The foreman and a laborer were laying an 8"-diameter, 20'-long plastic sewer pipe in the bottom of a trench that was 36" wide, 9' deep, and approximately 50' long. The trench was neither sloped nor shored, and there was water entering it along a rock seam near the bottom. The west side of the trench caved in near the bottom, burying one employee to his chest and completely covering the other. The rescue operation took two hours for the first man and four hours for the second. Both men died.

The Bottom Line: Working in an excavation can be fatal. Always ensure your trench is safe.

Source: The Occupational Safety and Health Administration (OSHA)

of trench protection system will be used. A single project can include several depth or width requirements and varying soil conditions. This may require that several different protective systems be used for the same site. A registered engineer must certify shields for the existing soil conditions. The certification must be available at the job site during assembly of the trench box. After assembly, certifications must be made available upon request.

6.2.1 Trench-Box Safety

If used correctly, trench boxes protect workers from the dangers of a cave-in. All safety guidelines for excavations also apply to trench boxes. Follow these safety guidelines when using a trench box:

- Be sure that the vertical walls of a trench box extend at least 18" above the lowest point where the excavation walls begin to slope. *Figure 15* shows proper trench box placement.
- Never enter the trench box during installation or removal.
- Never enter an excavation behind a trench box after the trench box has been moved.
- Backfill the excavation as soon as the trench box has been removed.
- If a trench box is to be used in a pit or a hole, all four sides of the trench box must be protected.
- An exit from the trench box and the excavation must be located within 25' of each worker.

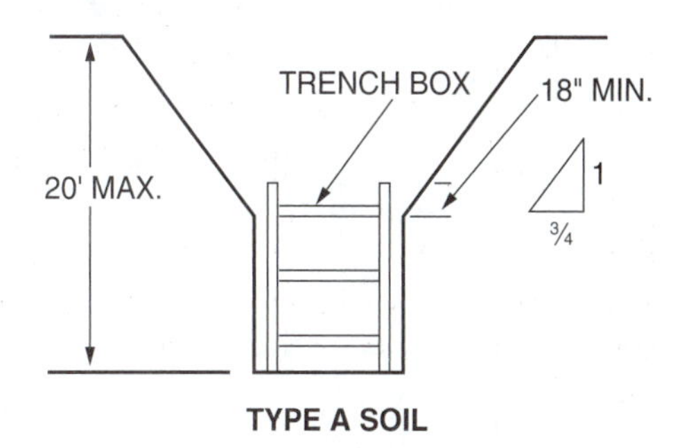

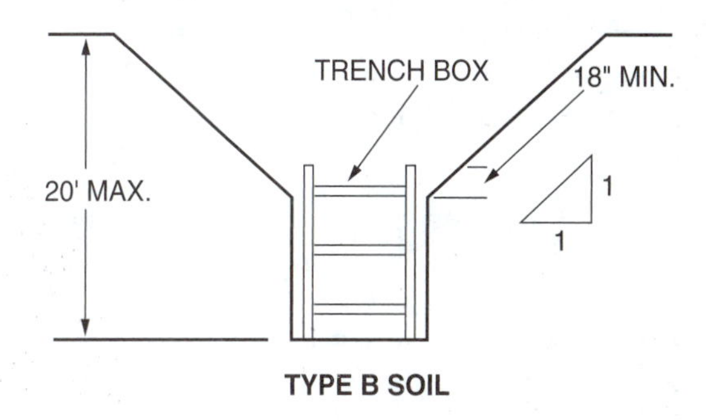

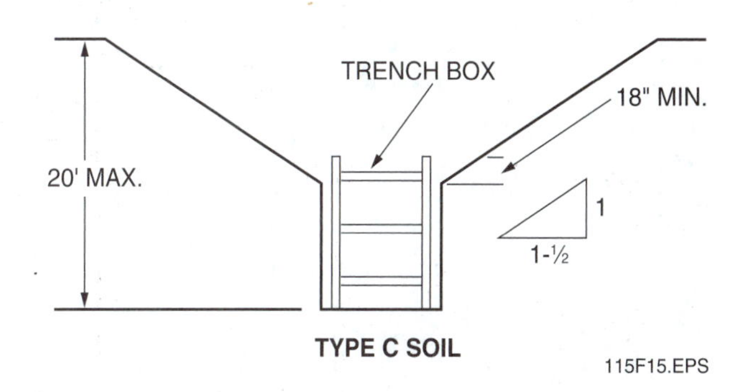

Figure 15 ◆ Proper trench-box placement.

6.3.0 Sloping Systems

A sloping system is a method of protecting workers from cave-ins. Sloping is accomplished by inclining the sides of an excavation. The angle of incline varies according to such factors as the soil type, environmental conditions of exposure, and nearby loads. There are three general classifications of soil types, and one type of rock. For each classification of soil type, OSHA defines maximum angles for the slope of the walls, as shown in *Table 3*. The designation and selection of the proper sloping system is far more complex than described in this section. Factors such as the depth of the trench, the amount of time the trench is to remain open, and other factors will affect the maximum allowable slope.

Table 3 Maximum Allowable Slopes

Soil or Rock Type	Maximum Allowable Slopes for Excavations Less Than 20' Deep
Stable rock	Vertical (90 degrees)
Type A	¾:1 (53 degrees)
Type B	1:1 (45 degrees)
Type C	1½:1 (34 degrees)

Step-back, also known as benching, is another method of sloping the excavation walls without the use of a support system. Step-back uses a series of steps that must rise on the approximate angle of repose for the type of soil being used. The same safety rules apply to step-back excavations that apply to sloped excavations.

Trencher Rollover

While operating a trencher, an operator was changing the equipment position by backing down a side slope leading to the edge of an empty ditch. The trenching bar was in an elevated position. As the operator shifted gears from reverse to forward, the equipment became top heavy, tipped over sideways, rolled one-half turn into the ditch, and came to rest on its top. Silt fences were being installed as a method to intercept and detain sediment from areas disturbed during grading operations in order to prevent the sediment from leaving the site or entering storm drains. During the installation, it was necessary to cut a shallow trench in which the barrier fabric and posts were placed. An operator was selected to dig the trench and the Activity Hazard Analysis (AHA) was reviewed with the operator; however, the AHA did not include precautions for operation on sloped surfaces. The accident occurred after the operator was finished with the excavation and was attempting to return to the beginning of the cut. Stability of the equipment was verified and the operator was removed. The operator was not injured.

The Bottom Line: Accidents can happen at any time. Use extreme caution when operating equipment near the edge of an excavation.

Source: The Occupational Safety and Health Administration (OSHA)

6.4.0 Sloping Requirements for Different Types of Soils

Many excavations are started with a vertical cut. Although some soils will stand to considerable depths when cut vertically, most will not. When vertical slopes fall to a more stable angle, large amounts of material usually fall into the excavation. *Figure 16* shows methods of failure in excavation walls.

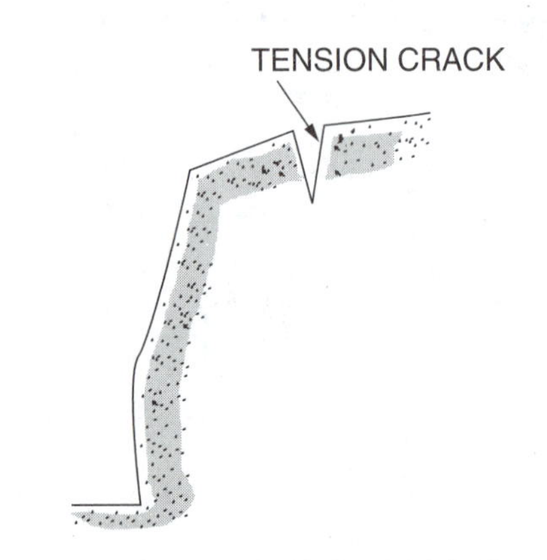

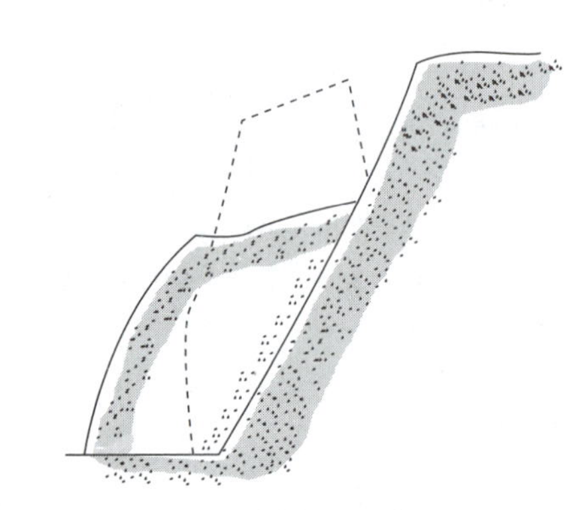

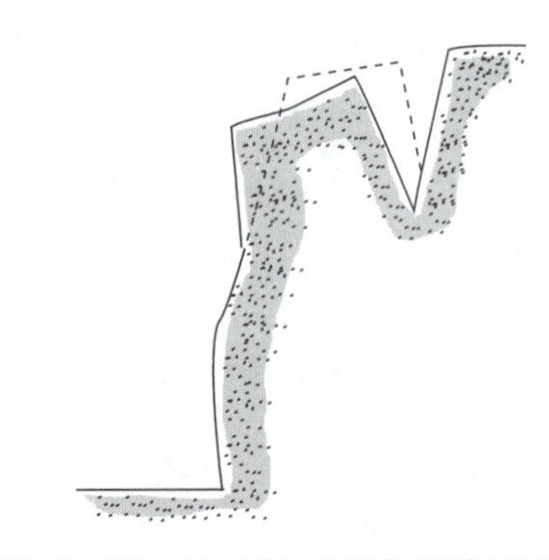

Figure 16 ◆ Methods of failure in excavation walls.

The slope of the excavation walls must be angled so that material will not fall into the excavation. The slope is figured by angle from the horizontal plane and by horizontal run to vertical rise. For example, a 45-degree slope would be a 1:1 slope, meaning that for each foot measured horizontally, the slope rises 1' vertically. The slope of the excavation walls varies according to the type and characteristics of the soil being used. Four basic classifications of soil have been identified. The classification is based on the stability of the soil and the maximum allowable slopes for excavations in each of the four types of soils. These four classifications, in decreasing order of stability, are stable rock, Type A, Type B, and Type C. *Figure 17* shows maximum allowable slopes for each soil type.

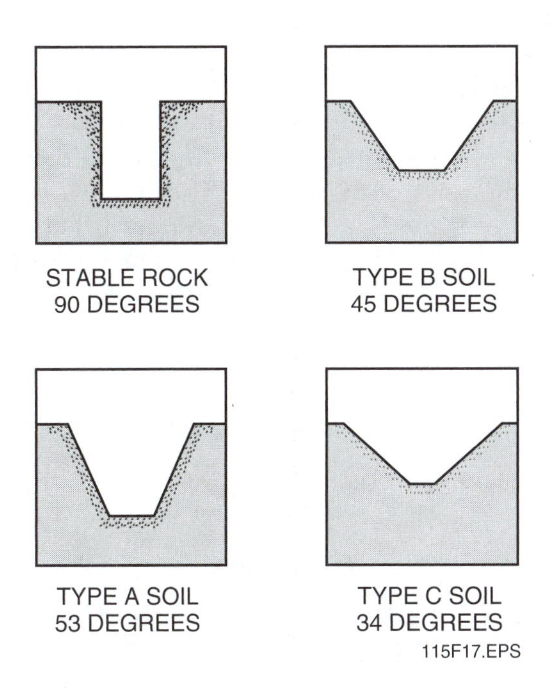

Figure 17 ◆ Maximum allowable slopes for soil types.

6.4.1 Stable Rock

Stable rock refers to natural solid mineral matter that can be excavated with vertical sides and remain intact while exposed. Stable rock includes solid rock, shale, or cemented sands and gravels. *Figure 18* shows an excavation cut in stable rock.

6.4.2 Type A Soil

Type A soil refers to solid soil with a **compression strength** of at least 1.5 tons per square foot. Cohesive soils are soils that do not crumble, are plastic when moist, and are hard to break up when dry.

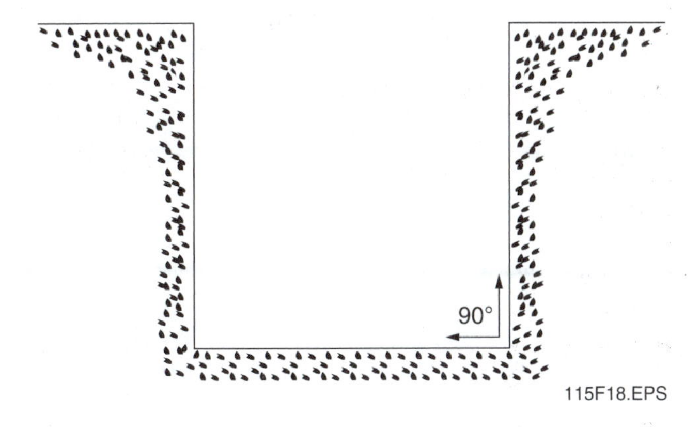

Figure 18 ◆ Excavation cut in stable rock.

Examples of cohesive soils are clay, silty clay, sandy clay, and clay loam. Cemented soils, such as caliche and hardpan, are also considered Type A. No soil can be considered Type A if any of the following conditions exist:

- The soil is fissured. Fissured means a soil material that has a tendency to break along definite planes of fracture with little resistance, or a material that has cracks, such as tension cracks, in an exposed surface.
- The soil can be affected by vibration from heavy traffic, pile driving, or other similar effects.
- The soil has been previously disturbed.

All sloped excavations that are 20' deep or less must have a maximum allowable slope of ¾ horizontal to 1 vertical. *Figure 19* shows a simple slope excavation in Type A soil.

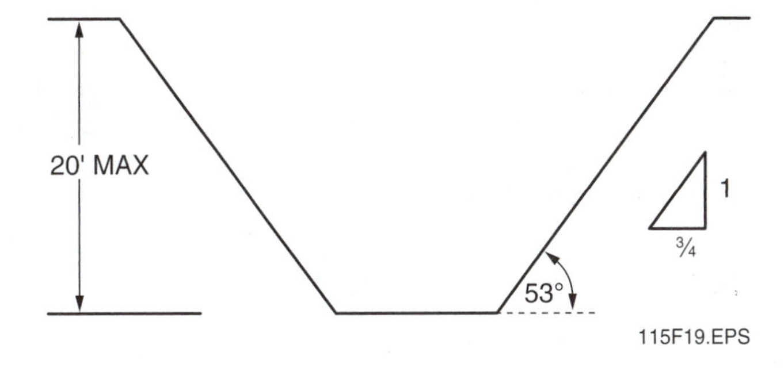

Figure 19 ◆ Simple slope excavation in Type A soil.

An exception to this rule occurs when the excavation is 12' deep or less and will remain open 24 hours or less. These excavations in Type A soil can have a maximum allowable slope of ½ horizontal to 1 vertical.

Figure 20 shows a simple slope, short-term excavation in Type A soil.

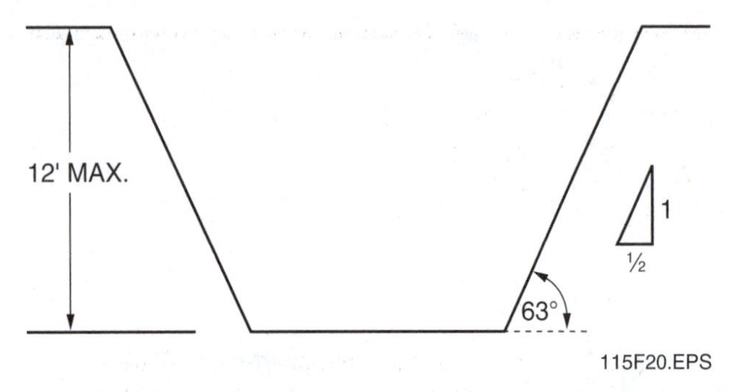

Figure 20 ◆ Simple slope, short-term excavation in Type A soil.

6.4.3 Type B Soil

Type B soil refers to cohesive soils with a compression strength of greater than 0.5 ton per square foot but less than 1.5 tons per square foot. It also refers to granular soils, including angular gravel, which are similar to crushed rock, silt, sandy loam, unstable rock, and any unstable or fissured Type A soils. Type B soils also include previously disturbed soils, except those that would fall into the Type C classification. Excavations made in Type B soils have a maximum allowable slope of 1 horizontal to 1 vertical. *Figure 21* shows simple slope excavations in Type B soil.

6.4.4 Type C Soil

Type C soil is the most unstable soil type. Type C soil refers to cohesive soil with a compression strength of 0.5 ton per square foot or less. Gravel, loamy soil, sand, any submerged soil or soil from which water is freely seeping, and unstable submerged rock are considered Type C soils. Excavations made in Type C soils have a maximum allowable slope of 1½ horizontal to 1 vertical. *Figure 22* shows a simple slope excavation in Type C soil.

6.5.0 Combined Systems

Slide-rail systems can be considered a cross between trench boxes and steel sheeting. These systems are designed to be used in shallow pits, tunnel pits, and trenches, virtually anywhere that a trench box or sheeting systems can be used. The slide-rail system shown in *Figure 23* consists of the following components:

- *Slide rails* – The slide rails are the horizontal components of the system. They have a cavity to accept the lining plates and are fitted with cross braces. Slide rails are available in single-

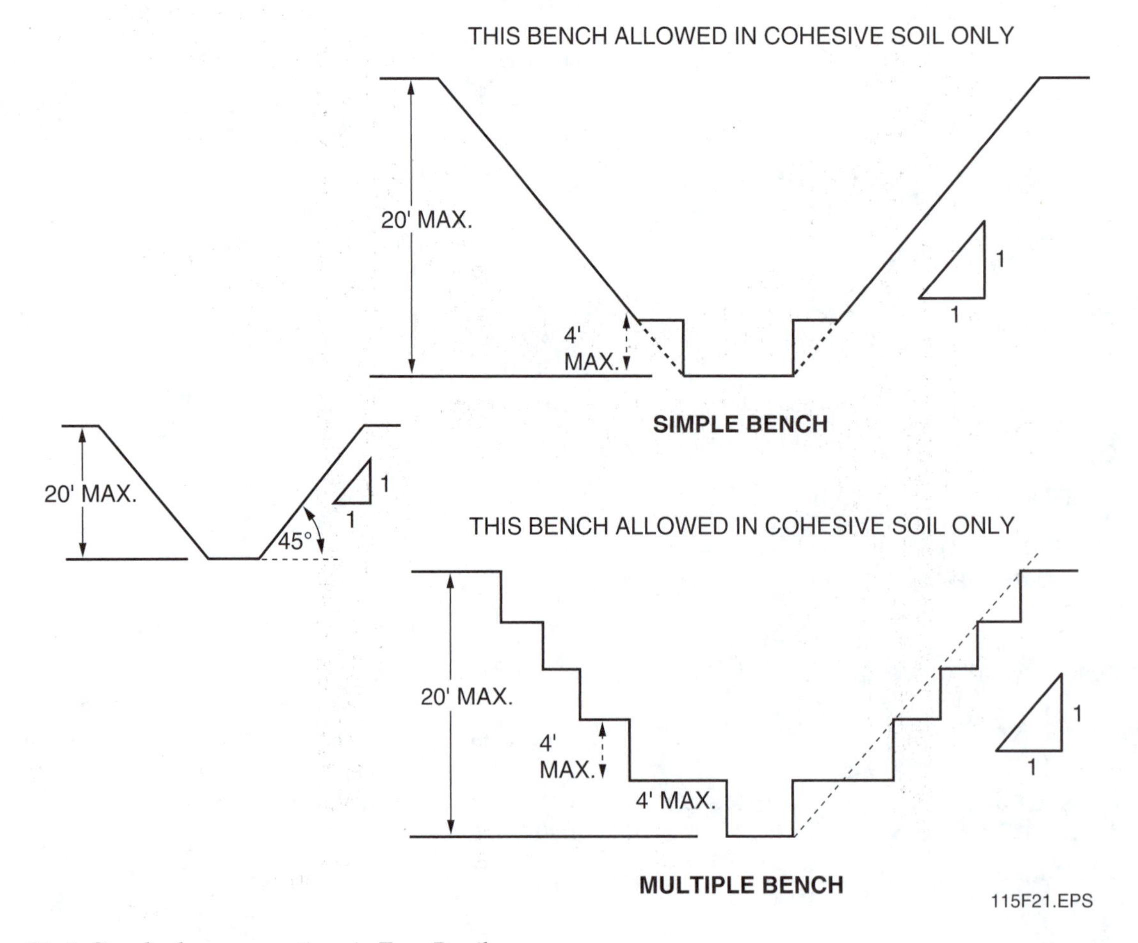

Figure 21 ◆ Simple slope excavations in Type B soil.

double-triple rail configurations. The triple-rail system is used at the greatest depth.

- *Lining plates* – Lining plates are used to provide the trench sidewall support. Lining plates are installed in the slide rails and are pushed into the ground as the excavation proceeds.
- *Cross braces* – Cross braces connect between the slide rails to provide lateral support to the system, usually at the end of the slide rail.

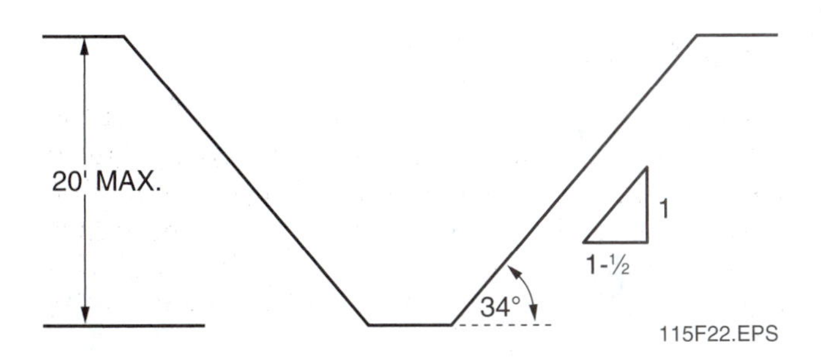

Figure 22 ◆ Simple slope excavation in Type C soil.

Exit Ladder

An employee was working in a trench that was 4' wide and 7' deep. About 30' away, a backhoe was straddling the trench. The backhoe operator noticed a large chunk of dirt falling from the side wall behind the worker in the trench. He called out a warning, but before the worker could climb out, 6' to 8' of the trench wall had collapsed on him and covered his body up to his neck. He suffocated before the backhoe operator could dig him out. No sloping or shoring had been used in the trench.

The Bottom Line: Excavations can collapse. Make sure proper shoring is in place before working in a trench.

Source: The Occupational Safety and Health Administration (OSHA)

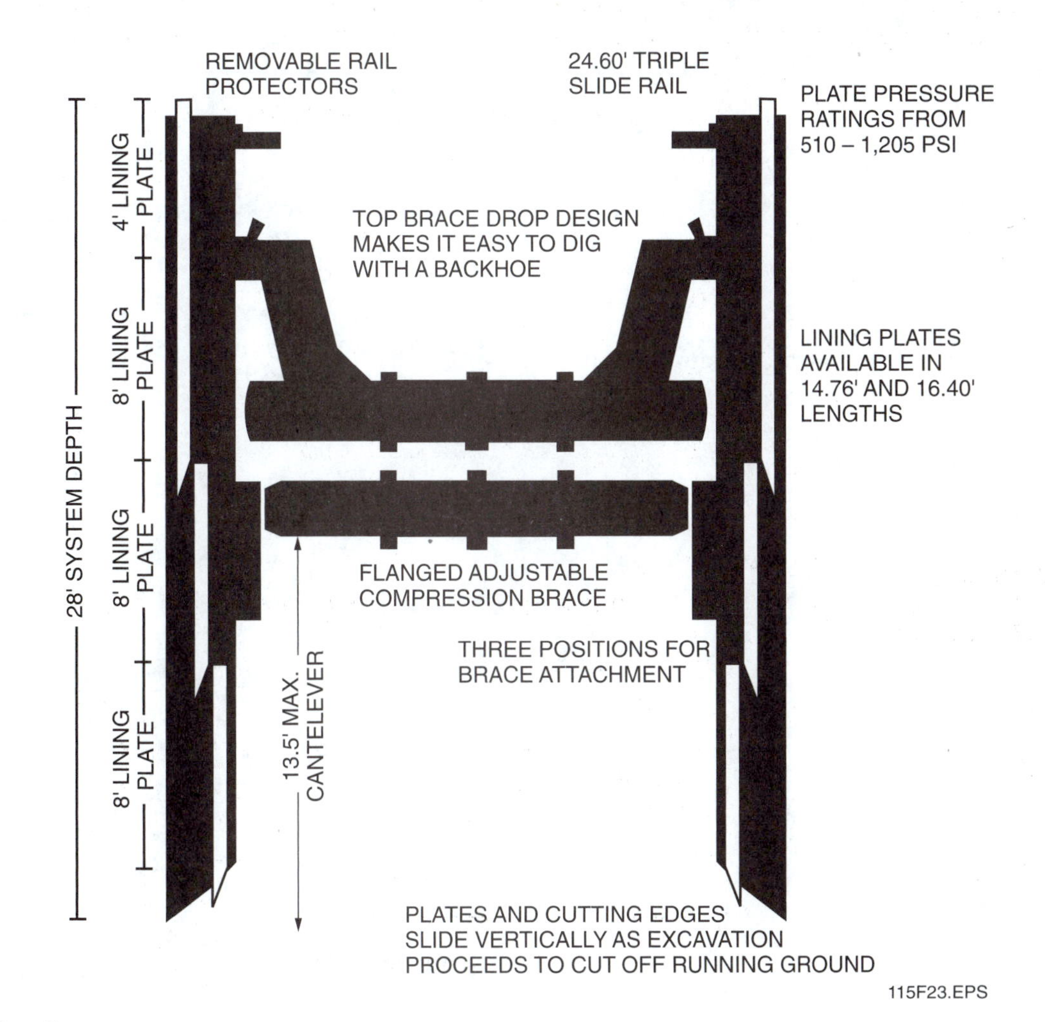

Figure 23 ◆ Slide-rail system.

Summary

Death and injuries from trenching accidents are common. Cave-ins, falls, flooding, and electrical shock are examples of the safety hazards related to trenches and excavations. Most accidents associated with trenches and excavations occur because the proper protective procedures were not followed and inspections by a competent person did not occur. OSHA has very specific procedures that must be followed for work in trenches and excavations. When these procedures are properly followed, the risk of accidents is greatly reduced.

Review Questions

1. One cubic yard of earth weighs approximately _____ pounds.

a. 1,000
b. 2,000
c. 3,000
d. 4,000

2. Flooding is considered a trenching hazard.

a. True
b. False

3. Tools and equipment must be kept at least _____ from the edge of a trench.

a. 1'
b. 2'
c. 3'
d. 4'

4. Ladders and ramps used as exits in a trench must be located every _____ in any trench that is over 4' deep.

a. 10'
b. 15'
c. 20'
d. 25'

5. The most common hazard during an excavation is _____.

a. a cave-in
b. an electrical strike
c. a fall
d. flooding

For Questions 6 through 8, match the correct protective system to its corresponding description.

6. _____ Shoring

7. _____ Trench shields

8. _____ Sloping

a. Cutting the walls of the excavation back at an angle to its floor
b. Supports the walls of the excavation and prevents their movement and collapse
c. Placed in unshored excavations to protect personnel from excavation wall collapse

9. The slope of the excavation walls must be angled so that material will not fall into the excavation.

a. True
b. False

10. As the excavation proceeds, _____ are installed in the slide rails and are pushed into the ground.

a. cross braces
b. lining plates
c. steel sheeting
d. trench boxes

GLOSSARY

Trade Terms Introduced in This Module

Angle of repose: The greatest angle above the horizontal place at which a material will adjust without sliding.

Benching: A method of protecting workers from cave-ins by excavating the sides of an excavation to form one or a series of horizontal levels or steps, usually with vertical or near-vertical surfaces between levels.

Carbonation: A chemical process in which carbon accumulates and causes a breakdown of surrounding minerals, such as dirt.

Competent person: A person who is capable of identifying existing and predictable hazards in the area or working conditions that are unsanitary, hazardous, or dangerous to employees, and who has the authority to take prompt corrective measures to fix the problem.

Compression strength: The ability of soil to hold a heavy weight.

Cross braces: The horizontal members of a shoring system installed perpendicular to the sides of the excavation, the ends of which bear against either uprights or wales.

Hydration: A chemical process in which water or liquids accumulate and cause a breakdown of surrounding minerals, such as dirt.

Oxidation: A chemical process in which oxygen accumulates and causes a breakdown of surrounding minerals, such as dirt.

Protective system: A method of protecting employees from cave-ins, from material that could fall or roll from an excavation face or into an excavation, or from the collapse of adjacent structures. Protective systems include support systems, sloping and benching systems, and shielding systems.

Shield: A structure that is able to withstand the forces imposed on it by a cave-in and thereby protect employees within the structure. Shields can be permanent structures or can be designed to be portable and moved along as work progresses. Additionally, shields can be either premanufactured or job-built in accordance with *29 CFR 1926.652 (c)(3) or (c)(4).*

Shoring: A structure such as a metal hydraulic, mechanical, or timber shoring system that supports the sides of an excavation and is designed to prevent cave-ins.

Skeleton: A condition that occurs when individual timber uprights or individual hydraulic shores are not placed in contact with the adjacent member.

Subsidence: A depression in the earth that is caused by unbalanced stresses in the soil surrounding an excavation.

Tight sheeting: The use of specially edged timber planks, such as tongue-and-groove planks, that are at least 3" thick. Steel sheet piling or similar construction that resists the lateral pressure of water and prevents loss of backfill is called tight sheeting.

Uprights: The vertical members of a trench shoring system placed in contact with the earth and usually positioned so that individual members do not contact each other. Uprights placed so that individual members are closely spaced, in contact with, or interconnected to each other, are often called sheeting.

Wales: Horizontal members of a shoring system placed parallel to the excavation face whose sides bear against the vertical members of the shoring system or the earth.

Figure Credits

Trenching Shoring Services, Inc.	115F01, 115F04, 115F05, 115F06, 115F08
Gary Wilson	115SA01

NCCER CURRICULA — USER UPDATE

NCCER makes every effort to keep its textbooks up-to-date and free of technical errors. We appreciate your help in this process. If you find an error, a typographical mistake, or an inaccuracy in NCCER's curricula, please fill out this form (or a photocopy), or complete the online form at **www.nccer.org/olf**. Be sure to include the exact module ID number, page number, a detailed description, and your recommended correction. Your input will be brought to the attention of the Authoring Team. Thank you for your assistance.

Instructors – If you have an idea for improving this textbook, or have found that additional materials were necessary to teach this module effectively, please let us know so that we may present your suggestions to the Authoring Team.

NCCER Product Development and Revision

13614 Progress Blvd., Alachua, FL 32615

Email: curriculum@nccer.org

Online: www.nccer.org/olf

❑ Trainee Guide ❑ AIG ❑ Exam ❑ PowerPoints Other ______________________________

Craft / Level: Copyright Date:

Module ID Number / Title:

Section Number(s):

Description:

Recommended Correction:

Your Name:

Address:

Email: Phone:

Module 75116-03

Forklift Safety

COURSE MAP

This course map shows all of the modules in Field Safety. The suggested training order begins at the bottom and proceeds up. The local Training Program Sponsor may adjust the training order.

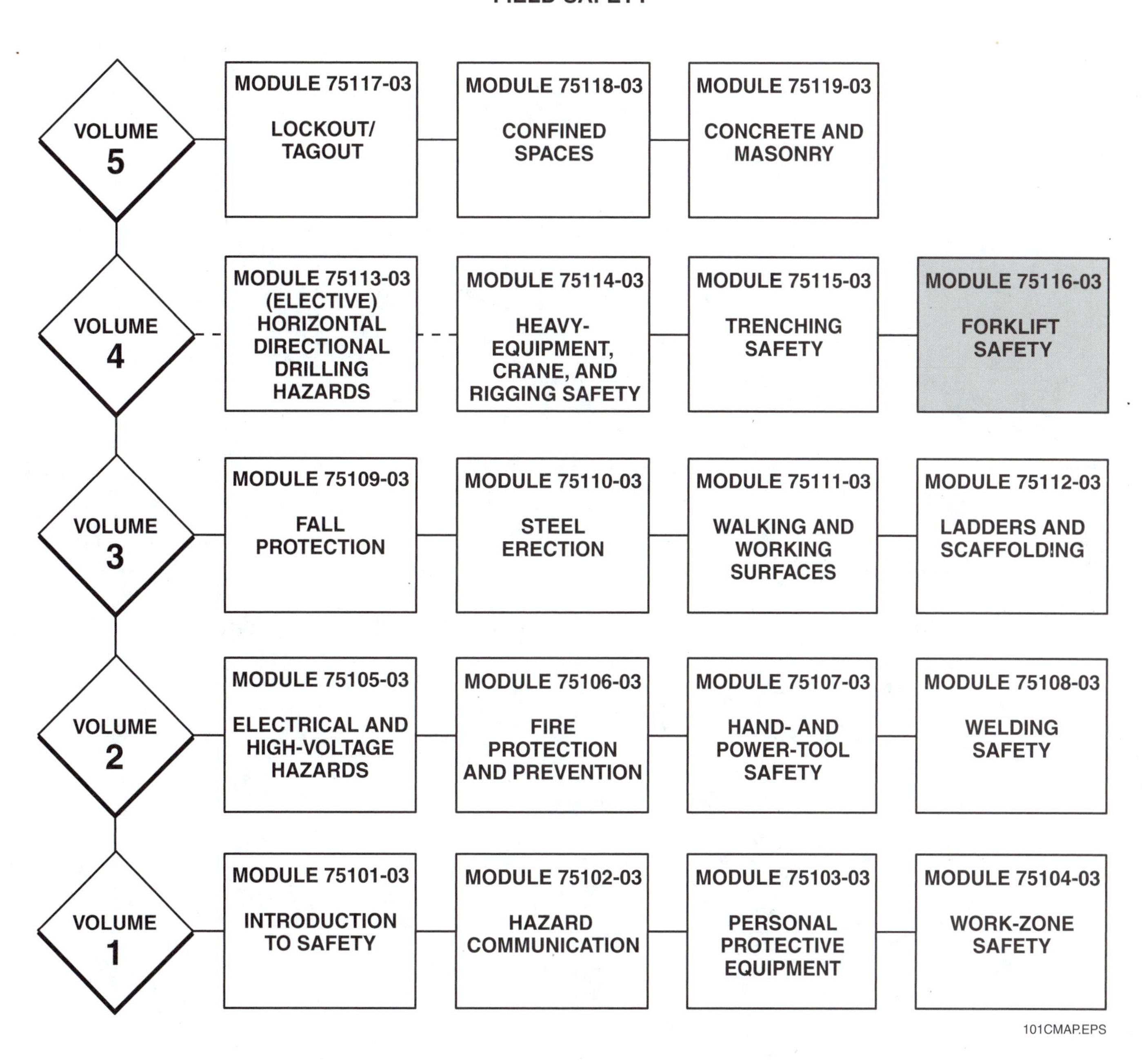

MODULE 75116-03 CONTENTS

Figures

Tables

MODULE 75116-03

Forklift Safety

Objectives

When you have completed this module, you will be able to do the following:

1. Explain the elements of a pre-shift inspection.
2. Describe the practices for safe traveling.
3. Describe the practices for safe load handling.
4. Understand how to operate a forklift safely on ramps and docks.
5. Explain how to work safely around a forklift.

Prerequisites

Before you begin this module, it is recommended that you successfully complete the following: Field Safety, Modules 75101-03 through 75112-03, 75114-03 and 75115-03. *Horizontal Directional Drilling Hazards* (module 75113-03) is an elective and is not required for successful completion of this course.

Required Materials

1. Pencil and paper
2. Appropriate personal protective equipment

1.0.0 ◆ INTRODUCTION

Forklifts are common on many work sites. They are useful for lifting and moving heavy or awkward loads of materials, supplies, and equipment (*Figure 1*). While extremely useful and relatively easy to operate, these machines can be very dangerous. They present several risks including hitting other workers, dropping loads, tipping over, and causing fires and explosions.

116F01.EPS

Figure 1 ◆ Forklift.

While mechanical and hydraulic problems can cause accidents, the most common cause of forklift accidents is human error. That means most forklift accidents can be avoided if the operator and other workers in the area stay alert and use caution and common sense. In fact, research by Liberty Mutual Insurance Company shows that drivers with more than a year of experience operating a forklift are more likely to have an accident than someone with little experience. This is because operators tend to become too comfortable and less attentive after they gain experience on the equipment. The same study showed that the most common type of forklift accident is one in which a pedestrian is hit by the truck.

Table 1 shows the most common types of forklift accidents and their percentage in relation to the total number of accidents, according to a Liberty Mutual Insurance Company study.

Table 1 Common Types of Forklift Accidents

Forklift Accidents	Percentage of the Number of all Workplace Accidents
Pedestrian struck	20%
Forklift tipped over	18%
Falling load	13%
Workers falling from forks	10%
Operator struck by own unsecured truck	5%

2.0.0 ◆ BEFORE YOU OPERATE A FORKLIFT

Before you can begin operating a forklift, you must be trained and certified on that particular piece of equipment. Once you are trained and certified, you must thoroughly inspect your forklift before you begin each shift.

2.1.0 Training and Certification

It is a common misconception that if you can drive a car, truck, or piece of heavy equipment, you can just hop on a forklift and start operating it. However, the Occupational Safety and Health Administration (OSHA) requires forklift operators to be trained and certified on each piece of equipment before they operate it on the job site. The operator's card only applies to the specific piece of equipment on which they are trained. Operators of powered forklifts must be qualified as to sight, hearing, physical, and mental ability to operate the equipment safely. Personnel who have not been trained in the operation of forklifts may only operate them for the purpose of training. The training must be conducted under the direct supervision of a qualified trainer.

2.2.0 Pre-Shift Inspection

A pre-shift inspection is required before operation of a forklift. The more thorough you are when inspecting the forklift, the safer and more productive you will be during your shift. *Figure 2* shows a sample checklist covering the basic items that need to be checked during a pre-shift inspection.

Your company's checklist and your supervisor can provide you with specific information about what you should check before you begin each shift on a forklift. Your training and the forklift operator's manual will help you to understand what to look for when you inspect the forklift. If you find any problems during your pre-shift inspection, notify your supervisor or maintenance manager immediately. The forklift should be locked out and tagged. It cannot be used until all problems are corrected.

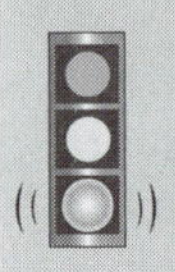

NOTE
For more information on the lockout/tagout process, refer to the *Lockout/Tagout* module in Volume Five.

2.3.0 General Safety Precautions

Safe operation is the operator's responsibility. Operators must develop safe working habits and be able to recognize hazardous conditions in order to protect themselves and others from death and injury. They must always be aware of unsafe conditions to protect the load and the forklift from damage. They must also understand the operation and function of all controls and instruments before operating any forklift. Operators must read and fully understand the operator's manual for each piece of equipment being used.

The following safety rules are specific to forklift operation.

- Always check the capacity chart mounted on the machine before operating any forklift.
- Never put any part of the body into the mast structure or between the mast and the forklift.
- Never put any part of the body within the reach mechanism.
- Understand the limitations of the forklift.
- Do not permit passengers to ride in the forklift unless a safe place to ride has been provided by the manufacturer.
- Never leave the forklift running unattended.
- Never carry passengers on the forks.

Forklift operators must pay special attention to the safety of any pedestrians on the job site. Safeguard pedestrians at all times by observing the following rules.

- Always look in the direction of travel.
- Do not drive the forklift up to anyone standing in front of an object or load.
- Make sure that personnel stand clear of the rear swing area before turning.
- Exercise particular care at cross aisles, doorways, and other locations where pedestrians may step into the travel path.
- Always use a spotter when moving an elevated load with a telescoping-boom forklift.

OPERATOR'S DAILY CHECKLIST

Check Each Item Before Start Of Each Shift   Date: ____________

Check One: Gas/LGP/Diesel Truck ☐ Electric Sit-down ☐ Electric Stand-up ☐ Electric Pallet

Truck Serial Number: ________________ Operator: __________________ Supervisor's OK: ________________

Hour Meter Reading: ________________

Check each of the following items before the start of each shift. Let your supervisor and/or maintenance department know of any problem. DO NOT OPERATE A FAULTY TRUCK. Your safety is at risk.

After checking, mark each item accordingly. Explain below as necessary.

Check boxes as follows: ☐ OK   ☐ NG, needs attention or repair. Circle problem and explain below.

OK	NG	Visual Checks
		Tires/Wheels: wear, damage, nuts tight
		Head/Tail/Working Lights: damage, mounting, operation
		Gauges/Instruments: damage, operation
		Operator Restraint: damage, mounting, operation oily, dirty
		Warning Decals/Operator's Manual: missing, not readable
		Data Plate: not readable, missing adjustment
		Overhead Guard: bent, cracked, loose, missing
		Load Back Rest: bent, cracked, loose, missing
		Forks: bent, worn, stops OK
		Engine Oil: level, dirty, leaks
		Hydraulic Oil: level, dirty, leaks
		Radiator: level, dirty, leaks
		Fuel: level, leaks
		Battery: connections loose, charge, electrolyte low
		Covers/Sheet Metal: damaged, missing
		Brakes: linkage, reservoir fluid level, leaks
		Engine: runs rough, noisy, leaks

OK	NG	Visual Checks
		Steering: loose/binding, leaks, operation
		Service Brake: linkage loose/binding, stops OK, grab
		Parking Brake: loose/binding, operational, adjustment
		Seat Brake (if equipped): loose/binding, operational, adjustment
		Horn: operation
		Backup Alarm (if equipped): mounting, operation
		Warning Lights (if equipped): mounting, operation
		Lift/Lower: loose/binding, excessive drift, leaks
		Tilt: loose/binding, excessive drift, "chatters," leaks
		Attachments: mounting, damaged operation, leaks
		Battery Test (electric trucks only): indicator in green
		Battery: connections loose, charge, electrolyte low while holding full forward tilt
		Control Levers: loose/binding, freely return to neutral
		Directional Controls: loose/binding, find neutral OK

Explanation of problems marked above: ____

____

____

____

____

____

116F02.EPS

Figure 2 ◆ Example of a forklift operator's daily checklist.

3.0.0 ◆ TRAVELING

Traveling refers to driving your forklift both with and without a load. To move either the forklift or a load of materials, you must travel. Sometimes the job requires only short travel distances, such as from a flatbed truck or rail car to a storage or **staging area** (*Figure 3*). Other times you must travel longer distances on the forklift. For example, you may need to move a load of roofing materials from a storage area by the receiving dock all the way across the work site to a building being roofed.

116F03.EPS

Figure 3 ◆ Forklift working in a storage area.

3.1.0 Stay Inside

You are safest when traveling on a forklift if you keep your whole body inside the vehicle. Many experienced drivers get into the unsafe habit of hanging an elbow outside of the truck, sliding a foot off the platform, or resting one hand with the fingers hanging over the edge of the truck. This often results in crushing injuries and amputation. The **operator's compartment**, along with the use of seatbelts, is designed to protect the operator from falling objects, impact from collisions, contact with electrical utilities, and tipping accidents. For example, if you allow your elbow to hang over the edge of the truck, and then accidentally back into a support beam, you could easily crush your arm between the forklift and the beam.

3.2.0 Pedestrians

Always yield the right-of-way to pedestrians. Forklifts are heavy machines that typically require a distance equal to the length of the forklift in order to stop. Pedestrians are usually not aware of this and walk around the site expecting these large and cumbersome machines to be able to stop quickly. Because of this, it is important to look out for any pedestrians.

3.3.0 Passengers

Forklifts are designed to carry one person: the operator. No one else should ever ride in or stand on a forklift. There are a few specially designed and certified attachments that allow forklifts to be converted to personnel lifts. Other than those few situations, no one other than the operator should be on the forklift.

3.4.0 Blind Corners and Intersections

As you approach a blind corner or intersection, always assume that a pedestrian or another piece of equipment is coming the other way. Stop at the intersection. Sound your horn. Then proceed slowly through the intersection or around the corner. Be prepared to stop if necessary.

3.5.0 Keep the Forks Low

Whether traveling with or without a load, keep the forks as low as possible (*Figure 4*). As a general rule of thumb, the forks should never be higher

Deadly Overload

A forklift operator was carrying a load that was stacked too high. The load obstructed his view. To make up for it, he stuck his head out the side of the operator's compartment to see around the load. Unfortunately, as he was preparing to drive forward, another forklift was backing up and sideswiped his machine, decapitating him.

The Bottom Line: Never carry a load that obstructs your view. Always keep all of your body parts inside the operator's compartment.

Source: The Occupational Safety and Health Administration (OSHA)

116F04.EPS

Figure 4 ◆ Travel with your forks low.

than 6" from the travel surface while traveling, unless you are moving over an extremely rough surface. The forks are strong and pointed. Ramming into something or someone can cause serious damage. If the forks are low, the chance of critically or fatally injuring someone is greatly reduced.

3.6.0 Absolutely No Horseplay

Driving a forklift is a serious operation that requires maturity and attention. Never drive a forklift toward another person as a joke. This is especially important if they are in front of a solid object or another piece of equipment. Doing so could easily lead to a very serious crushing accident. Accidents caused by horseplay are the most avoidable problem on the job site. Working with heavy equipment is dangerous and your behavior on the job should reflect that. Your safety and that of your co-workers depends on it.

3.7.0 Travel Surface

Whenever possible, take the smoothest and driest route when traveling. Rough or bumpy surfaces can cause a lot of bouncing, which may destabilize a forklift and make the forklift or its load tip. Wet or slippery surfaces can cause the tires to lose traction, resulting in loss of control of the forklift.

4.0.0 ◆ HANDLING LOADS

A forklift's main use is to transport large, heavy, or awkward loads. If not handled correctly, loads can fall from the forks, obstruct the operator's view, or cause the forklift to tip. The most important factor to consider when using any forklift is its capacity. Each forklift is designed with an intended capacity, and this capacity must never be exceeded. Exceeding the capacity jeopardizes not only the machine and the load, but also the safety of everyone on or near the forklift. Every manufacturer supplies a capacity chart for each forklift. The operator must be aware of the capacity of the machine before being allowed to operate the forklift.

4.1.0 Picking Up Loads

Some forklifts are equipped with a sideshift device that allows the operator to shift the load sideways several inches in either direction with respect to the mast. A sideshift device enables more precise placing of loads, but it also changes the forklift's **center of gravity** and must be used with caution. If the forklift being used is equipped with a sideshift device, be sure to return the fork carriage to the center position before attempting to pick up a load. The following is a typical procedure used when picking up a load.

Step 1 Check the position of the forks with respect to each other. They should be centered on the carriage. If the forks have to be moved, check the operator's manual for the proper procedure. Usually, there is a pin at the top of each fork that, when lifted, allows each fork to be slid along the upper backing plate until the fork centers over the desired notch.

Step 2 Travel at a safe rate of speed. Always keep the forks lowered when traveling.

Step 3 Before picking up a load with a forklift, make sure the load is stable. If it looks like the load might shift when picked up, secure the load. Knowing the center of gravity is crucial, especially when picking up tapered sections. Make a trial lift, if necessary, to determine and adjust the center of gravity.

Step 4 Approach the load so that the forks straddle the load evenly. It is important that the weight of all loads be distributed evenly on the forks. Overloading one fork at the expense of the other can damage the forks. In some cases, it may be necessary to measure the load and mark its center.

Step 5 Drive up to the load with the forks straight and level. If the load being picked up is on a pallet, be sure the forks are low enough to clear the pallet boards.

Step 6 Move forward until the leading edge of the load rests squarely against the back of both forks. If you cannot see the forks engage the load, ask someone to signal for you.

Step 7 Tilt the mast rearward until the forks contact the load. Raise the carriage until the load safely clears the ground. Then, tilt the mast fully rearward to cradle the load. This ensures that the load will not slip during travel.

4.2.0 Traveling With Loads

Always travel at a safe rate of speed with a load. Never travel with a raised load. Keep the load as low as possible and be sure the mast is tilted rearward to cradle the load.

As you travel, stay alert and pay attention. Watch the load and the conditions ahead of you, and alert others to your presence. Avoid sudden stops and abrupt changes in direction. Be careful when downshifting because sudden deceleration can cause the load to shift or topple. Watch the machine's rear clearance when turning.

If you are traveling with a telescoping-boom forklift, be sure the boom is fully retracted. If you have to drive on a slope, keep the load as low as possible. Do not drive across steep slopes. If you have to turn on an incline, make the turn wide and slow.

4.3.0 Traveling With Long Loads

Traveling with long loads presents special hazards, particularly if the load is flexible and subject to damage. Traveling multiplies the effect of bumps over the length of the load. A stiffener may be added to the load to give it extra rigidity.

To prevent slippage, secure long loads to the forks. This may be done in one of several ways. A field-fabricated cradle may be used to support the load. While this is an effective method, it requires that the load be jacked up.

The forklift may be used to carry pieces of rigging equipment. This method requires the use of slings and a spreader bar.

In some cases, long loads may be snaked through openings that are narrower than the load itself. This is done by approaching the opening at an angle and carefully maneuvering one end of the load through the opening first. Avoid making quick turns because abrupt maneuvers will cause the load to shift.

4.4.0 Placing Loads

Position the forklift at the landing point so that the load can be placed where you want it. Be sure everyone is clear of the load. The area under the load must be clear of obstructions and must be able to support the weight of the load. If you cannot see the placement, use a signaler to guide you.

With the forklift in the unloading position, lower the load and tilt the forks to the horizontal position. When the load has been placed and the forks are clear from the underside of the load, back away carefully to disengage the forks.

4.5.0 Placing Elevated Loads

Special care needs to be taken when placing elevated loads. Some forklifts are equipped with a leveling device that allows the operator to rotate the fork carriage to keep the load level during travel. When placing elevated loads, it is extremely important to level the machine before lifting the load.

One of the biggest potential safety hazards during elevated load placement is poor visibility. There may be workers in the immediate area who cannot be seen. The landing point itself may not be visible. Your depth perception decreases as the height of the lift increases. To be safe, use a signal person to help you position the load.

Use tag lines to tie off long loads. Tie off loads to the mast of the forklift. Drive the forklift as closely as possible to the landing point with the load kept low. Set the parking brake. Raise the load slowly and carefully while maintaining a slight rearward tilt to keep the load cradled. Under no circumstances should the load be tilted forward until the load is over the landing point and ready to be set down.

If the forks start to move, sway, or lean, stop immediately but not abruptly. Lower the load slowly; reposition it, or break it down into smaller components if necessary. If ground conditions are poor at the unloading site, it may be necessary to reinforce the ground with planks to provide greater stability. As the load approaches the landing point, slow the lift speed to a minimum. Continue lifting until the load is slightly higher than the landing point.

4.6.0 Using a Forklift to Rig Loads

A forklift can be a very useful piece of rigging equipment if it is properly and safely used. Loads can be suspended from the forks with slings, moved around the job site, and placed. All the rules of careful and safe rigging apply when using a forklift to rig loads. Be sure not to drag the load or let it swing freely. Use tag lines to control the load.

Never attempt to rig an unstable load with a forklift. Be especially mindful of the load's center of gravity when rigging loads with a forklift.

When carrying cylindrical objects, such as oil drums, keep the mast tilted rearward to cradle the load. If necessary, secure the load to keep it from rolling off the forks.

4.7.0 Dropping Loads

Momentum is a physical force that makes objects in motion tend to stay in motion, even if the mode of transportation stops. Have you ever carried a meal on a cafeteria tray? In your experience, what happened to the items on your tray if you had to stop suddenly or if you turned too quickly? The items on your tray probably started to topple or slide. They might have fallen right off the tray or they may have shifted the load so that the tray tipped, spilling everything.

The same principle applies to the load on your forklift. If you turn or stop too quickly, the load will keep going in the direction you were originally headed, causing it to slide off the forks. At the very least, this may be inconvenient, as you have to stop everything to restack your load. It can be expensive if you drop fragile materials or equipment. It can be deadly if the load falls on a co-worker or causes your truck to tip over. Avoid sudden maneuvers when operating a forklift.

4.8.0 Tipping

There are three main causes for a forklift tipping:

- The load is too heavy.
- The load is placed too far forward on the forks.
- The operator is not driving safely.

To avoid tipping, you need to understand what the center of gravity is and how it applies to forklifts. The center of gravity is the point around which all of an object's weight is evenly distributed (*Figure 5*). Your forklift has a center of gravity and the load you're moving will have its own center of gravity. When the forklift picks up the load, the center of gravity shifts to the **combined center of gravity** (*Figure 6*).

The forklift will tip if the center of gravity moves too far forward, backward, right, or left. Putting too heavy a load on the forklift, or placing the load too far forward on the forks, will cause the combined center of gravity to be too far forward, causing the forklift to tip forward. Turning too sharply or quickly can cause the forklift to sway or swing to the left or right, causing the combined center of gravity to veer far enough off center to tip the forklift over. These types of accidents can easily result in the operator or a bystander being crushed by the forklift.

4.9.0 Obstructing the View

It is common to try to move as much as possible in the fewest number of trips. Sometimes this causes forklift operators to stack loads too high. This may exceed the safe weight limit of the forklift, and it may also block the operator's vision. To operate a forklift safely, the operator must be able to clearly see what is in front of and behind the forklift. This should be done without the operator leaning outside of the operator's compartment. Leaning outside the operator's compartment can result in serious injuries.

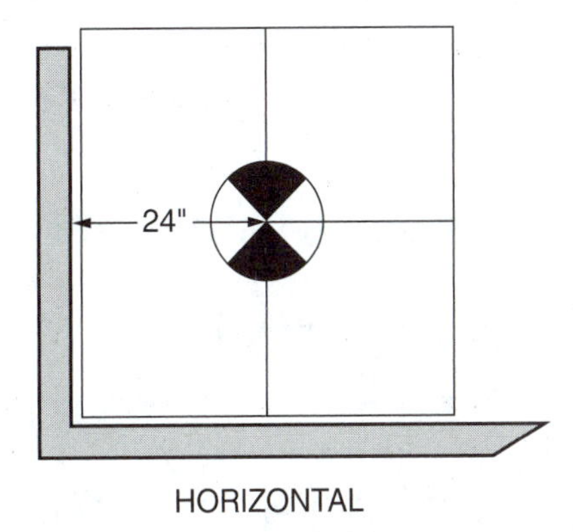

Figure 5 ◆ Center of gravity.

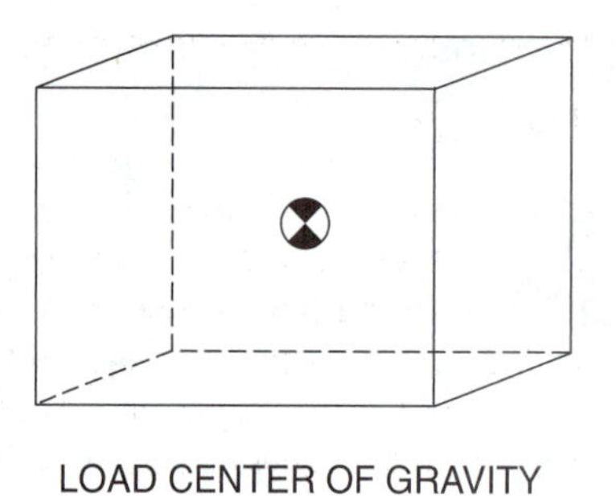

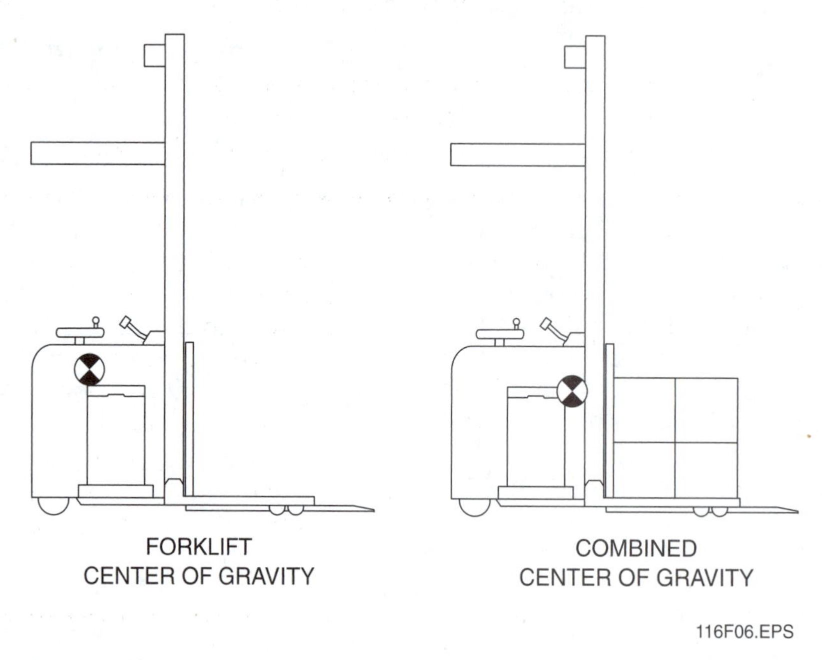

Figure 6 ◆ Combined center of gravity.

5.0.0 ◆ WORKING ON RAMPS AND DOCKS

Ramps and docks have special working conditions with specific safety requirements. Ramps allow wheeled vehicles to move easily from one level to the next. However, going up and down an angled surface has an impact on the forklift's center of gravity, making it easier for the forklift to tip. Operators can be crushed by or thrown from the forklift if this happens. Docks elevate the driving surface to a convenient height for the loading and unloading of over-the-road (OTR) trucks and rail cars. However, they also create a risk that the forklift might fall off the edge of the dock.

5.1.0 Ramps

The ramp's grade increases the tipping hazard for the forklift. Follow these rules when working on ramps:

- Keep the load as low as possible.
- Always keep the load uphill.
- When working on any graded surface, make sure your load is pushed as far back onto the forks as possible.
- If possible, tilt the forks so that the load is level with the graded surface.
- Do not turn or make quick starts or stops on a ramp.

5.2.0 Docks

It is possible to drive a forklift off the edge of a dock if you are not careful and attentive. Not only can this damage the equipment, it can also kill or injure the operator and anyone near the area of the falling load.

Forklift operators need to be aware of the edge of the dock. The edge is normally painted a bright color, such as yellow. Besides wheel chocks, devices called dock plates may be used to smoothly bridge the gap between a dock and the floor of a truck.

Chock the Wheels

As a forklift passes from a dock to an over-the-road truck and back, the force can slowly move the truck forward if the wheels are not properly chocked. In one case, a forklift operator incorrectly assumed that the truck driver had chocked the wheels of the truck. Every time the forklift passed from the dock onto the truck, the truck moved away from the dock. Because the operator was concentrating on his job, he did not notice that the gap between the dock and the truck was growing. After several trips in and out of the truck, the gap was wide enough for the forklift's front wheels to fall into the gap. This caused the forklift to tip forward into the truck and tip the truck trailer backward, crushing the operator between the truck and the forklift.

The Bottom Line: Never assume that the wheels are chocked properly. Always chock the wheels yourself or check to make sure it was done correctly.

Source: The Occupational Safety and Health Administration (OSHA)

6.0.0 ◆ FIRE AND EXPLOSION HAZARDS

Forklifts use fuels such as gasoline, liquid propane (LP) gas, and diesel fuel. All of these fuels are capable of causing a fire or explosion if not handled properly. In addition, LP gas is stored in cylinders under pressure, creating an explosion hazard if the cylinder is exposed to extreme heat or fire. It is extremely important to keep these fuels away from any source of fire, and to keep the areas in which the forklift is used free of any flammable materials. There are specific precautions that must be taken to avoid the possibility of a fire or explosion.

6.1.0 Fire Prevention

The best way to prevent a fire is to make sure that the three elements needed for fire—fuel, oxygen, and heat—are never present in the same place at the same time. Here are some basic safety guidelines for fire prevention:

- Always work in a well-ventilated area, especially when you are using flammable materials such as shellac, lacquer, paint stripper, or construction adhesives.
- Never smoke or light matches when you are working with flammable materials.
- Keep oily rags in approved, self-closing metal containers.
- Store combustible materials only in approved containers.
- Know where to find fire extinguishers, what kind of extinguisher to use for different kinds of fires, and how to use the extinguishers.
- Keep open fuel containers away from any sources of sparks, fire, or extreme heat.
- Make sure all extinguishers are fully charged. Never remove the tag from an extinguisher; it shows the date the extinguisher was last serviced and inspected.
- Don't fill a gasoline or diesel fuel container while it is resting on a truck bed liner or other ungrounded surface. The flow of fuel creates static electricity that can ignite the fuel if the container is not grounded.

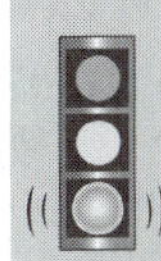

NOTE

For detailed information on fire prevention and protection, refer to the *Fire Prevention and Protection* module in Volume Two.

6.1.1 Flammable and Combustible Liquids

Liquids can be flammable or combustible. Flammable liquids have a flash point below 100°F. Combustible liquids have a flash point at or above 100°F. Fires can be prevented by doing the following things:

- *Removing the fuel* – Liquid does not burn. What burns are the gases (vapors) given off as the liquid evaporates. Keeping the liquid in an approved, sealed container prevents evaporation. If there is no evaporation, there is no fuel to burn.
- *Removing the heat* – If the liquid is stored or used away from a heat source, it cannot ignite.
- *Removing the oxygen* – The vapor from a liquid will not burn if oxygen is not present. Keeping safety containers tightly sealed prevents oxygen from coming into contact with the fuel.

There's No Such Thing as a Free Ride

It was the end of a hard day. Knowing the forklift could quickly and easily travel several hundred yards to the site office, a worker hitched a ride on the forks. The forklift began to bounce from front to back as it passed over a particularly rough road. The worker was thrown off in front of the moving forklift and crushed to death.

The Bottom Line: Never hitch a ride on a forklift.

Source: The Occupational Safety and Health Administration (OSHA)

6.1.2 Flammable Gases

Flammable gases used on construction sites include acetylene, hydrogen, ethane, and LP gas. To save space, these gases are compressed so that a large amount can be stored in a small cylinder or bottle. As long as the gas is kept in the cylinder, oxygen cannot get to it and start a fire. The cylinders must be handled carefully and stored away from sources of heat.

Oxygen is also classified as a flammable gas. If it is allowed to escape and mix with another flammable gas, the resulting mixture can explode.

6.2.0 Fire Fighting

You are not expected to be an expert firefighter, but you may have to deal with a fire to protect yourself and others. You need to know the location of fire-fighting equipment on your job site. You also need to know which equipment to use on different types of fires. However, only qualified personnel are authorized to fight fires.

Most companies tell new employees where fire extinguishers are kept. If you have not been told, be sure to ask. Also ask how to report fires. The telephone number of the nearest fire department should be clearly posted in your work area. If your company has a company fire brigade, learn how to contact them. Learn your company's fire-safety procedures.

7.0.0 ◆ PEDESTRIAN SAFETY

You may not be a forklift operator, but you will probably work on a site with forklifts. Here are some guidelines for working safely around forklifts.

Remember that it may be difficult for the operator to see you. The operator may be concentrating on the load and may not be paying attention to pedestrians. Therefore, always assume that the forklift operator doesn't see you. Remember, the most common type of forklift accident is hitting a pedestrian.

Forklifts are heavy and usually carry large loads. This results in a large amount of momentum, which means that it will probably take a distance equal to the length of the forklift to stop it. Never risk your safety on a forklift's ability to stop in time. Make sure that you're never positioned between a forklift and an immovable object.

Sometimes forklifts carry heavy loads high overhead. Never stand under the raised forks of a forklift. Objects may fall from the forks or, if there is a sudden loss of hydraulic pressure, the forks and load may drop suddenly, crushing any people or objects beneath them.

When working in a storage room or warehouse with racking, never work in the aisle on the other side of a racking unit where a forklift is working. Occasionally, the operator may push a load into the rack, causing it to fall off the other side. If you are unlucky enough to be on the other side, there is a good chance that you will be struck or crushed by a falling object.

Never hitch a ride on a forklift. Forklifts are designed for one operator and a load. Riding on the forks creates a high risk of falling and being run over by the forklift. Riding on the tractor of the forklift is also not permitted. It is easy to fall and/or be crushed between the forklift and other objects.

Summary

Forklifts are the workhorses of a construction site. They can move, lift, and lower large, heavy, and awkward loads. They can get equipment off rail cars and out of trucks. They are heavy and powerful machines that take special training to operate. The best way to work safely around forklifts is to follow all safety procedures and stay alert to any possible hazards.

Review Questions

1. The most common type of forklift accident is the _____.
   a. forklift tipping over
   b. load falling from a forklift
   c. forklift falling off a dock
   d. forklift hitting a pedestrian

2. In order to operate a forklift you must _____.
   a. have a commercial driver's license
   b. be trained on how to operate it
   c. be trained in the use of other heavy equipment
   d. supply your employer with a copy of your driver's license

3. If you find any problems with the forklift during your pre-shift inspection, you should _____.
   a. tell the next driver
   b. leave a note for your supervisor, then use the truck until it can be repaired
   c. report the problem to your supervisor, then lock out and tag the forklift
   d. look for another forklift without any problems so you can get started

4. Leaning your head out of the operator's compartment _____.
   a. is often necessary to see around a tall load
   b. requires the proper personal protective equipment
   c. makes you vulnerable to crushing and amputation injuries
   d. is only acceptable when working outdoors

5. Pedestrians always have the right-of-way over a forklift.
   a. True
   b. False

6. It is safe to carry passengers on a forklift _____.
   a. if they wear safety harnesses
   b. if a bench is bolted to a pallet for transport
   c. when a certified personnel-lift attachment is used
   d. under no circumstances

7. A forklift may tip if its center of gravity moves too far forward.
   a. True
   b. False

8. A load can fall off the forks of a forklift for any of the following reasons *except* _____.
   a. if the forklift turns or stops too quickly
   b. if the load's center of gravity shifts too far
   c. if the load is not stacked high enough
   d. if the forklift tips too far forward

9. Internal combustion engines on a forklift increase the risk of _____.
   a. amputation
   b. getting crushed by equipment
   c. falling from a height
   d. fire and explosion

10. It is safe to work in an aisle on the other side of a racking unit where a forklift is in use.
    a. True
    b. False

GLOSSARY

Trade Terms Introduced in This Module

Center of gravity: The point around which all of an object's weight is evenly distributed.

Combined center of gravity: When the weight of two items is combined, the center of gravity shifts to one point for both items.

Momentum: A physical force that causes an object in motion to stay in motion.

Operator's compartment: The portion of a forklift where the operator is positioned to control the forklift.

Staging area: An area where materials are stored in preparation for delivery. Staging areas often have a high volume of traffic.

ACKNOWLEDGMENTS

Figure Credits

Sellick Equipment Limited	116F01, 116F03
Manitou North America, Inc.	116F04

NCCER CURRICULA — USER UPDATE

NCCER makes every effort to keep its textbooks up-to-date and free of technical errors. We appreciate your help in this process. If you find an error, a typographical mistake, or an inaccuracy in NCCER's curricula, please fill out this form (or a photocopy), or complete the online form at **www.nccer.org/olf**. Be sure to include the exact module ID number, page number, a detailed description, and your recommended correction. Your input will be brought to the attention of the Authoring Team. Thank you for your assistance.

Instructors – If you have an idea for improving this textbook, or have found that additional materials were necessary to teach this module effectively, please let us know so that we may present your suggestions to the Authoring Team.

NCCER Product Development and Revision

13614 Progress Blvd., Alachua, FL 32615

Email: curriculum@nccer.org
Online: www.nccer.org/olf

❑ Trainee Guide ❑ AIG ❑ Exam ❑ PowerPoints Other ______________________

Craft / Level: Copyright Date:

Module ID Number / Title:

Section Number(s):

Description:

Recommended Correction:

Your Name:

Address:

Email: Phone:

Module 75117-03

Lockout/Tagout

COURSE MAP

This course map shows all of the modules in Field Safety. The suggested training order begins at the bottom and proceeds up. The local Training Program Sponsor may adjust the training order.

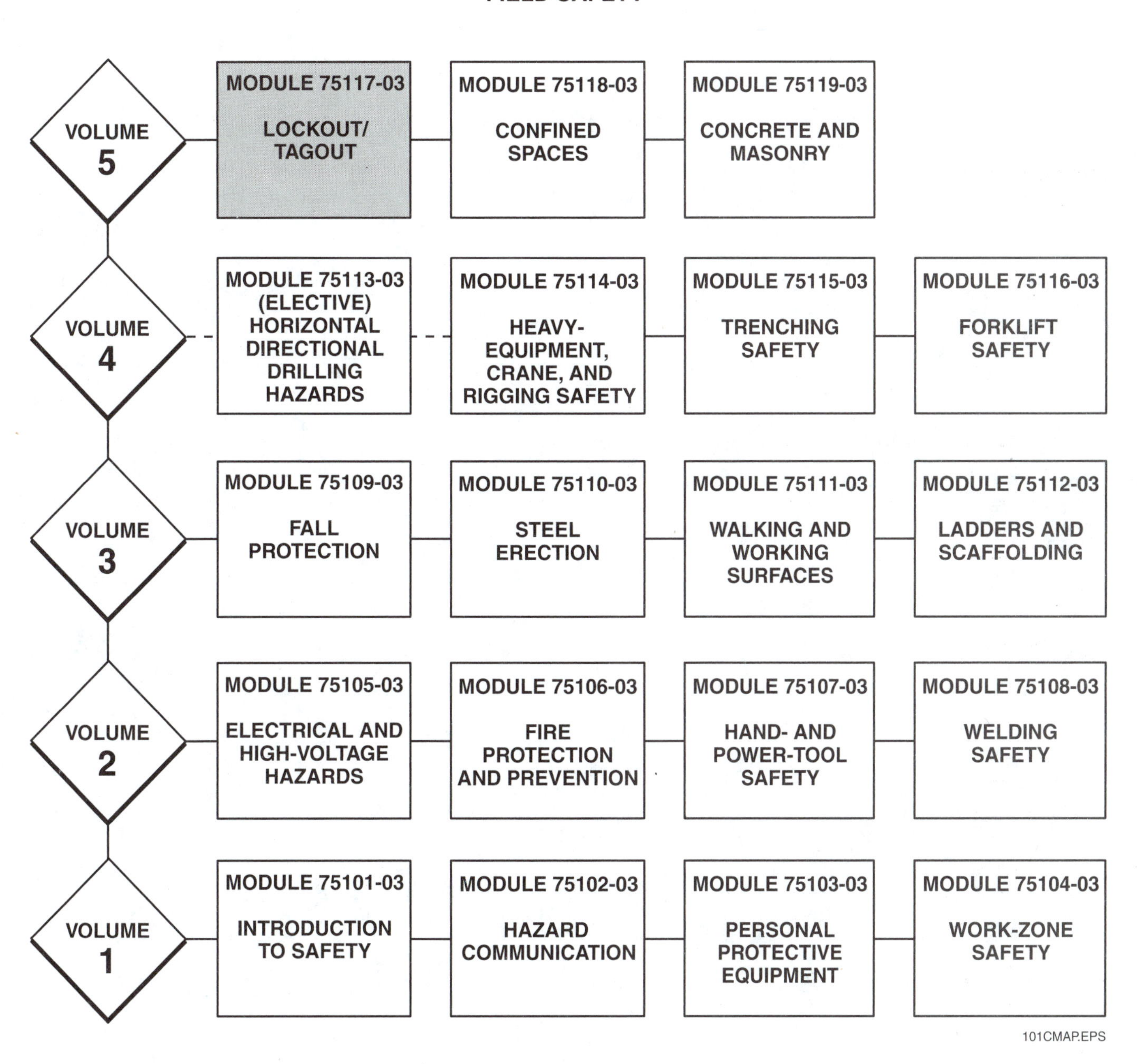

MODULE 75117-03 CONTENTS

Figures

MODULE 75117-03

Lockout/Tagout

Objectives

When you have completed this module, you will be able to do the following:

1. Describe the lockout/tagout process.
2. Identify common safety hazards associated with lockout/tagout.
3. Describe the safeguards associated with lockout/tagout.

Prerequisites

Before you begin this module, it is recommended that you successfully complete the following: Field Safety, Modules 75101-03 through 75112-03, and 75114-03 through 75116-03. *Horizontal Directional Drilling Hazards* (module 75113-03) is an elective and is not required for successful completion of this course.

Required Materials

1. Pencil and paper
2. Appropriate personal protective equipment

1.0.0 ◆ INTRODUCTION

Many accidents on a job site involve machinery and equipment. These accidents often happen because there is an uncontrolled release of energy. Uncontrolled releases of energy can cause machinery to start up while being serviced or maintained. Failure to **lockout** and **tagout** machinery before working on it is a major cause of injury and death on job sites. Some of those injuries include electrocution, amputation of body parts, and severe crushing injuries. In one instance, a maintenance worker was killed when he was knocked off balance by equipment that moved unexpectedly, knocking him into the spinning blade of a 36" saw.

The proper implementation of an effective lockout/tagout system can eliminate this type of hazard. It safeguards workers from unexpected releases from various **energy sources**. An energy source is any source of electrical, mechanical, hydraulic, pneumatic, chemical, thermal, or other energy. Energy can be stored in a system or machine even after the power is shut off. Stored energy in a system must be released or grounded to prevent unexpected operation of the machine while it is being serviced. For example, although the power has been shut off, residual air pressure remains in a compressor, which can activate pneumatic tools.

There is an additional danger if the machine contains acids, chemicals, flammable liquids, or high-temperature liquids or gases. Stored energy may cause the explosive release of these hazardous materials, resulting in serious injuries.

When anyone is working on or around any of these hazards, mechanical and other systems must be shut down, drained, or de-energized. **Energy-isolating devices,** such as switches, circuit breakers, valves, or other components, are used with tags and locks (*Figure 1*) to make sure that motors aren't started, valves aren't opened or closed, and any other changes that would endanger workers cannot be made.

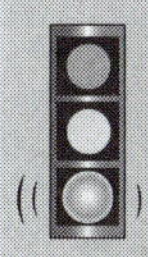

NOTE

Electrical hazards are discussed in more detail in the *Electrical and High-Voltage Hazards* module in Volume Two.

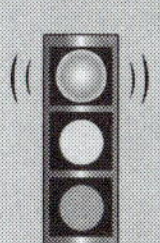

WARNING!

Electrical equipment is not the only source of lockout/tagout accidents.

117F01.EPS

Figure 1 ◆ Lockout/tagout device.

NOTE

Some disconnect devices are equipped with keyed interlocks for protection during operation. These locks are called kirklocks and are relied upon to ensure proper sequence of operation only. They are not to be used for the purpose of locking out a circuit or system. Where disconnects are installed for use in isolation, they should never be opened under load. When opening a disconnect manually, it should be done quickly using a positive force. Again, lockouts should be used when the disconnects are open.

Generally, each lock has its own key, and the person who puts the lock on keeps the key. That person is the only one who can remove the lock. Tags typically have the word *DANGER* on them (*Figure 2*).

2.0.0 ◆ LOCKOUT/TAGOUT

Lockouts and tagouts protect workers from all possible sources of energy. In a lockout, an energy-isolating device such as a disconnect switch or circuit breaker is placed in the OFF position and locked. *Figure 3* shows some examples of the types of **lockout devices** that can be used with key or combination locks. Multiple lockout

117F02.EPS

Figure 2 ◆ Typical safety tags.

CASE HISTORY

Near Miss

While maintenance workers were de-coupling an actuator from a 100-psi steam control valve, the packing gland was removed from the valve. The valve was not under lockout/tagout control. As a result, the wrong fasteners were removed. This caused the packing gland follower fasteners and the packing bore components to be expelled from the valve due to residual steam pressure. Fortunately, no one was injured, but the valve was destroyed and had to be replaced.

The Bottom Line: A lockout/tagout program can prevent accidents from the unexpected release of steam, spring, or hydraulic energy. Lockout/tagout is not limited to electrical safety.

Source: U.S. Department of Energy

devices (*Figure 4*) are used when more than one person has access to the equipment. In some instances, as with work involving valves, a chain and lock can be used to hold the valve in place and keep it from being turned.

In a tagout, components that power up equipment and machinery, such as switches, are set in a safe position and a written warning or **tagout device** is attached to them (*Figure 5*).

An effective lockout/tagout program should include the following:

- An inspection of equipment by a trained individual who is thoroughly familiar with the equipment operation and associated hazards.
- Identification and labeling of lockout devices.
- The purchase of locks, tags, and blocks.
- A standard written operating procedure that is followed by all employees.

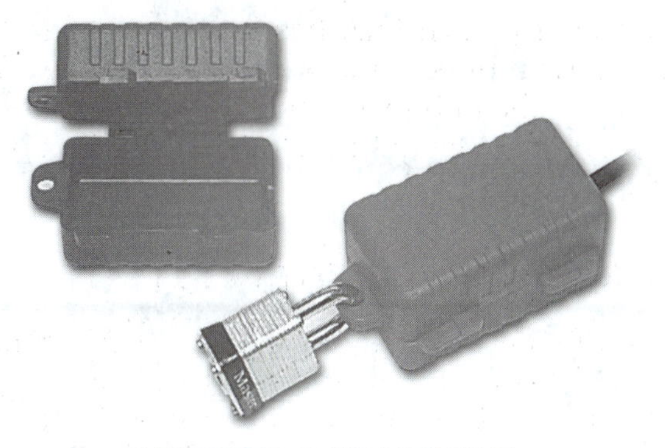

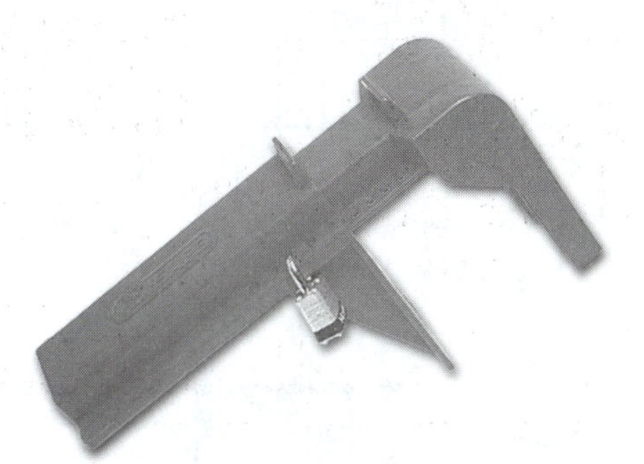

Figure 3 ◆ Lockout devices.

Figure 4 ◆ Multiple lockout/tagout device.

117F05.EPS

Figure 5 ◆ Placing a lockout/tagout device.

3.0.0 ◆ LOCKOUT/TAGOUT PROCEDURES

The exact procedures for lockout/tagout may vary at different companies and job sites. Ask your supervisor to explain the lockout/tagout procedure on your job site. You must know this procedure and follow it. This is for your safety and the safety of your co-workers. If you ever have questions about lockout/tagout procedures, ask your supervisor.

The following is an example of a typical lockout/tagout procedure. It is made up of these four components.

- Preparation for lockout/tagout
- Sequence for lockout/tagout
- Restoration of energy
- Emergency removal authorization

3.1.0 Preparation for Lockout/Tagout

When preparing for a lockout/tagout, it is important to follow these steps.

Step 1 Check the procedures to ensure that no changes have been made since you last used a lockout/tagout.

Step 2 Identify all authorized and affected employees involved with the pending lockout/tagout.

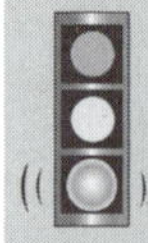

NOTE

There is a difference between turning off a machine and actually disengaging a piece of equipment. When turning off a control switch, you are opening a circuit; however, there is still electrical energy at the switch. A short in the switch or someone turning on the machine may start it running again.

Flood Warning

The threaded male end of a ½" pressure-relief valve (PRV) on a sprinkler system riser broke off during an attempt to re-pipe the discharge side of the valve. This caused approximately 550 gallons of water to be discharged into a utility closet and surrounding office areas. There were no injuries as a result of this event, but there was significant property damage.

The Bottom Line: A lockout/tagout should have been performed on the sprinkler system.

Source: U.S. Department of Energy

Harmful Energy

An employee was attempting to correct an electrical problem involving two non-operational lamps. He did not shut off the power at the circuit breaker panel nor did he test the wires to see if they were live. He was killed instantly when he fell from the ladder after grabbing the two live wires with his left hand.

The Bottom Line: Do not work on electrical circuits unless they are positively de-energized and tagged out.

Source: The Occupational Safety and Health Administration (OSHA)

P.R.O.P.E.R. Lockout/Tagout Procedures

Using P.R.O.P.E.R lockout/tagout procedures is the best way to prevent accidents, injury, and death.

P – Process shut down.
R – Recognize energy sources.
O – Off (shut energy sources off).
P – Place locks and tags.
E – Release stored energy.
R – Return controls to neutral.

3.2.0 Sequence for Lockout/Tagout

To perform a lockout/tagout, follow these steps.

Step 1 Notify all authorized and affected personnel that a lockout/tagout is to be used and explain why it is necessary.

Step 2 Shut down the equipment or system using the normal OFF or STOP procedures.

Step 3 Lock out all energy sources and test disconnects to be sure they cannot be moved to the ON position. Next, open the control cutout switch. If there is no cutout switch, block the magnet in the switch open position before working on electrically operated equipment or apparatus such as motors or relays. Remove the control wire.

Step 4 Lock and tag the required switches or valves in the open position. Each authorized employee must affix a separate lock and tag, if necessary.

Step 5 Dissipate any stored energy by attaching the equipment or system to ground.

Step 6 Verify that the test equipment is functional using a known power source.

Step 7 Confirm that all switches are in the open position and use test equipment to verify that all components are de-energized.

Step 8 If it is necessary to temporarily leave the area, upon returning, retest to ensure that the equipment or system is still de-energized.

3.3.0 Restoration of Energy

After work is done on the machinery or equipment, use these steps to restore energy.

Step 1 Completely reassemble and secure the equipment or system.

Step 2 Confirm that all equipment and tools, including shorting probes, are accounted for and removed from the equipment or system.

Step 3 Replace and reactivate all of the safety controls.

Step 4 Remove the locks and tags from the isolation switches. Each employee must remove his or her own lock and tag.

Step 5 Notify all affected personnel that the lockout/tagout has ended and the equipment or system will be re-energized.

Step 6 Operate or close the isolation switches to restore energy.

3.4.0 Emergency Removal Authorization

There are times when emergency removal of a lockout/tagout device is required. The device can only be removed by an authorized supervisor. Follow these guidelines if a lockout/tagout device

Lockout Method

Which lockout method would you rather use if your life was at stake?

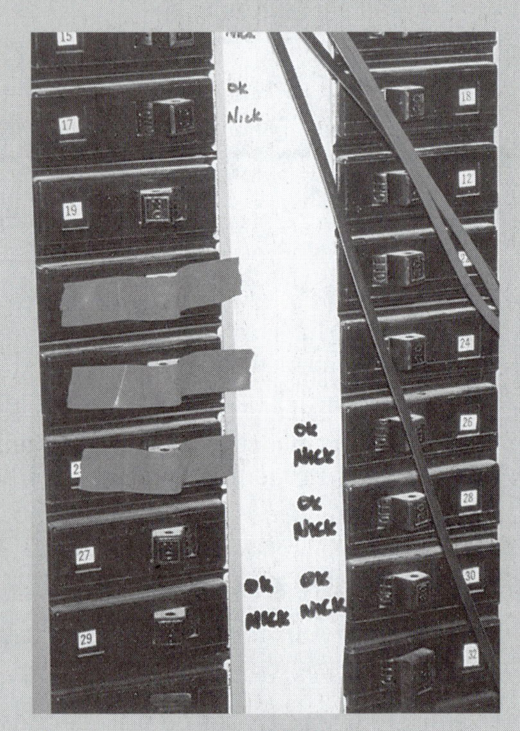

(A) TAPE

(B) CIRCUIT BREAKER LOCK

117SA01.EPS

is left secured, and the authorized employee is absent, or the key is lost.

- The authorized employee must be informed that the lockout/tagout device has been removed.
- Written verification of the action taken, including informing the authorized employee of the removal, must be recorded in the job journal.

4.0.0 ◆ SAFEGUARDS

It is important to know how to protect yourself and your co-workers on the job. Everyone must be aware of what activities are being done on the site and understand how to perform them safely. These are the most common safeguards that should be used to keep the job site safe.

- Never operate any device, valve, switch, or piece of equipment that has a lock or a tag attached to it.
- Use only tags that have been approved for your job site.
- If a device, valve, switch, or piece of equipment is locked out, make sure the proper tag is attached.
- Lock out and tag all electrical systems.
- Lock out and tag pipelines containing acids, explosive fluids, and high-pressure steam.
- Lock and tag motorized vehicles and equipment when they require repair or are being replaced. Also, disconnect or disable any starting devices.

Testing Can Save Your Life

On a site in Georgia, electricians found energized switches after the lockout of a circuit panel in an older system that had been upgraded several times. The existing wiring did not match the current site drawings. A subsequent investigation found many such situations in older facilities.

The Bottom Line: Never rely solely on drawings. The circuit must be tested after lockout to verify that it is de-energized.

Source: The Occupational Safety and Health Administration (OSHA)

Summary

Lockouts and tagouts protect workers from all possible sources of energy. In a lockout, an energy-isolating device such as a disconnect switch or circuit breaker is placed in the OFF position and locked.

In a tagout, components that power up equipment and machinery, such as switches, are set in a safe position and a written warning or tagout device is attached to them

Failure to follow your company's lockout/tagout procedures can result in serious injury or death.

Review Questions

1. An uncontrolled release of energy can be prevented by using _____.

a. circuit breakers
b. locks and tags
c. switches
d. valves

2. Lockout/tagout systems only protect workers from electrical hazards.

a. True
b. False

3. All of the following are hazards associated with lockout/tagout *except* _____.

a. acids
b. hydraulics
c. cold temperatures
d. compressed air

4. Disconnect switches and circuit breakers are examples of _____ devices.

a. energy-isolating
b. energy-removal
c. lockout/tagout
d. multiple lockout

5. A chain and a lock can be used as a lockout device for some valves.

a. True
b. False

6. All of the following are components of a lockout/tagout system *except* ________.

a. preparation for lockout/tagout
b. sequence for lockout/tagout
c. restoration of energy
d. distribution of keys

7. The first step in the lockout/tagout sequence is to _____.

a. shut down the equipment
b. dissipate any stored energy
c. verify that the test equipment is functional
d. notify all authorized and affected personnel

8. Energy can be restored to repaired equipment or machinery _____.

a. at the beginning of each day
b. when workers come back after lunch
c. after repair work has been completed
d. when a new shift begins

9. The only person authorized to perform an emergency removal of a lockout/tagout device is a(n) _____.

a. authorized contractor
b. authorized supervisor
c. rescue worker
d. OSHA inspector

10. All of the following are common lockout/tagout safeguards *except* _____.

a. lock and tag all electrical equipment
b. never operate any device, valve, switch, or piece of equipment that has a lock or tag attached to it
c. use only tags that have been approved for your job site
d. tag motorized vehicles only when there will be multiple users

PROFILE IN SUCCESS

Dennis Truitt, The TIMEC Group of Companies

Vice President,
Corporate Health and Safety

"My biggest contribution to health and safety, which is both a personal and company-endorsed accomplishment, is the annual reduction in the number of injuries our employees experience. By utilizing proven loss-control techniques, we have been able to make a positive impact on health and safety.... These efforts directly impact and play an important role in the success of our company."

How did you choose a career in the Safety Industry?
After earning a degree in psychology, I decided to get into something that was behavior based, but also business driven.

What types of training have you been through?
I have been involved with a variety of curricula, with my main focus being on health and safety. As well, I have earned regulatory and standards-driven education-based certifications. I have also been involved in many emergency-response and business-based curricula.

What kinds of work have you done in your career?
I have had the pleasure of working for both contractors and clients in the health and safety field. I have done everything from field inspecting and auditing to developing, implementing, and maintaining comprehensive loss-control programs from both the field side and the management side of the business.

Tell us about your present job.
I currently manage and provide a corporate-level advisory service to all of the organizations within The TIMEC Group of Companies, throughout our multiple regional offices and field working sites. My responsibilities include all aspects of the loss-control model and our companies' business strategies. I also maintain adherence to all regulatory agency and client-driven business practices, standards, and requirements.

What factors have contributed most to your success?
Training; management commitment to safety; a positive, proactive approach to new ideas; and utilization of proven, field-tested loss-control policies and procedures with an emphasis on identification and mitigation of substandard acts/conditions have contributed the most to my success.

What advice would you give to those new to Safety industry?
Approach the industry with as much knowledge as you can hold on to. Never stop educating and re-educating yourself. Realize that standards can only be adhered to if a person understands the responsibility he/she plays in safety. Teach your people to all be safety representatives for the sakes of your company, themselves, and their families. Safety is unnerving and reactive when folks are injured or ill, but when your personnel return home in exactly the same way they came to work, the rewards are priceless.

GLOSSARY

Trade Terms Introduced in This Module

Energy-isolating device: Any mechanical device that physically prevents the transmission or release of energy. These include, but are not limited to, manually operated electrical circuit breakers, disconnect switches, line valves, and blocks.

Energy source: Any source of electrical, mechanical, hydraulic, pneumatic, chemical, thermal, or other energy.

Lockout: The placement of a lockout device on an energy-isolating device, in accordance with an established procedure, ensuring that the energy-isolating device and the equipment being controlled cannot be operated until the lockout device is removed.

Lockout device: Any device that uses positive means such as a lock to hold an energy-isolating device in a safe position, thereby preventing the energizing of machinery or equipment.

Tagout: The placement of a tagout device on an energy-isolating device, in accordance with an established procedure, to indicate that the energy-isolating device and the equipment being controlled may not be operated until the tagout device is removed.

Tagout device: Any prominent warning device, such as a tag and a means of attachment, that can be securely fastened to an energy-isolating device in accordance with an established procedure. The tag indicates that the machine or equipment to which it is attached is not to be operated until the tagout device is removed in accordance with the energy-control procedure.

Figure Credits

Veronica Westfall	117F01
Marzetta Signs	117F02
North Safety	117F03, 117F05, 117SA01 (B)
Mike Powers	117SA01 (A)

NCCER CURRICULA — USER UPDATE

NCCER makes every effort to keep its textbooks up-to-date and free of technical errors. We appreciate your help in this process. If you find an error, a typographical mistake, or an inaccuracy in NCCER's curricula, please fill out this form (or a photocopy), or complete the online form at **www.nccer.org/olf**. Be sure to include the exact module ID number, page number, a detailed description, and your recommended correction. Your input will be brought to the attention of the Authoring Team. Thank you for your assistance.

Instructors – If you have an idea for improving this textbook, or have found that additional materials were necessary to teach this module effectively, please let us know so that we may present your suggestions to the Authoring Team.

NCCER Product Development and Revision

13614 Progress Blvd., Alachua, FL 32615

Email: curriculum@nccer.org
Online: www.nccer.org/olf

❑ Trainee Guide ❑ AIG ❑ Exam ❑ PowerPoints Other ______________________________

Craft / Level: Copyright Date:

Module ID Number / Title:

Section Number(s):

Description:

Recommended Correction:

Your Name:

Address:

Email: Phone:

Module 75118-03

Confined Spaces

COURSE MAP

This course map shows all of the modules in Field Safety. The suggested training order begins at the bottom and proceeds up. The local Training Program Sponsor may adjust the training order.

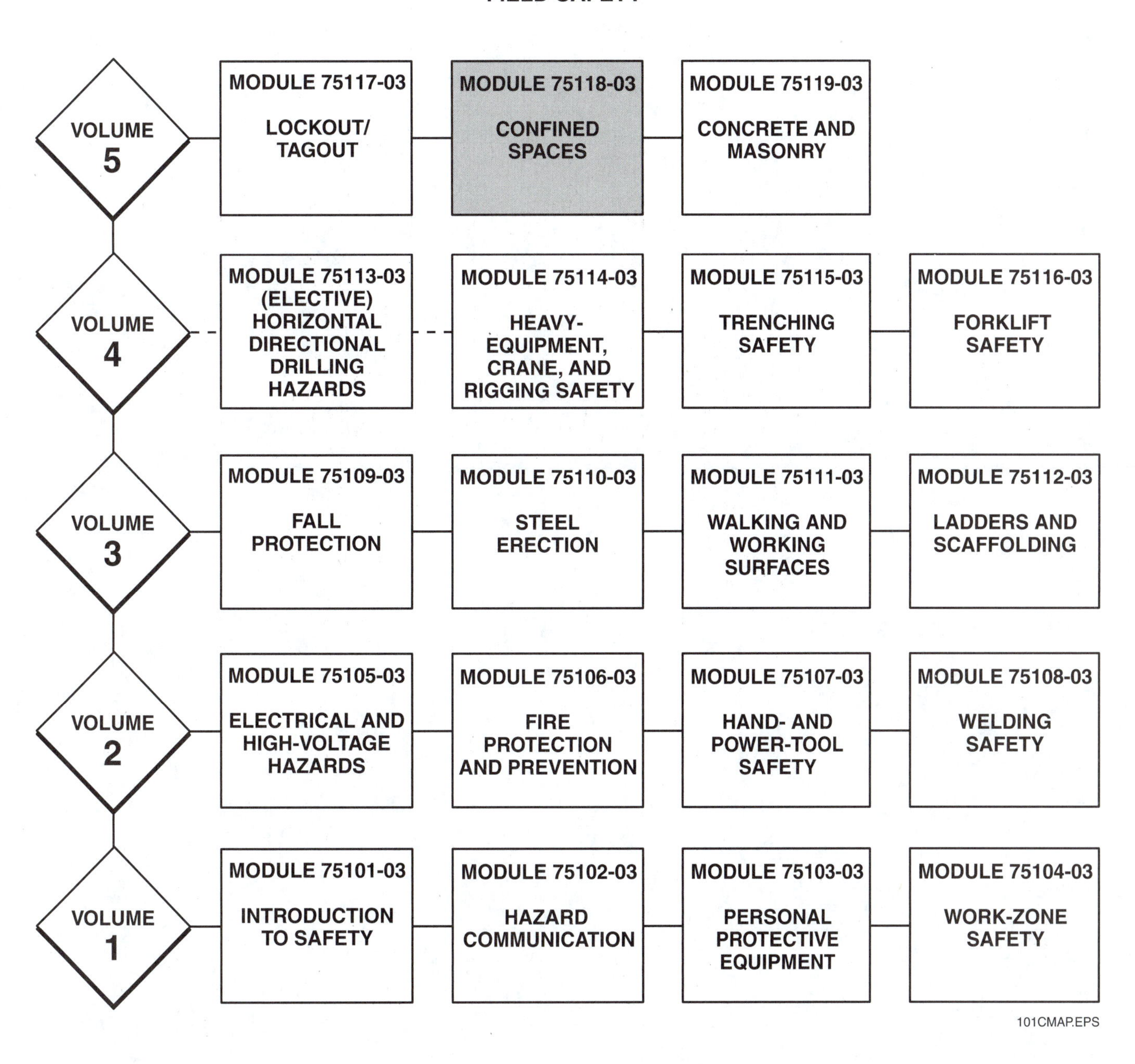

MODULE 75118-03 CONTENTS

Figures

MODULE 75118-03

Confined Spaces

Objectives

When you have completed this module, you will be able to do the following:

1. Describe the difference between a permit-required confined space and a nonpermit-required confined space.
2. Explain the purpose of an entry permit.
3. Explain the hazards associated with confined spaces.
4. Describe the responsibilities of all workers on the site.
5. Demonstrate and explain proper on-site safety and emergency-response procedures.

Prerequisites

Before you begin this module, it is recommended that you successfully complete the following: Field Safety, Modules 75101-03 through 75112-03, and 75114-03 through 75117-03. *Horizontal Directional Drilling Hazards* (module 75113-03) is an elective and is not required for successful completion of this course.

Required Materials

1. Pencil and paper
2. Appropriate personal protective equipment

1.0.0 ◆ INTRODUCTION

Spaces on a job site are considered confined when their size and shape restrict the movement of anyone who must enter, work in, and exit the space. **Confined spaces** often have poor ventilation and are difficult to enter and exit. For example, employees who work in process vessels generally must squeeze in and out through narrow openings and perform their tasks while in a cramped or awkward position. In some cases, confinement itself creates a hazard.

Culvert Check Saves a Life

One lucky worker checked out a culvert before entering it and found a nest of rattlesnakes.

The Bottom Line: Always be sure to look inside the confined space before entering.

In a confined space, hazards such as poor air quality, toxins, explosions, fire, and moving machinery parts tend to be far more deadly. In fact, over an 8-year period in the 1990s an average of 38 workers per year were killed in confined spaces.

Confined spaces may contain unknown hazards. In one instance, a worker was lowered into a 21'-deep manhole on a looped chain seat. Twenty seconds after entering the manhole, he started gasping for air and fell. He landed face down in the water at the bottom of the manhole. An autopsy determined that he died from lack of oxygen.

Most confined spaces have restricted entrances and exits. Workers are often injured as they enter or exit through small doors and hatches. It can also be difficult to move around in a confined space and workers can be struck by moving equipment. Escapes and rescues are much more difficult in confined spaces than they are elsewhere.

Confined spaces are entered for inspection, equipment testing, repair, cleaning, or emergencies. They should only be entered for short periods of time.

Some of the confined spaces you may work in include:

- Manholes
- Boilers
- Trenches
- Tunnels
- Sewers
- Underground utility vaults
- Pipelines
- Pits
- Air ducts
- Process vessels (*Figure 1*)

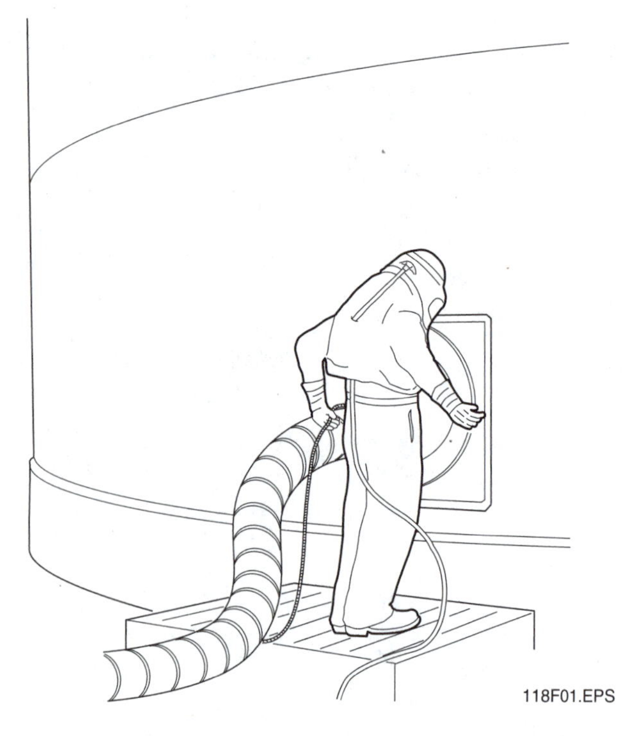

Figure 1 ◆ Entry into a process vessel.

A written confined-space entry program can protect you. It will identify the hazards and specify the equipment or support that is needed to avoid injury. All industrial and some construction sites have written confined space entry programs. You need to know and follow your company's policy.

2.0.0 ◆ CONFINED-SPACE CLASSIFICATION

Confined spaces must be inspected before work can begin. This helps to identify possible hazards. After an inspection by a company-authorized person, the confined space is classified based on any hazards that are present. The two classifications are nonpermit-required and permit-required. See *Figure 2*.

WARNING!

No one may enter a confined space unless they have been properly trained and have a valid entry permit.

Figure 2 ◆ Nonpermit-required and permit-required confined spaces.

2.1.0 Nonpermit-Required Confined Space

A **nonpermit-required confined space** is a work space free of any mechanical, physical, electrical, and **atmospheric hazards** that can cause death or injury. After a space has been classified as a nonpermit-required space, workers can enter using the appropriate personal protective equipment for the type of work to be performed. Always check with your supervisor if it is unclear what personal protective equipment is required. A space is considered confined if it:

- Is large enough and so configured that an employee can bodily enter and perform assigned work
- Has a limited or restricted means of entry or exit (such as tanks, vessels, silos, storage bins, hoppers, vaults, and pits)
- Is not designed for continuous employee occupancy

2.2.0 Permit-Required Confined Space

A **permit-required confined space** is a confined space that has real or possible hazards. These hazards can be atmospheric, physical, electrical, or mechanical. *OSHA CFR 1910.146* defines a permit-required confined space as a confined space that:

- Contains or has the potential to contain a hazardous atmosphere
- Contains a material that has the potential for engulfing an entrant
- Has an internal configuration such that an entrant could be trapped or asphyxiated by inwardly converging walls or by a floor that slopes downward and tapers to a small cross-section
- Contains any other recognized serious safety or health hazard

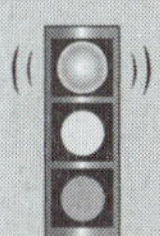

WARNING!

All permit-required confined spaces must be identified, and the associated permit must be posted.

An entry permit must be issued and signed by the job-site supervisor before the confined space is entered. No one is allowed to enter a confined space unless there is a valid entry permit. The permit is to be kept at the confined space while work is being performed. Always check with your supervisor if it is unclear whether or not you need a permit to enter a confined space.

3.0.0 ◆ ENTRY PERMITS

Confined spaces can be extremely dangerous. Entry into the space begins when any part of your body passes the entrance or opening of a confined space. Before entering a permit-required confined space, you must have an entry permit (*Figure 3*).

An entry permit is a job checklist that verifies that the space has been inspected. It also lets everyone on the site know about the hazards of the job. All entry permits must be filled out and signed by the supervisor before anyone enters the space. The permit must also be posted at the entrance to the site and be available for workers to review. Entry permits must include the following information:

- A description of the space and the type of work that will be done
- The date the permit is valid and how long it lasts
- Test results for all atmospheric testing including oxygen, toxin, and flammable material levels
- The name and signature of the person who did the tests
- The name and signature of the entry supervisor
- A list of all workers, including supervisors, who are authorized to enter the site
- The means by which workers and supervisors will communicate with each other
- Special equipment and procedures that are to be used during the job
- Other permits needed for work done in the space, such as welding
- The contact information for the emergency response rescue team

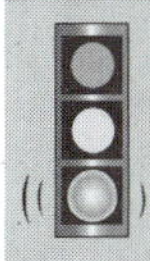

NOTE

Some confined spaces require both a permit to enter and a permit to work in the space.

4.0.0 ◆ HAZARDS

Confined spaces are dangerous. The main hazards include poor airflow and restricted movement. Poor ventilation can allow toxic gases to build up. Physical hazards are more dangerous because escape is limited.

Attachment 16 – 2
Confined-Space Entry Permit

Master Card / Safe Work Ticket No. ______________

1. **Work Description:** ______________________________
   Equip. Name / Number & Location or Area ______________________________
   Purpose of Entry ______________________________
   Valid Start Date ______________ Duration Time ________ to ________

2. **Hazardous Materials:**
   What did the equipment last contain? ______________________________
   Will the work generate a hazardous atmosphere? ☐ Yes ☐ No If yes, specify hazards and controls.
   ____

3. **Rescue Requirements:**
   ☐ External, by attendant ☐ Complex Rescue, by rescue team at point of entry
   ☐ Non-IDLH and/or Simple Rescue, by rescue team on-site ☐ IDLH, by rescue team at point of entry
   Has the rescue team been notified of the entry? ☐ Yes ☐ N/A Time of notification ____________
   How will the rescue team be summonsed for an emergency? ☐ Radio Channel: ____ ☐ Other: ________

4. **Gas Test Requirements:**
   LEL/O_2 - Instrument Mfg./No. ______/________ Bump Check Time/Gas Tester - ______/______
   Toxicity - Instrument Mfg./No. ______/________ Bump Check Time/Gas Tester - ______/______
   Frequency of Testing: ☐ Continuous ☐ Other - Specify - ______________________

- Continuous monitoring results must be recorded every three hours

Acceptable Levels	Results								
Oxygen: 19.5% - 23.5%									
Combustible Gas: %LEL - <10%									
Other ________ < PEL* ______									
Other ________ < PEL* ______									
Other ________ < PEL* ______									

- Entry in excess of the PEL will require appropriate PPE.

5. **Ventilation / Exhaust Equipment:**
   ☐ None required, natural ventilation adequate ☐ Forced air ventilation ☐ Exhaust ventilation
   Equipment Type: ☐ Air powered horn ☐ Electric blower Volume Required - ________ cfm

6. **Personal Protection:**
   ☐ Gloves (type) ______________ ☐ Respirator (type) ______________________
   ☐ Goggles or face shield ☐ Self Contained Breathing Equipment
   ☐ Lifelines Attached to Harness ☐ Other, specify: ______________________
   ☐ Chemical Resistant Suit, Specify Type ______________________

7. **Fire Protection:** ☐ None required ☐ Portable Fire Extinguisher – type and size: ____________
   ☐ Fire Watch ☐ Other, specify: ______________________

118F03A.EPS

Figure 3 ◆ Entry permit (1 of 2).

8. Condition of Area and Equipment:

Required Yes	N/A	THESE KEY POINTS MUST BE CHECKED
		a. Equipment locked and tagged out?
		b. Piping is disconnected, capped or plugged and/or blinded.
		c. Equipment emptied, washed, purged & ventilated?
		d. Low voltage or GFCI protected equipment provided?
		e. Explosion proof electrical equipment provided?
		f. Provisions are made to barricade or post signs at entry points when attendant is not on duty.

Other Requirements:

9. Special Instructions: ☐ None ☐ Check with issuer before starting work

10. Approval	Permit			Permit Acceptance		
	Supt. / Area Supv.	Date	Time	Maint. Supv. / Engineer / Contractor Supv.	Date	
Issued by						
Endorsed by						
Endorsed by						

11. Individual Review / Entrant Roster: I have been instructed in the proper Work Permit, Confined Space Entry, Lockout/Tagout Procedures, associated physical and atmospheric hazards and have reviewed the gas testing results.

Entrants	Date	Time In / Out		Time In / Out		Time In / Out		Time In / Out		Time In / Out	

I have been informed of the duties and responsibilities for an attendant, the associated physical and atmospheric hazards, and have reviewed the gas testing results.

Attendants	Date	Time On / Off		Time On / Off		Time On / Off	

12. Job Completion:

☐ Yes ☐ N/A Has the rescue team been notified?
☐ Yes ☐ No Is the work on equipment complete & the confined space ready to return to service?
☐ Yes ☐ No Has the worksite been cleaned and made safe?
Workers answering above questions: ______________________________

13. Post Job Review: Were any hazards encountered or created during entry operations?
☐ Yes ☐ No If yes, describe: ______________________________
Possible solutions: ______________________________

Forward to job file within 7 days of job completion.

118F03B.EPS

Figure 3 ◆ Entry permit (2 of 2).

4.1.0 Atmospheric Hazards

Atmospheric hazards are the most common hazards in a confined space. In a hazardous atmosphere, the air can have either too little or too much oxygen, be explosive or flammable, or contain toxic gases. Special meters are used to detect these atmospheric hazards (*Figure 4*).

> **Think About It**
>
> ### *Confined-Space Entry*
>
> What safe and unsafe behaviors do you see in these photographs? How could they be corrected?
>
> 118SA01.EPS
>
> 118SA02.EPS

4.1.1 Oxygen-Deficient or Oxygen-Enriched Atmospheres

A confined space that does not have enough oxygen is called an **oxygen-deficient** atmosphere; a confined space that has too much oxygen is called an **oxygen-enriched** atmosphere. For safe working conditions, the oxygen level in a confined space

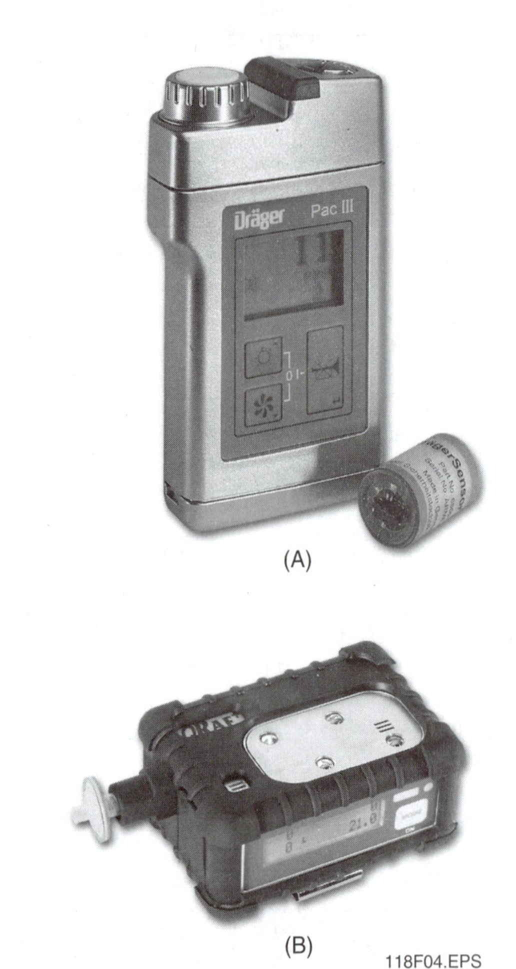

Figure 4 ◆ Detection meters.

Added Oxygen Causes Lost Life

A welder entered a 24"-diameter steel pipe to grind a bad weld on a valve about 30' from the entry point. Before he entered, other crew members decided to add oxygen to the pipe near the bad weld to make sure the air was safe. The welder had been grinding off and on for about five minutes when a fire broke out. The fire covered his clothing. He was pulled from the pipe and the fire was put out. The burns were so serious that the welder died the next day.

The Bottom Line: This accident could have been avoided. It happened because of poor communication between workers and unsafe work practices.

Source: The Occupational Safety and Health Administration (OSHA)

must range between 19.5% and 23.5% by volume, with 21% being considered the normal level. Oxygen concentrations below 19.5% by volume are considered oxygen-deficient; those above 23.5% by volume are considered oxygen-enriched.

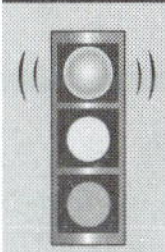

WARNING!
Never enter a confined space if you are not sure how safe it is.

Many of the processes that occur in a confined space use oxygen and may reduce the percentage of oxygen to an unsafe level. These processes include:

- Burning
- Rusting of metal
- Breaking down of plants or garbage
- Oxygen mixing with other gases

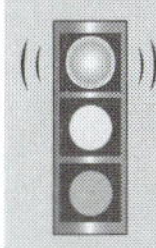

WARNING!
Normal breathing in a confined space can also create an oxygen deficiency.

When the oxygen in a confined space is reduced, it becomes harder to breathe. The symptoms of insufficient oxygen happen in this order.

1. Fast breathing and heartbeat
2. Impaired mental judgment
3. Extreme emotional reaction
4. Unusual fatigue
5. Nausea and vomiting
6. Inability to move your body freely
7. Loss of consciousness
8. Death

Too much oxygen in a confined space is a fire hazard and can cause explosions. Materials like clothing and hair are highly flammable and will burn rapidly in oxygen-enriched atmospheres. Fires can start easily in a confined space with oxygen-enriched air.

4.1.2 Combustible Atmospheres

Air in a confined space becomes **combustible** when chemicals or gases reach a certain concentration. Flammable gases can be trapped in confined spaces. These include acetylene, butane, propane, methane, and others. Dust and work by-products from spray painting or welding can also form a combustible atmosphere.

Some flammable gases are lighter than air and have a higher concentration at the top of a confined space. Vapors from fuels are generally heavier than air and will form a greater concentration at the bottom of the space.

A spark or flame will cause an explosion in a combustible atmosphere.

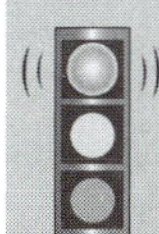

WARNING!
Explosion-proof lights, motors, exhaust fans, and other equipment must be used to prevent fires and explosions in any combustible area.

4.1.3 Toxic Atmospheres

Toxic gases and vapors come from many sources. They can be deadly when they are inhaled or absorbed through the skin above certain concentration levels. In spaces with no ventilation, high concentrations can gather and quickly become toxic. Even in lower doses, some chemicals can seriously affect your breathing and brain functions.

Four Die in Confined Space

A project involved the upgrade/replacement of a sewer pumping station and the contractor prepared a confined-space entry permit for the work. One employee was disconnecting a sewer bypass connection in a manhole while three others were at the manhole entrance. The manhole filled with sewage and gases from the sewer line, and the employee was overcome by a lack of oxygen. The other three employees tried to help. Each entered the manhole one at a time, apparently to attempt rescue. Each was overcome by the sewer gases and died.

The Bottom Line: Asphyxiation (lack of oxygen) is not like what you see in the movies. You can't go in unprotected, stagger around awhile, save a couple of lives, then exit coughing and unharmed. Reality is quite different. Upon your first breath you pass out. If a retrieval system is not in place, anyone who enters the space to rescue you will die. Many workers have died simply by putting their heads in manholes to assess a situation. Even without fully entering the confined space, these workers were immediately incapacitated.

Source: The Occupational Safety and Health Administration (OSHA)

The harmful effects of toxic gases and vapors vary. Many toxic gases, such as carbon monoxide, cannot be detected by sight or smell. Some toxic gases have harmful effects that may not show up until years after contact. Others, such as nitric oxide, can kill quickly.

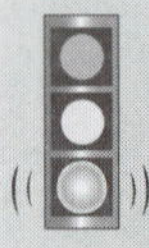

NOTE

The material safety data sheet (MSDS) attached to the entry permit will have information about any toxins you may encounter.

4.1.4 Monitoring the Atmosphere in Confined Spaces

The air inside a permit-required confined space must be tested before anyone is allowed to enter. Atmospheric tests must be done in this order:

1. Oxygen content
2. Flammable gases and vapors
3. Potential toxic contaminants

The test for oxygen must be performed first because most combustible gas meters are oxygen dependent and will not provide reliable readings in an oxygen-deficient atmosphere. The test for combustible gases is performed next because the threat of fire or explosion is usually more urgent and more life-threatening than exposure to toxic gases and vapors.

Various instruments are used to test and monitor confined-space atmospheres. Portable, battery-operated gas detection meters can measure oxygen levels by changing the sensor in the detection meter. Gas detection meters must be calibrated and operated according to the manufacturer's instructions. The meter must be able to detect oxygen and combustible gases at the levels specified in *OSHA 29 CFR 1910.146*.

4.2.0 Additional Hazards

In addition to atmospheric hazards, there are several other physical and environmental hazards in confined spaces. These hazards include:

- Electric shock
- Purging
- Falling objects
- Engulfment
- Extreme temperatures
- Noise
- Slick or wet surfaces
- Moving parts

4.2.1 Electric Shock

Electric shock can occur when power tools and line cords are used in an area where there are wet floors or surfaces. Tools and equipment should be grounded or a ground fault circuit interrupter should be used when working in a confined space.

4.2.2 Purging

Purging happens when toxic, corrosive, or natural gases enter and mix with the air in a confined space. Purging is most often experienced when working in pipes. Purging gases are used to clean pipelines. These gases can create an oxygen imbalance in the space that will suffocate workers almost immediately. Once purging is complete, appropriate ventilation must be established to render the atmosphere safe. Air monitoring is necessary to verify air quality.

4.2.3 Falling Objects

Materials or equipment can fall into confined spaces and strike workers. This usually happens when another worker enters or exits a confined space with a top-side opening, such as a manhole. Vibrations can also cause materials or tools to fall and strike workers.

4.2.4 Engulfment

Engulfment occurs when a worker is buried alive by a liquid or material that enters a confined space (*Figure 5*). Small, loose material stored in bins and hoppers, such as grain, sand, or coal, can engulf and suffocate a worker. Heavier materials or liquids that enter a confined space can crush or strangle a worker.

118F05.EPS

Figure 5 ◆ Buried alive.

4.2.5 *Extreme Temperatures*

Confined spaces that are too hot or too cold can be hazardous to workers. Spaces that are too hot can cause heat stroke or heat exhaustion, while spaces that are too cold can cause **hypothermia**. Another temperature-related hazard in a confined space is steam. Steam is extremely hot and can cause serious or deadly burns.

4.2.6 *Noise*

Noise in a confined space can be very loud. This happens because the size of the space is small and sound bounces off the walls. Too much noise, or noise that is too loud, can permanently damage hearing. It can also prevent workers from communicating. If this happens, an evacuation warning could be missed.

4.2.7 *Slick or Wet Surfaces*

Workers can get seriously hurt by slips and falls on slick or wet surfaces. Wet surfaces also add to the chance of electrocution from electrical circuits, equipment, and power tools.

4.2.8 *Moving Parts*

Workers can get struck or trapped by moving parts such as augers or belts. This usually happens when a worker slips or falls. It also can happen if the operator of the moving machinery is wearing loose clothing or jewelry.

5.0.0 ◆ RESPONSIBILITIES AND DUTIES

Everyone involved in confined space work must have special training. This includes entrants doing work in a confined space, attendants at the opening of the space, entry supervisors, and rescue workers.

5.1.0 Entrants

Entrants are people who enter a confined space to do the work. They must be aware of the dangers of the job and know how to protect themselves. The duties of an entrant are to:

- Make sure they have a valid entry permit.
- Know the atmospheric, fire, and toxic contamination hazards.
- Use the specified personal protective equipment, including face and eye protection, gloves, aprons, and coveralls as required.
- Stay in contact with the attendant to make sure they are being monitored and can be told to evacuate, if needed.
- Alert the attendant when warning signs or symptoms of exposure exist.
- Know how to escape when necessary.
- Exit the space immediately upon hearing an evacuation alarm or if an uncontrolled hazard is detected within the space.

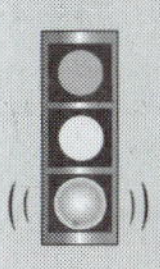

NOTE

Retrieval methods should be part of an emergency procedure. However, exercise caution when using retrieval devices in situations where the device may cause worker entanglement.

5.2.0 Attendants

An attendant stays outside of the confined space and communicates directly with the entrant. The attendant has constant contact with the entrant though telephone, radio, visual, or other means. An attendant's basic responsibility is to protect the entrant. To do this, the attendant must do the following:

- Set up a station at the exit of the confined space.
- Keep count of the personnel inside.
- Know what work is allowed or disallowed in the space.
- Know how to monitor for safety.
- Test the atmosphere remotely using a probe (*Figure 6*) and recognize safe and unsafe levels.
- Maintain contact with entrants and be able to recognize symptoms of physical distress.
- Order evacuation when problems occur.
- Know how to call rescuers and use the alarm system.
- Refuse entry to unauthorized personnel.
- Never enter the space for rescue attempts.

5.3.0 Supervisors

The entry supervisor is the person responsible for safe confined-space entry operations. This means that he or she is responsible for the lives of all workers involved in the operation. The entry supervisor must do the following:

- Authorize and oversee entry.
- Be well trained in entry procedures.
- Know the hazards of the confined space.

- Make sure entry permits are correct and complete.
- Make sure that proper lockout/tagout procedures are performed.
- Make sure that all equipment required for entry is available and understand how it works.
- Make sure that workers understand the job.
- Know rescue plans and rescue workers.
- Be responsible for canceling entry authorization and terminating entry when unacceptable conditions are present or when the job is completed.
- Evaluate problems and recognize when reentry is safe.
- Ensure that explosion-proof equipment is used.
- Notify all workers in the area or facility that work is being done in a confined space.

5.4.0 Rescue Workers

There are two types of rescues: non-entry rescues and entry rescues.

A non-entry rescue can be done by anyone outside of the confined space. It is usually done by the attendant. In this type of rescue, winches and tripods can be used to pull the entrant out of the space without endangering other workers (*Figure 7*).

Rescue workers have a responsibility to do the following:

- Know the hazards of the confined space.
- Know the proper use of personal protective and rescue equipment, including respirators.
- Practice rescue operations and know how to perform rescues.
- Be trained in first aid and CPR.
- Know the company's confined-space rescue program.

Entry rescues can be extremely hazardous. Only trained rescue workers are allowed to enter a confined space during an emergency. More rescue workers are killed in confined-space accidents than the workers they are trying to rescue.

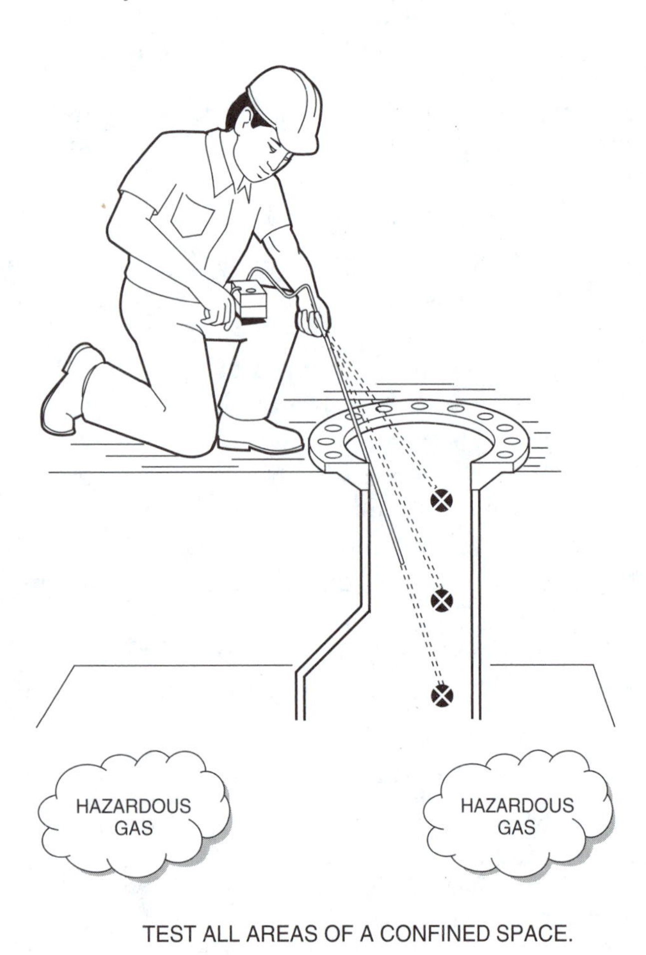

Figure 6 ◆ Attendant checks the atmosphere.

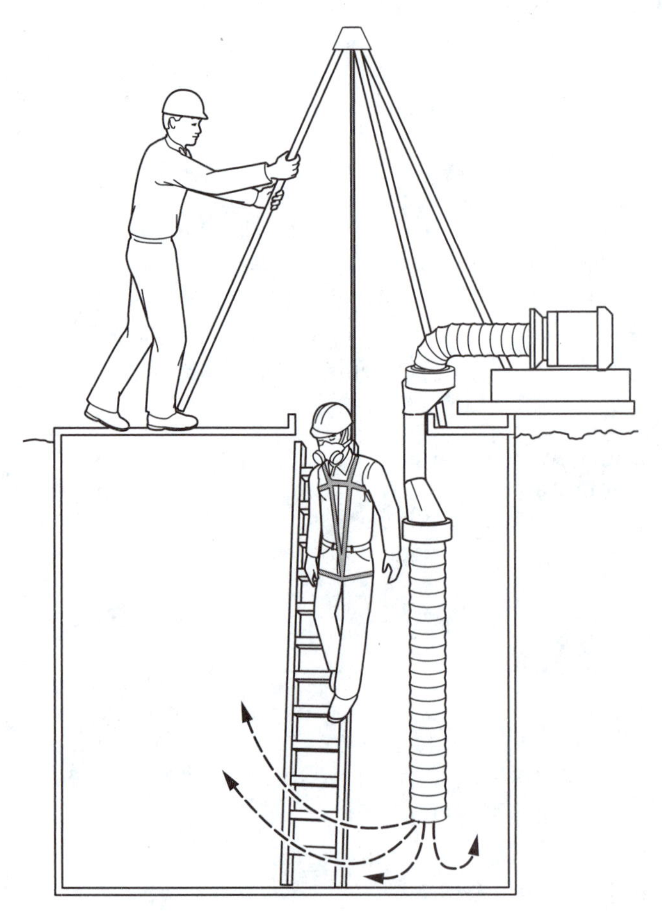

Figure 7 ◆ Non-entry rescue.

6.0.0 ◆ SAFEGUARDS

It is important to understand how to protect yourself and your co-workers when working in a confined space. In order to do this, everyone must be aware of what is happening on the site and understand how to work safely. These are the most common safeguards that should be used during confined-space operations.

- Monitoring and testing
- Ventilation
- Personal protective equipment
- Communications
- Training

6.1.0 Monitoring and Testing

The air in a confined space must be tested before any workers enter it. Testing must be done by a properly trained, qualified person. This person must be a company-apporoved or otherwise designated individual. He or she is referred to as the confined-space attendant. The air is tested for oxygen content, explosive gases or vapors, and toxic chemicals. This can be done by inserting a wand attached to a gas meter into the confined space.

> **NOTE**
> Entrants and attendants have the right to witness or review the gas testing results and may request additional tests if deemed necessary.

The atmosphere in a confined space may need to be monitored during the entire job. This is done by attaching monitors to entrants or by using outside devices. When the atmosphere is monitored, workers can be assured that the air quality is good and that they will immediately know about changes in the atmosphere that would require them to leave.

6.2.0 Ventilation

If the air in a confined space is hazardous or has the potential of becoming dangerous, the space must be ventilated immediately to remove toxic gases or vapors and replace lost oxygen. Ventilators blow clean air into the space, as shown in *Figure 8*. They must stay on as long as workers are in the space.

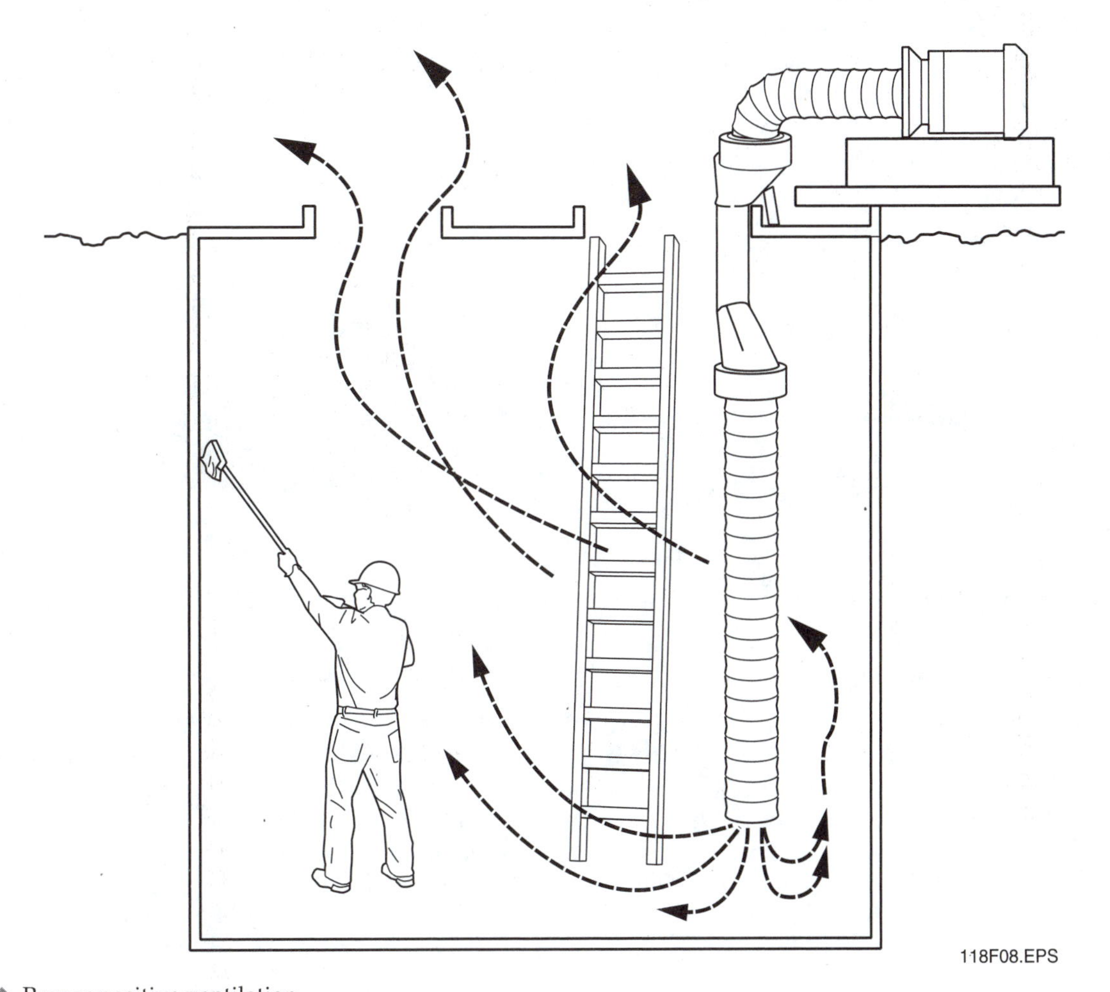

Figure 8 ◆ Proper positive ventilation.

DID YOU KNOW?

Digital Cameras for Safety

Industrial hygiene personnel, supervisors, and planners can use digital cameras with a 20X zoom and flash to capture data on equipment located inside confined spaces. The camera can be used from outside the confined space to capture information such as data on tags, part numbers, and equipment configurations. In some cases, this eliminates the need for entry altogether, which in turn eliminates potential accidents and injuries.

Source: The Occupational Safety and Health Administration (OSHA)

It's important to remember that just because there is air being blown into or being removed from the space, it doesn't mean that the air is being ventilated. Toxic gases can hide in confined spaces. Make sure the attendant has carefully tested the entire space before entering it.

6.3.0 Personal Protective Equipment

Every job requires some type of personal protective equipment. Standard personal protective equipment (*Figure 9*) includes hard hats, safety goggles and glasses, boots, and gloves. On a confined-space job site, the following items may be needed in addition to the standard equipment:

- Full body harness
- Lifelines
- Air-purifying respirator
- Air-supplying respirator

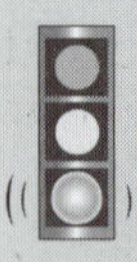

NOTE

Always check the entry permit to make sure you have the personal protective equipment needed to enter a confined space.

6.4.0 Communication

All workers on a site must be able to communicate with one another. It is especially important for attendants and entrants to be able to communicate. This allows attendants to warn entrants about dangers and order an evacuation when necessary. Communication between workers is another way to monitor the confined space.

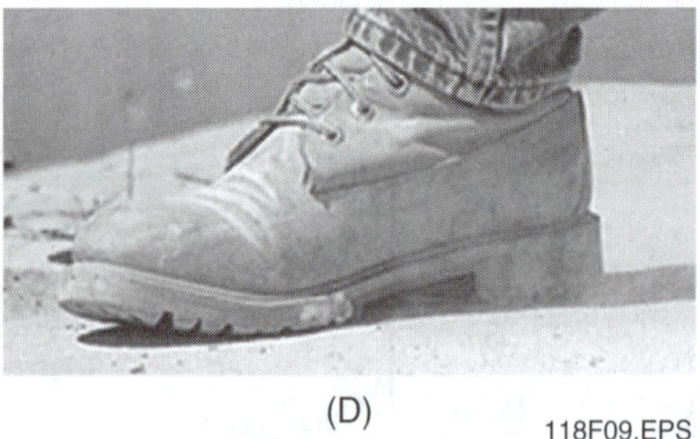

Figure 9 ◆ Standard personal protective equipment.

6.5.0 Training

Entering a confined space requires specialized training. In fact, no one is allowed to enter a confined space unless they have been properly trained and authorized by the site supervisor to enter the space. Training gives workers the knowledge needed to complete their jobs safely and efficiently. If you have not been properly trained, do not enter any confined space.

Summary

Each confined-space job is different. Whether you are working in a sewer, a sawdust bin, or an oil refinery tank, you will experience danger. The hazards associated with confined-space work include entry and exit hazards, atmospheric instability, toxic gases, electric shock, and being buried alive. Many of the accidents that occur in confined spaces are due to unsafe work practices.

It is important for everyone on the job, from entrants to entry supervisors, to know the hazards of confined-space jobs and follow all established safety rules and rescue plans.

Review Questions

1. The two classifications for confined spaces are nonpermit-required and permit-required.

a. True
b. False

2. The type of permit that must be issued before entering a permit-required confined space is called a(n) _____ permit.

a. hot work
b. entry
c. excavation
d. manhole

3. Entry into a confined space begins when _____.

a. the permit is issued
b. all of the entrants have been assigned a job
c. any part of your body passes through the entrance of a confined space
d. the first entrant enters the confined space

4. The most common type of hazard in a confined space is _____.

a. atmospheric
b. electric
c. temperature
d. noise

5. A confined space that does not have enough oxygen for a person to survive is called oxygen-deficient.

a. True
b. False

6. An explosion in a combustible atmosphere can be caused by _____.

a. flammable clothing or materials
b. a source of ignition such as a flame or spark
c. toxic vapors
d. vibrations from tools outside of the confined space

7. The order in which atmospheric testing must be done is _____.

a. flammable gases and vapors, potential toxic contaminants, and oxygen content
b. oxygen content, potential toxic contaminants, and flammable gases and vapors
c. oxygen content, flammable gases and vapors, and potential toxic contaminants
d. potential toxic contaminants, oxygen content, and flammable gases and vapors

8. Staying in a confined space that is too cold for a long period of time can cause hypothermia.

a. True
b. False

9. An attendant is responsible for all of the following *except* _____.

a. entering the space to perform a rescue
b. ordering an evacuation if required
c. knowing how to monitor for safety
d. refusing entry to unauthorized personnel

10. Rescue workers are the only workers allowed to enter a permit-required confined space during an emergency.

a. True
b. False

GLOSSARY

Trade Terms Introduced in This Module

Atmospheric hazard: A potential danger in the air or a condition of poor air quality.

Combustible: Air or materials that can explode and cause a fire.

Confined space: Spaces on a job site are considered confined when their size and shape get in the way of anyone who must enter, work in, and exit the space.

Hypothermia: A life-threatening condition caused by exposure to very cold temperatures.

Nonpermit-required confined space: A work space free of any atmospheric, physical, electrical, and mechanical hazards that can cause death or injury.

Oxygen-deficient: An atmosphere in which there is not enough oxygen. This is usually considered less than 19.5% oxygen by volume.

Oxygen-enriched: An atmosphere in which there is too much oxygen. This is usually considered more than 23.5% oxygen by volume.

Permit-required confined space: A confined space that has real or possible hazards. These hazards can be atmospheric, physical, electrical, or mechanical.

ACKNOWLEDGMENTS

Figure Credits

Gary Wilson	118SA01, 118SA02
Draeger Safety, Inc.	118F04 (A)
RAE Systems, Inc.	118F04 (B)
Bacou-Dalloz	118F09 (A)
Bullard Classic Head Protection	118F09 (B)
North Safety Products	118F09 (C)
Milwaukee Electric Tool Corporation	118F09 (D)

NCCER CURRICULA — USER UPDATE

NCCER makes every effort to keep its textbooks up-to-date and free of technical errors. We appreciate your help in this process. If you find an error, a typographical mistake, or an inaccuracy in NCCER's curricula, please fill out this form (or a photocopy), or complete the online form at **www.nccer.org/olf**. Be sure to include the exact module ID number, page number, a detailed description, and your recommended correction. Your input will be brought to the attention of the Authoring Team. Thank you for your assistance.

Instructors – If you have an idea for improving this textbook, or have found that additional materials were necessary to teach this module effectively, please let us know so that we may present your suggestions to the Authoring Team.

NCCER Product Development and Revision

13614 Progress Blvd., Alachua, FL 32615

Email: curriculum@nccer.org
Online: www.nccer.org/olf

❑ Trainee Guide ❑ AIG ❑ Exam ❑ PowerPoints Other ______________________________

Craft / Level: Copyright Date:

Module ID Number / Title:

Section Number(s):

Description:

Recommended Correction:

Your Name:

Address:

Email: Phone:

Module 75119-03

Concrete and Masonry

COURSE MAP

This course map shows all of the modules in Field Safety. The suggested training order begins at the bottom and proceeds up. The local Training Program Sponsor may adjust the training order.

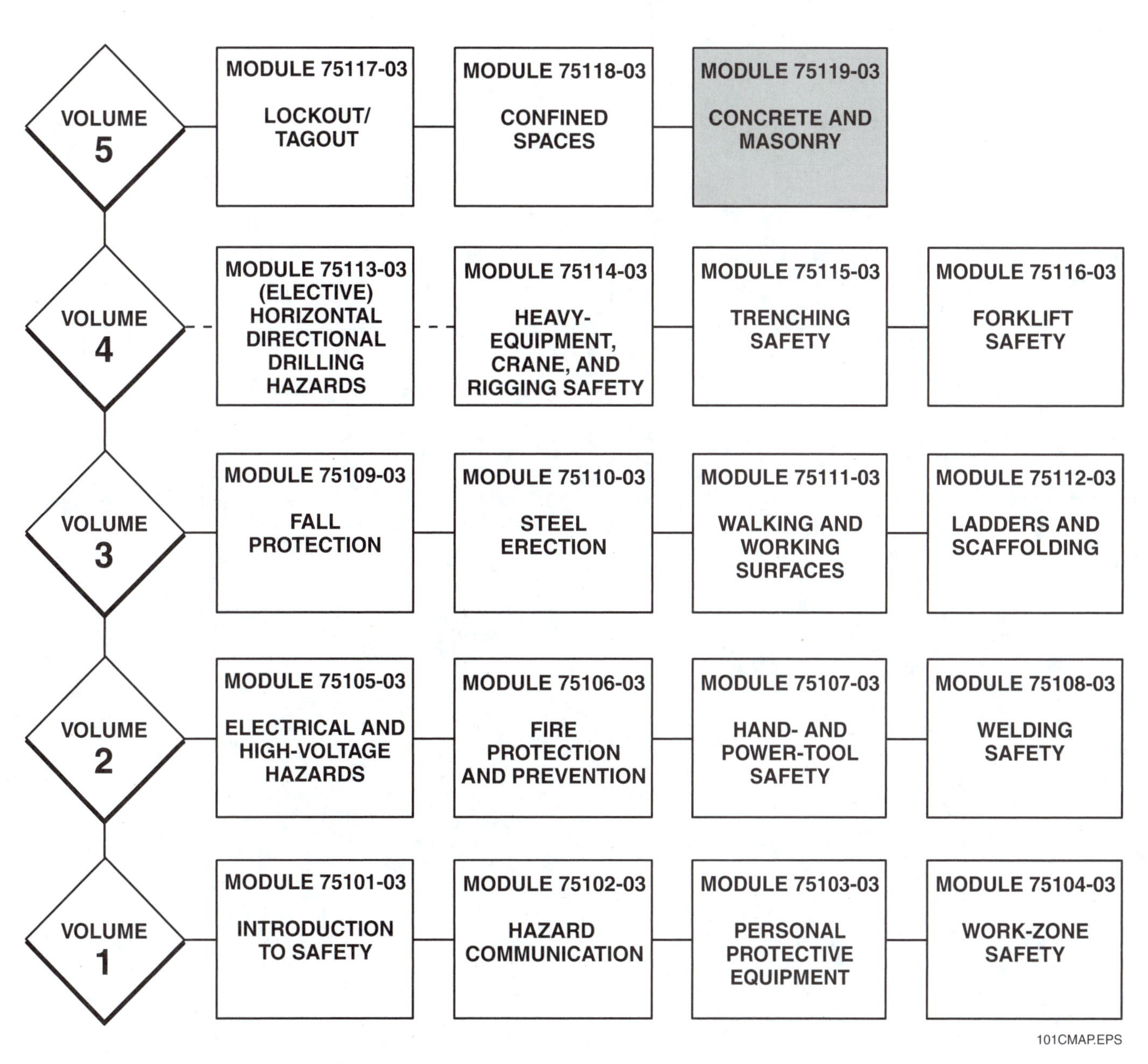

MODULE 75119-03 CONTENTS

Figures

MODULE 75119-03

Concrete and Masonry

Objectives

When you have completed this module, you will be able to do the following:

1. Explain and identify safety hazards associated with concrete construction and masonry work.
2. Demonstrate and explain proper on-site safety, including the use of personal protective equipment (PPE).

Prerequisites

Before you begin this module, it is recommended that you successfully complete the following: Field Safety, Modules 75101-03 through 75112-03, and 75114-03 through 75118-03. *Horizontal Directional Drilling Hazards* (module 75113-03) is an elective and is not required for successful completion of this course.

Required Materials

1. Pencil and paper
2. Appropriate personal protective equipment

1.0.0 ◆ INTRODUCTION

In order to stay safe, you must be aware of all hazardous materials on the job site. This is especially true with concrete and masonry jobs. Anyone working with dry **cement** or wet concrete should be aware that these are **toxic** materials. Dry cement dust can enter open wounds and cause blood poisoning. When it comes in contact with body fluids, cement dust can cause chemical burns to the membranes of the eyes, nose, mouth, throat, and lungs. It can also cause a fatal lung disease known as **silicosis**. Wet cement or concrete can also cause chemical burns to the skin. Repeated contact with cement or wet concrete can result in an allergic skin reaction known as cement dermatitis.

Some of the structures that are built during concrete construction and masonry work include walls, buildings (*Figure 1*), bridges, street curbs, and stairs. The greatest hazard on concrete and masonry jobs is falling. In fact, each year more than 100,000 workers are injured as a result of falls.

119F01.EPS

Figure 1 ◆ Concrete building under construction.

Working responsibly and safely can help prevent harm to you and your co-workers. The inexperienced worker is arguably the most feared person on a construction site. Experience has shown that new workers cause many of the accidents on the job, although they themselves may not be injured. This is why it is important to encourage everyone, especially new workers, to be safety conscious.

2.0.0 ◆ CONCRETE AND MASONRY JOB-SITE HAZARDS

While the actual work of concrete and masonry construction is not the same, workers on both of these jobs are exposed to many of the same hazards. For example, falling is a major risk in both types of work. Workers in both crafts may perform their jobs at extremely high elevations (*Figure 2*). Concrete and masonry workers both use ladders and **scaffolding**. They are also in danger of being struck by falling objects and injured by tools that aren't working properly.

119F02.EPS

Figure 2 ◆ Modern concrete structure.

2.1.0 Falls

More people are injured in falls than in any other type of accident in the construction industry. On average, between 150 and 200 construction workers are killed in falls each year. In one instance, a 27-year-old cement finisher died when he lost his balance and fell off an unguarded end of a **scaffold**. He fell 160' from a suspension scaffold and died when his safety **lanyard** snapped, causing him to fall to the ground.

Falls are classified into two groups: falls from an elevation and falls on the same level. Falls from an elevation can happen when someone is doing work from scaffolding, work platforms, decking, concrete forms, ladders, or excavations. Falls from elevations are almost always fatal. This is not to say that falls on the same level aren't also extremely dangerous. When a worker falls on the same level, usually from tripping or slipping, sharp edges and pointed objects can cut or stab the worker.

The following safe practices can help prevent slips and falls:

- Wear safe, strong work boots that are in good repair.
- Watch where you step. Be sure your footing is secure.
- Do not allow yourself to get in an awkward position. Stay in control of your movements at all times.
- Maintain clean, smooth walking and working surfaces. Fill holes, ruts, and cracks. Be sure to clean up slippery material.
- Pick up litter.
- Install cables, extension cords, and hoses so that they will not become tripping hazards.
- If you must climb to reach something, use a sound ladder that has been safely set in position and properly secured at both the top and bottom (*Figure 3*).
- When climbing a ladder, always face the ladder and use both hands.
- When working from a ladder, keep your shoulders inside the vertical stringers. Do not overreach. Move the ladder if you cannot safely reach the working surface.
- Do not run on scaffolding, work platforms, decking, ladders, or other elevated work areas.

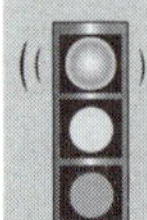

WARNING!

When it is impossible to provide safe platforms and railings, you need to protect yourself from falls by using fall-protection equipment such as harnesses and lifelines, shock-absorbing lanyards, and safety nets. For more information on fall protection, refer to the *Fall Protection* module in Volume Three.

2.1.1 Ladders

Ladders are some of the most important tools on a job site. When selecting a ladder for a job, it is important to choose one that will extend at least 36" above the landing surface you are trying to reach. Always place the base of the ladder so that the distance between the base and the wall is one-quarter of the ladder length from the base elevation to the point where the ladder touches the wall, as illustrated in *Figure 4*.

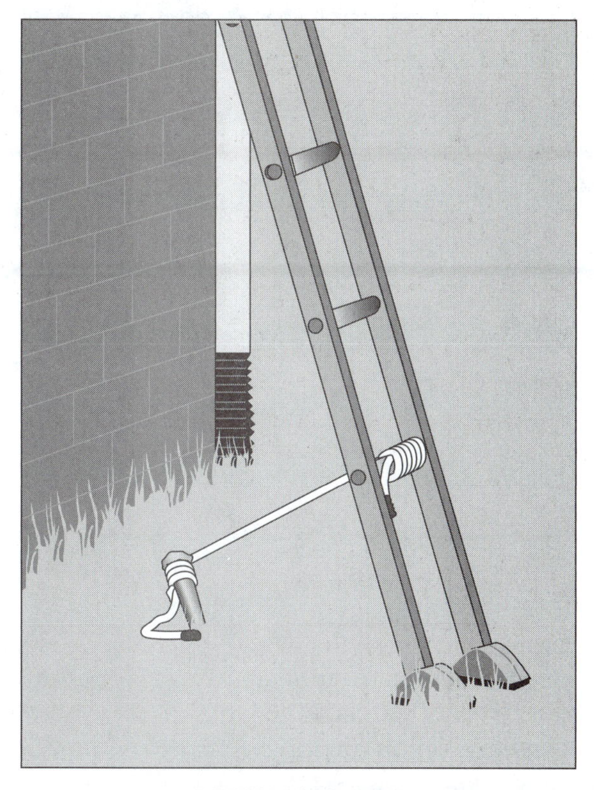

Figure 3 ◆ Properly secured ladder.

Figure 4 ◆ Proper positioning of a ladder.

The following precautions should be kept in mind when setting up and using a ladder:

- Do not use a metal ladder when performing any type of electrical work or whenever there is a possibility that you might come into contact with electrical conductors.
- Portable and extension ladders should be placed on a solid base in such a way that the base of the ladder is one-quarter the distance of its length from its vertical support.

Training is Important

An employee was building the third level of a tubular welded frame scaffold while standing on the second level. The scaffold was built on a poured concrete floor and had been leveled. Each section of the framework measured 6'-5" high. The working surface was solidly planked.

When the employee tried to set the third level frame into the pins of the second level, the frame he was trying to position flipped to one side. The momentum of the frame thrust the employee backward off the second level. He fell to the ground, sustaining a fatal blow to his head.

Following an inspection, the employer was sited for failure to provide specific employee training and failure to implement an effective safety program.

The Bottom Line: Make sure you have been properly trained before doing any work on an elevated work site.

Source: The Occupational Safety and Health Administration (OSHA)

- Place the ladder so that it leaves 6" of clearance in back of the ladder and 30" of clearance in front of the ladder.
- Place the ladder so that it leans against a solid and immovable surface. Never place a ladder against a window, door, doorway, sash, loose or movable wall, or box.
- Face the ladder when climbing up or down.
- Climb or descend the ladder one rung at a time. Never run up or slide down a ladder.
- Do not use ladders during high winds. If you must use a ladder in windy conditions, make sure you lash the ladder securely in order to prevent slippage.
- Make sure the soles of your shoes are free of oil, mud, and grease.
- Never rest any tools or materials on the top of a ladder.
- Keep both hands free so that you can hold the ladder securely while climbing. Use a rope to raise and lower any tools and materials that you might need.
- Move the ladder in line with the work to be done. Never lean sideways away from the ladder in order to reach the work area.
- Never stand on the top two rungs of a ladder.
- Use ladders only for short periods of elevated work. If you must work from a ladder for extended periods, use a lifeline fastened to a safety belt.
- Lay the ladder down on the ground when you have finished using it, unless it is anchored securely at the top and bottom where it is being used.
- Never use makeshift substitutes for ladders.
- Never use stepladders for straight ladder work.

2.1.2 Scaffolds

Scaffolds are elevated working platforms that support workers and materials. One general safety rule for all types of scaffolds is that every scaffold should have a minimum safety factor ratio of 4:1. This means that the scaffolding should

Pay Attention

Late in the afternoon, a maintenance worker was descending from a fixed ladder and fell approximately 5' to the floor. He was not paying attention to what he was doing. The worker sustained injuries to the left ankle and right knee. He was hospitalized and required surgery. The worker was not able to return to his regular work activity for several months.

The Bottom Line: Even the most simple task, such as climbing a fixed ladder, requires your maximum attention. Tasks that are often considered low hazard and routine may have the potential to result in significant injuries.

Source: U.S. Department of Energy

be able to support at least four times the weight that will be placed on it.

Typical scaffolding is shown in *Figure 5.* The main part of the scaffolding is the working platform. A working platform should have a guardrail system that includes a top rail, midrail, toeboard, and screening. To be safe and effective, the top rail should be approximately 42" high, the midrail should be located halfway between the toeboard and the top rail, and the toeboard should be a minimum of 4" high. If workers will be passing or working under the scaffold, the area between the top rail and the toeboard must be screened. Finally, the platform planks must be laid closely together. For safety purposes, the ends of the planks must overlap at least 6" and no more than 12".

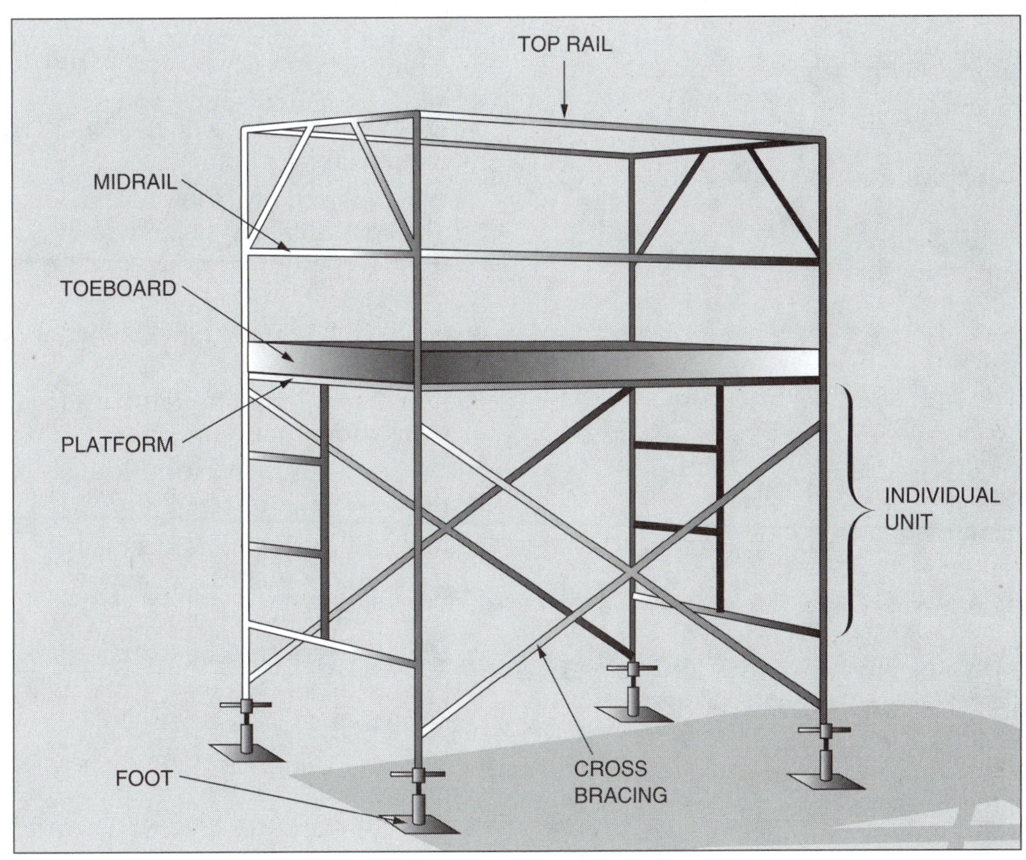

Figure 5 ◆ Typical scaffolding.

Experienced Worker Killed

A construction crew was preparing to pour concrete into forms. A laborer climbed up a ladder on one side of the forms and stepped over the form to stand on an unguarded scaffold on the opposite side. He was carrying two hand trowels and a brush to be used by other workers after the concrete was poured. He sustained fatal injuries when he fell and hit his head on a concrete slab at ground level.

The employee had previously worked for his employer on several different occasions and had been performing this type of work for 21 years. The employer unwisely believed that no training was necessary for this employee, so none was provided.

The Bottom Line: Safety training should be a part of regular operations. It benefits everyone, even the most experienced workers.

Source: The Occupational Safety and Health Administration (OSHA)

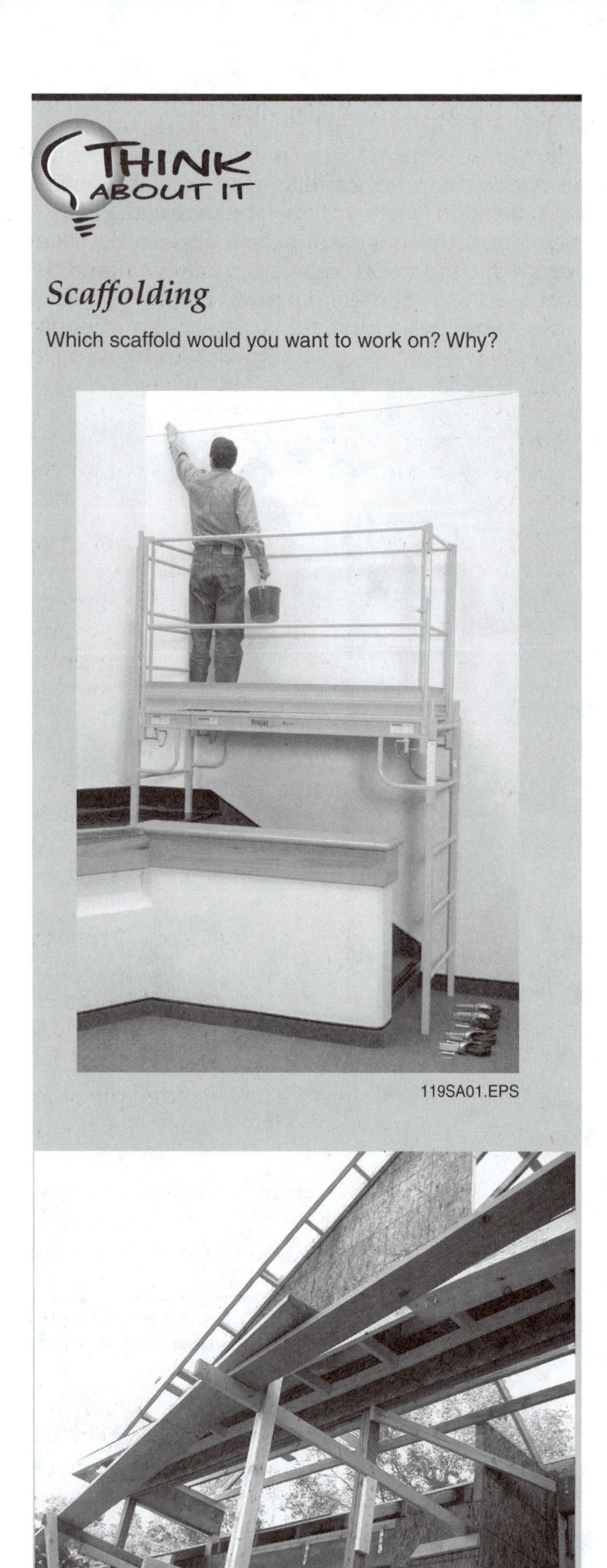

Scaffolding

Which scaffold would you want to work on? Why?

119SA01.EPS

119SA02.EPS

The following guidelines will help to ensure safe scaffold use.

- Always raise the scaffold to the desired height. Do not try to increase the height by using ladders, boxes, or sawhorses on top of a scaffold.
- Always keep scaffolding planks clear of extra tools and materials. Clean up any slippery substances that get spilled on scaffolding.
- Use extreme caution when working near power lines.
- Anchor freestanding scaffolding with guy wires to prevent tipping or sliding.
- Locate and read the posted safety rules and regulations for scaffolding use.
- When objects and tools are to be removed from the work area, lower them to the ground with a rope. Never throw or drop them from the scaffold.
- Never move scaffolding when someone is on it.
- Keep tools and materials back from the edge of work platforms.
- Use only approved scaffolding.
- Inspect the scaffolding regularly.
- Install all braces and accessories according to the manufacturer's recommendations.
- Make sure the scaffolding is plumb and level.
- Do not overload scaffolding.
- Be sure all safety rails are installed according to regulations.
- Make sure the wheels are locked before ascending or disassembling rolling scaffolding.
- Always check for electrical wires before erecting any scaffolding.
- Do not work on scaffolding during storms or high winds, or when the scaffolding is covered with ice or snow.
- If the scaffolding is wobbly, bouncy, or can be pulled down easily, it is not safe. Scaffolding must have a sound structure and include toeboards and handrails. Wooden floor planks must be free of knots. Broken members must be repaired immediately.

2.2.0 Falling Objects

Falling and flying objects are a common hazard on a construction site. You are at risk from falling objects when you are beneath cranes, scaffolds, concrete forms, or where overhead work is being performed. There is a danger from flying objects when power tools or activities like pushing, pulling, or prying cause objects to become airborne. Injuries can range from minor abrasions to concussions, blindness, or death.

Follow these safeguards to avoid being struck by falling or flying objects.

- Always wear a hard hat.
- Use toeboards, screens, and guardrails on scaffolds to stop objects from falling.
- Secure tools and materials to prevent them from falling on anyone working below you.
- Stack materials properly to prevent sliding, falling, or collapse (*Figure 6*).
- Use debris nets, catch platforms, or canopies to catch or deflect falling objects.
- Use safety glasses, goggles, and face shields around machines or tools that can cause flying particles.
- Inspect tools to ensure that protective guards are in good condition.
- Make sure you are trained in the proper operation of powder-actuated tools.
- Avoid working underneath loads being moved.
- Barricade hazard areas and post warning signs.
- Inspect cranes and hoists to see that all components, such as wire rope, lifting hooks, and chains are in good condition.
- Do not exceed the lifting capacity of cranes and hoists.

119F06.EPS

Figure 6 ◆ Stacks of bricks.

2.3.0 Tools and Equipment

Many injuries result from improper or unsafe use of hand tools and small power tools. Becoming familiar with the basic tools and materials of your trade and understanding how to handle them safely are important steps toward working safely. The need to keep tools clean and in good working order cannot be overstressed. Safe and proper use of tools will not only help preserve the tools in good working condition but can also help to prevent accidents. All tools are capable of inflicting injury, even hand tools.

Some rules for the care and safe use of power tools are:

- Do not attempt to operate any power tool without being certified on that particular tool.
- Always wear the appropriate safety equipment and protective clothing for the job. For example, wear safety glasses and close-fitting clothing that cannot become caught in the moving parts of the tool. Roll up long sleeves, tuck in your shirt tail, and tie back long hair.
- Never leave a power tool running unattended.
- Assume a safe and comfortable position before starting a power tool.
- Do not distract others or let anyone distract you while operating a power tool.
- Be sure that any electric power tool is properly grounded before using it.
- Be sure that power is not connected before performing maintenance or changing accessories.
- Do not use dull or broken accessories.
- Use a power tool only for its intended purpose.
- Do not use a power tool with any guards or safety devices removed.
- Use a proper extension cord of sufficient size to service the particular electric tool you are using.
- Do not operate an electric power tool if your hands or feet are wet.
- Become familiar with the correct operation and adjustments of a power tool before attempting to use it.
- Keep a firm grip on the power-float or trowel-machine handle while operating the machine.
- Be sure there is proper ventilation before operating gasoline-powered equipment indoors.
- Keep a fire extinguisher nearby when filling and operating gasoline-powered equipment.
- Keep hands and feet away from cutting tools such as concrete saws, grinders, and trowel-machine blades.
- Store tools properly when not in use.

2.4.0 Personal Protective Equipment

One of the first steps to ensuring a safe operation is wearing appropriate safety equipment and clothing. Safety gear and clothing, when worn correctly, can help protect you from harm. The type of clothing and equipment required varies with job responsibilities and daily activities. It is the responsibility of all workers to wear the appropriate clothing for the tasks they are performing. Your supervisor will provide a specific list of safety clothing and equipment required for your job.

The following safety precautions, in addition to your personal protective equipment, will also help to keep you and your co-workers safe.

- Remove all jewelry, including wedding rings, bracelets, necklaces, and earrings. Jewelry can get caught on or in equipment, which could result in a lost finger, ear, or other appendage.
- Tie back long hair in a ponytail and tuck it under your hard hat. Flying hair can obscure your view or get caught in machinery.
- Wear close-fitting clothing that is appropriate for the job. Clothing should be comfortable and should not interfere with the free movement of your body. Clothing or accessories that are too loose or torn may get caught in tools, materials, or scaffolding.
- Wear face and eye protection as required or if there is a risk from flying particles, debris, or other hazards such as brick dust or chemicals.
- Wear a long-sleeved shirt to provide extra protection for your skin.
- Protect any exposed skin by applying skin cream, body lotion, or petroleum jelly.
- Wear sturdy work boots or work shoes with thick soles. Never show up for work dressed in sneakers or gym shoes.
- Wear fall-protection equipment as required.

3.0.0 ◆ CONCRETE CONSTRUCTION

Concrete is one of the most widely used construction materials (*Figure 7*). It is strong, durable, and can be readily formed into many shapes and sizes. It is also economical compared to other materials because the raw materials from which concrete is made are usually inexpensive and readily available.

Because concrete is manufactured at or near the job site, concrete workers need to understand the safety requirements for mixing and placing concrete. They should also understand that the success of concrete construction depends largely on the worker's skill, knowledge, and craftsmanship.

119F07.EPS

Figure 7 ◆ Bridge.

Concrete used in construction can take several forms: plain, **reinforced**, **prestressed**, and **precast.** Plain concrete has no reinforcement of any type. Reinforced concrete, such as **post-tensioned** concrete, has some type of reinforcement to increase its tensile strength. Prestressed concrete uses a technique that places the concrete under compression so that it can resist greater amounts of tension. Precast concrete is concrete that is cast and cured away from the construction site. It may or may not be reinforced.

Different types of construction require different types of concrete. Plain and reinforced concrete, for example, are normally used in **slab-on-grade** construction. Reinforced, prestressed, post-tensioned and precast concrete are usually associated with elevated construction.

3.1.0 Forms

A concrete form is a temporary structure or mold used to support concrete until the hardening process gives it enough strength to be self-supporting. Structures such as **slabs**, beams, walls, columns, and stairs are made with forms.

Forms are an essential part of concrete construction because they control the position, alignment, size, and shape of the concrete structure (*Figure 8*).

119F08.EPS

Figure 8 ◆ Examples of forms.

A concrete form is subjected to tremendous pressure from wet concrete. The pressure increases as the height of the structure increases. The rate at which the concrete is poured also affects the pressure on the form. The faster the concrete is poured, the greater the pressure. Workers are at risk of being crushed if a form collapses.

Some common causes of form failure include:

- Form ties incorrectly placed or improperly fastened
- Exceeding the design working pressure of the form as a result of excessive rate of concrete placement, concrete mix design not being taken into account, improper vibration of the concrete, or temperature not being taken into account
- Improper form construction, especially when layout plans are not provided
- Form fillers, corners, and bulkheads not adequately designed and/or constructed

- Connecting hardware not installed
- Not being inspected by authorized qualified personnel to see if the form layout has been interpreted correctly

If the forms are not properly built and supported, they will not be able to withstand the pressure of the wet concrete. They will fracture, releasing the concrete. This can also happen if the concrete is poured too quickly. It can be very dangerous to workers and create a mess that will make the work area unsafe. It is therefore very important that the form builder construct the form in accordance with established standards and safety practices. Basic formwork safety guidelines include:

- The structure must be capable of supporting the load created by the concrete, forms, machines, and people. Forms and **shoring** (*Figure 9*) must be designed by qualified engineers.
- Exposed reinforcing steel on which people could fall must be guarded.
- Workers are not permitted behind jacks during tensioning operations.
- Workers are not permitted to work under concrete buckets. Buckets must be routed to avoid workers (*Figure 10*).
- Appropriate personal protective equipment must be worn.
- Formwork must be erected and braced in a manner that ensures it will support all vertical and lateral loads.
- Shoring equipment must be inspected and properly maintained. Damaged or defective shoring equipment must be repaired or replaced.
- Forms and shoring equipment may not be removed until the concrete has been verified as strong enough to support its weight and that of its loads.
- Reinforcing steel must be adequately supported to prevent it from falling or collapsing.
- Measures must be taken to prevent uncoiled wire mesh from recoiling.
- Measures must be taken to avoid skin contact with concrete, which is a hazardous material.

3.2.0 Reinforcing Bar (Rebar)

Reinforced concrete is a combination of concrete and steel. The steel that is used to reinforce concrete is called reinforcing bar or rebar (*Figure 11*). Rebar comes in different shapes and sizes. It is placed in the form where concrete is poured.

119F09.EPS

Figure 9 ◆ Steel shoring.

119F10.EPS

Figure 10 ◆ Worker standing under a concrete bucket.

119F11.EPS

Figure 11 ◆ Using rebar.

Working with rebar can be dangerous. Anyone working with rebar can be cut, crushed by dropped loads of rebar, or impaled by landing on vertical rebar. The following safety precautions will help prevent accidents and injury.

- Wear hard hats, safety boots, and leather gloves.
- Do not wear loose or ragged clothing.
- Make sure your footing is always solid.
- Wear approved fall-protection equipment.
- Keep the work area clean.
- Block piles of reinforcing steel to prevent sideways movement.
- Wear goggles with the proper lens when cutting with a torch.
- Know and observe the proper hand signals for hoisting equipment.
- Lift bars or bundles properly to avoid strain.
- Always know the location of your co-workers.
- Bend down projecting nails.
- When carrying bars with a partner, make sure the load is balanced. Each person should be about one-quarter the length of the rods from each end. Lift and drop all bars in unison.
- Stay alert for concrete buggies or any hoisted material that is swinging nearby.
- Never land hoisted bundles or drop carried bars on a form without checking that the form is strong enough to carry the additional weight.
- Do not hoist bundles by the wire wrappings that tie the bars together. Use proper slings and chokers.
- Reinforcing steel for vertical structures such as piers and columns must be adequately **guyed** and supported to prevent collapse.
- Report all unsafe conditions.
- Always use plastic caps on exposed ends of bars (*Figure 12*).
- Bend down loose ends of tie wires with pliers after each tie. If this is not done, the wire may puncture gloves, boots, or skin.

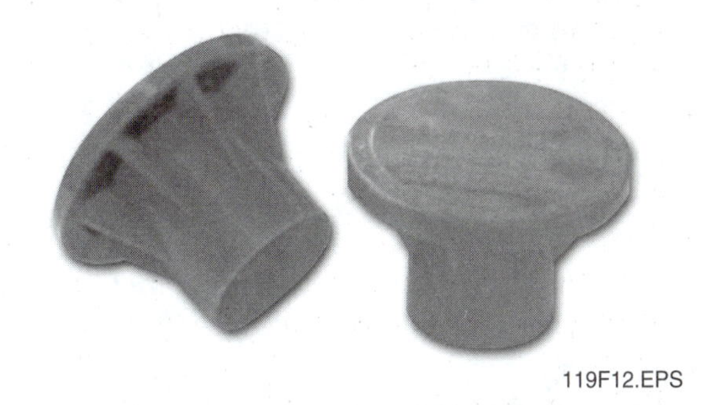

119F12.EPS

Figure 12 ◆ Capped rebar.

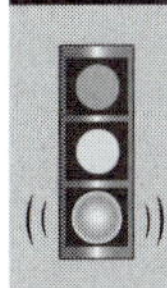

NOTE

If you or a co-worker experience an injury related to rebar, a supervisor must be contacted immediately and your company's established emergency-response plan must be followed.

3.3.0 Tilt-Up Panels

Tilt-up construction is a process in which concrete wall panels are poured and finished in forms on the floor of the building, or on a separate casting slab at the job site. After the panels have hardened, they are tilted up to a near-vertical position with a crane, and then moved to the desired location (*Figure 13*).

119F13.EPS

Figure 13 ◆ Panel being lifted.

Tilt-up panels are extremely heavy. The largest panel on record weighed over 300,000 pounds. Because of their weight, tilt-up panels present special hazards to workers at a tilt-up site. In one incident, a worker was directing a crane operator during the tilt-up process, while standing in front of the panel. The stabilizer legs of the crane sank into the ground, unbalancing the crane and causing the panel to fall on the worker, who was crushed and killed instantly.

There are some important safety precautions associated with tilt-up concrete work.

- Workers must stay clear of the panel while it is being moved.
- The temporary lifting inserts used to pick up the panels must be placed in accordance with the manufacturer's instructions. Only a qualified engineer can make changes.
- Once placed, the panels must be braced in accordance with the manufacturer's instructions until the roof is in place.
- Because concrete pouring and finishing is involved, the precautions associated with handling and finishing concrete must be observed.

3.4.0 Tools and Equipment Used in Concrete Construction

There are many special-purpose tools that are used only in concrete construction. Tools are important for concrete work because concrete cannot be handled with bare hands. Its elastic physical qualities also make it difficult to work with.

Because of the many stages involved in working concrete before it is finished, there are many special-purpose tools and equipment used. These tools range from hand tools like trowels and floats to powered saws and **screeds**.

To prepare the underlying earth and **sub-grade**, mechanical equipment such as a bulldozer, backhoe, or large earth mover may be used. Workers are in danger of being run over, crushed, and struck by these large machines.

> **NOTE**
> Heavy-equipment operations are covered in detail in the *Heavy-Equipment, Crane, and Rigging Safety* module in Volume Four.

Placing tools such as buggies and buckets (*Figure 14*) are used to guide and move concrete into the form or between the slab forms. If these tools are not used safely, workers can be struck or have a load of cement dumped on them.

CONCRETE BUCKET

BUGGY

119F14.EPS

Figure 14 ◆ Concrete bucket and buggy.

Screeding tools are used to level and to smooth concrete to a given elevation. Screeds can be manual or powered. Powered screeds (*Figure 15*) can be as dangerous as any other type of mechanical equipment if they are not used properly. Workers can get struck by or caught in the tool's machinery.

Finishing tools such as power saws, trowels, and single-rotary finishing machines (*Figure 16*) are used to further level the concrete and smooth the surface, or to finish it as specified. Finishing tools that use power can be particularly dangerous because there is a risk of amputation from the saw and being struck by moving finishing equipment.

Tool safety requires using tools properly and keeping them clean. A defective tool or piece of equipment can cause accidents, as well as slow down or stop work. Accidents can also happen when tools and equipment are left in the way of other workers.

Follow these general job-safety rules for working with tools and equipment to make sure everyone on the site is safe.

- Keep all tools sharp, clean, and in safe working order.
- Use retaining guards and safety devices on all equipment.
- Report all defective tools, machines, or other equipment to your supervisor and report all accidents and near-accidents.
- Disconnect power, lock out, and tag machines before performing maintenance.
- Never use compressed air to clean yourself or your clothing.
- Make sure others are at a safe distance.
- Use traffic-control devices where required.
- Follow the manufacturer's safety rules and limitations for all equipment.

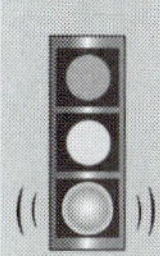

NOTE

Not every tool will be used on every job. Tool choices will depend on the type of concrete, the type of job, and other conditions.

STEEL TRUSS SCREED

ROLLER SCREED

119F15.EPS

Figure 15 ◆ Powered screeds.

POWER SAW

TROWEL

SINGLE-ROTARY FINISHING MACHINE

119F16.EPS

Figure 16 ◆ Power saw, trowel, and single-rotary finishing machine.

3.5.0 Cement Dermatitis

Working with dry cement or wet concrete can be harmful. Always use the appropriate personal protective equipment when you work with dry cement or wet concrete. Repeated contact with these materials can cause an allergic skin reaction known as cement dermatitis.

Cement dermatitis is a condition that irritates your skin and causes sores to appear. Nasal passages and the lining of your mouth can also be affected. You get cement dermatitis by touching cement or inhaling the fumes of fresh concrete or **mortar**. Some conditions can make cement dermatitis worse such as excessive sweating, preexisting dermatitis, and allergies. The following safeguards will help reduce the risk of cement dermatitis.

- Wear coveralls, long-sleeved shirts, and boots.
- Use rubber gloves, goggles, or face masks.
- Use a **respirator**.
- Wash frequently with a lanolin-based soap or use lotion.
- Bathe after each shift.

3.6.0 Personal Protective Equipment

Personal protective equipment (*Figure 17*) can protect you from illness, injury, and death. The general types of personal protective equipment used by concrete workers include:

- Hard hats
- Safety goggles
- Masks
- Gloves
- Knee pads
- Toeboards
- Safety boots

4.0.0 ◆ MASONRY

Masonry is used to build structures ranging from high-rise buildings to patios. In masonry work, cement blocks or clay bricks are assembled by hand using mortar, dry-stacking, or mechanical connectors. Once a brick is placed, a connecting element such as mortar is applied to the top surface and another brick is placed on top. Walls are built layer by layer up to surprisingly high elevations.

Masonry workers have a responsibility to learn all of the safety rules for a site. This includes:

- The safety gear and clothing required for each job
- The safety rules and procedures associated with each job
- General work-site safety procedures and awareness of potential hazards
- The proper use of the tools and equipment associated with each job

4.1.0 Masonry Tools and Equipment

There are many tools that masonry workers must know how to use (*Figure 18*). Skill in using them must be developed before they can work efficiently. Faulty, damaged, or improperly used tools and equipment pose a substantial threat to the safety and well being of all employees. One of the most important rules regarding tool use is to use quality tools. Quality tools will last longer if they are used and maintained properly. Next, it is important to use the correct tool for the job. If you use the proper tool, the job will not require as much muscle power. Here are some additional rules:

- Position all cutting edges away from your body.
- Keep your fingers away from all cutting edges.
- Hold cutting tools in such a way that you will not be hurt if there is a slip.
- When working around electrical equipment, use tools with insulated or wooden handles.
- Never use broken tools or tools with loose handles or dull blades.
- Maintain all tools in good working order and keep them clean, dry, and sharp.
- Store tools properly when they are not being used.
- Any defective tools, machines, or other equipment should be tagged and reported to your supervisor.

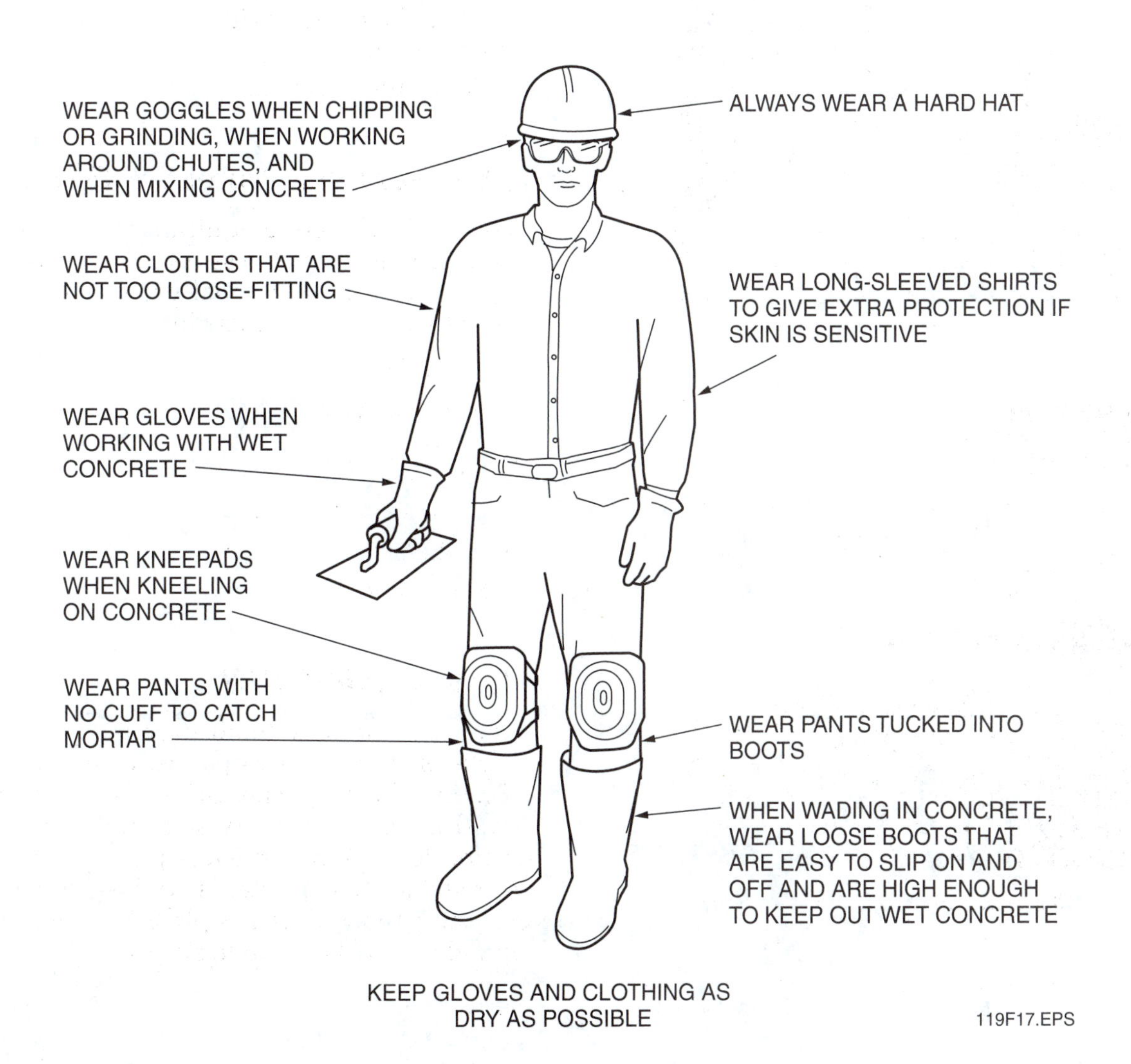

Figure 17 ◆ Appropriate personal protective equipment.

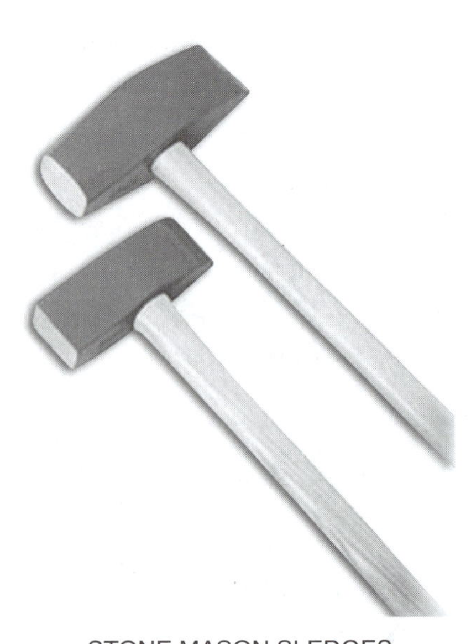

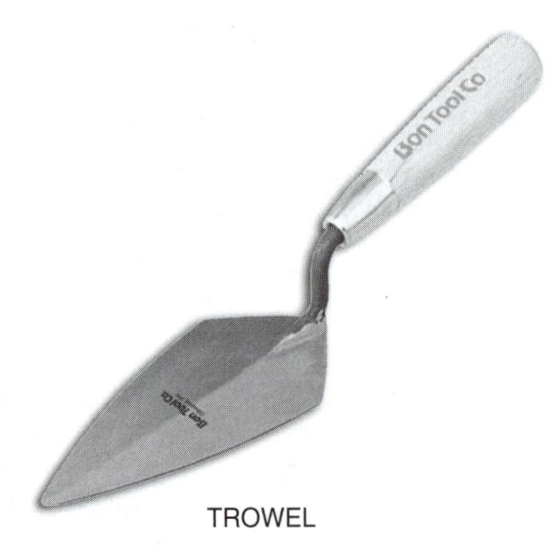

Figure 18 ◆ Masonry tools.

DID YOU KNOW?

Brick Upon Brick

For nearly 40 years, the Empire State building was the tallest building in the world. Approximately 10 million bricks were used in its construction, which was completed in 1931. Today, it is likely that curtain walls containing brick facades would be used in place of individual bricks.

Despite the enormity of the project, the construction of the Empire State Building was completed in about 15 months. One of the methods used to speed up construction was having trucks dump the bricks down a chute, instead of dumping them in the street. The chute led to a large hopper from which bricks were then dumped into carts and hoisted to the location where they were needed. This innovative technique eliminated the back-breaking work of moving the bricks from the pile to the bricklayer using a wheelbarrow.

Powder-Actuated Tools are Dangerous

A carpenter apprentice was killed when he was struck in the head by a nail that was fired from a powder-actuated tool in another room. The tool operator was attempting to anchor a plywood form in preparation for pouring a concrete wall. When he fired the gun, the nail passed through the hollow wall and traveled 27' before striking the victim. The tool operator had never received training in the proper use of the tool, and none of the employees in the area were wearing personal protective equipment.

The Bottom Line: Proper training and the use of personal protective equipment can save your life.

Source: The Occupational Safety and Health Administration (OSHA)

Worker Injured Using Safety Glasses

While on a tour of a facility, an individual removed his prescription glasses and put on safety glasses that distorted his depth perception. When he stepped up to enter a large pipe, his foot did not reach the entrance, resulting in a fall forward and laceration of the lower right leg.

The Bottom Line: Always use the type of safety glasses that can fit over prescription glasses. Abnormal distances for steps up or down may be difficult for touring individuals to maneuver in an unfamiliar environment, even without distorted vision.

Source: U.S. Department of Energy

Summary

Accidents on concrete and masonry jobs can be fatal. Falls are the most common types of accidents that happen on these jobs. Workers can also be injured by falling objects, direct contact with wet concrete and mortar, using unsafe tools and equipment, noise, and working on or around hazardous equipment. Most of these accidents happen due to carelessness. It is your responsibility to know all of the hazards and follow all safety procedures.

Review Questions

1. The greatest hazard on concrete and masonry jobs is _____.
   a. cement dermatitis
   b. falling
   c. falling objects
   d. tool malfunction

2. Dry cement can enter open wounds and cause blood poisoning.
   a. True
   b. False

3. Ladders should extend at least _____ above the surface you are trying to reach.
   a. 12"
   b. 24"
   c. 36"
   d. 48"

4. Elevated work platforms that support workers and materials are called_____.
   a. forms
   b. ladders
   c. scaffolds
   d. rebar

5. The ends of platform planks should overlap by at least _____.
   a. 3"
   b. 4"
   c. 5"
   d. 6"

6. Injuries from improper or unsafe use of tools are very rare on a construction site.
   a. True
   b. False

7. One of the most widely used and most economical construction materials is _____.
   a. brick
   b. concrete
   c. wood
   d. steel

8. A temporary structure or mold used to support concrete until it hardens is called a(n) _____.
   a. form
   b. rebar
   c. tilt-up panel
   d. wall

9. Repeated contact with cement or wet concrete can cause an allergic reaction known as silicosis.
   a. True
   b. False

10. The material that is used to connect bricks is _____.
   a. concrete
   b. mortar
   c. rebar
   d. sand

GLOSSARY

Trade Terms Introduced in This Module

Cement: A powder ground from lime, iron, and other minerals heated in a kiln that chemically reacts with water and serves as a binder of the ingredients in mortar and concrete.

Guyed: A steady or guided load that is laterally stable.

Lanyard: A short section of rope used to attach a worker's safety harness to a strong anchor point located above the work area.

Mortar: A mixture of portland cement, an aggregate or sand, lime, and water in certain combinations, that is used to bind masonry units, such as brick, block, or stone.

Post-tension: A method of reinforcing concrete with embedded cables or tendons tensioned after the concrete has hardened.

Precast: A concrete member that is cast and cured away from the construction site.

Prestressed: A method of reinforcing concrete with embedded cables or tendons tensioned before the concrete hardens.

Reinforced: A structure that resists load stress and gives additional working load support.

Respirator: A device that provides clean, filtered air for breathing when the surrounding air is unsafe.

Scaffold: An elevated work platform for both personnel and materials.

Scaffolding: A manufactured or job-built structure that supports a work platform.

Screed: A manual or motorized straightedge tool used to strike off the surface of placed concrete to a predetermined grade.

Silicosis: A disease of the lungs characterized by massive fibrosis of the lungs resulting in shortness of breath and caused by prolonged inhalation of cement dust.

Shoring: Props or posts of timber or other material in compression used for temporary support for excavations, formwork, or unsafe structures. Also, the process of erecting shores.

Slab: A section of concrete that is larger in its horizontal dimensions than in its thickness (such as a floor).

Slab-on-grade: A ground-supported concrete slab that is used as a foundation system. It combines concrete foundation walls with a concrete floor slab that rests directly on an approved base that has been placed over the ground.

Sub-grade: Soil prepared and compacted to support a structure or pavement system.

Toxic: Being harmful, poisonous, or containing a lethal chemical (toxin).

Figure Credits

PERI USA	119F01
Courtesy of the Construction Education Foundation	119F02, 119F06–119F16
Bil-Jax, Inc.	119SA01
Gary Wilson	119SA02
Bon Tool Company	119F18

NCCER CURRICULA — USER UPDATE

NCCER makes every effort to keep its textbooks up-to-date and free of technical errors. We appreciate your help in this process. If you find an error, a typographical mistake, or an inaccuracy in NCCER's curricula, please fill out this form (or a photocopy), or complete the online form at **www.nccer.org/olf**. Be sure to include the exact module ID number, page number, a detailed description, and your recommended correction. Your input will be brought to the attention of the Authoring Team. Thank you for your assistance.

Instructors – If you have an idea for improving this textbook, or have found that additional materials were necessary to teach this module effectively, please let us know so that we may present your suggestions to the Authoring Team.

NCCER Product Development and Revision

13614 Progress Blvd., Alachua, FL 32615

Email: curriculum@nccer.org
Online: www.nccer.org/olf

❑ Trainee Guide ❑ AIG ❑ Exam ❑ PowerPoints Other ________________

Craft / Level: Copyright Date:

Module ID Number / Title:

Section Number(s):

Description:

Recommended Correction:

Your Name:

Address:

Email: Phone:

Index

Index